Acousto-Optical Spectral Technologies

Acousto-Optical Spectral Technologies

Editors

Alexander S. Machikhin
Vitold Pozhar

MDPI

Basel • Beijing • Wuhan • Barcelona • Belgrade • Novi Sad • Cluj • Manchester

Editors
Alexander S. Machikhin
Acousto-optic Spectroscopy
Laboratory
Russian Academy of Sciences
Moscow
Russia

Vitold Pozhar
Department of Acousto-optic
Information Technologies
Russian Academy of Sciences
Moscow
Russia

Editorial Office
MDPI
St. Alban-Anlage 66
4052 Basel, Switzerland

This is a reprint of articles from the Special Issue published online in the open access journal *Materials* (ISSN 1996-1944) (available at: www.mdpi.com/journal/materials/special_issues/Acousto-Optical_Spectral_Technologies).

For citation purposes, cite each article independently as indicated on the article page online and as indicated below:

Lastname, A.A.; Lastname, B.B. Article Title. *Journal Name* **Year**, *Volume Number*, Page Range.

ISBN 978-3-0365-9519-1 (Hbk)
ISBN 978-3-0365-9518-4 (PDF)
doi.org/10.3390/books978-3-0365-9518-4

Cover image courtesy of Scientific and Technological Center of Unique Instrumentation of the Russian Academy of Sciences

Contents

Preface . . . . **vii**

Jinning Li, Yuhua Gui, Rui Xu, Zehong Zhang, Wei Liu and Gang Lv et al.
Applications of AOTF Spectrometers in In Situ Lunar Measurements
Reprinted from: *Materials* **2021**, *14*, 3454, doi:10.3390/ma14133454 . . . . **1**

Alexander Machikhin, Alina Beliaeva, Galina Romanova and Egor Ershov
Color Reproduction by Multi-Wavelength Bragg Diffraction of White Light
Reprinted from: *Materials* **2023**, *16*, 4382, doi:10.3390/ma16124382 . . . . **12**

Vladislav Batshev, Alexander Machikhin, Alexey Gorevoy, Grigoriy Martynov, Demid Khokhlov and Sergey Boritko et al.
Spectral Imaging Experiments with Various Optical Schemes Based on the Same AOTF
Reprinted from: *Materials* **2021**, *14*, 2984, doi:10.3390/ma14112984 . . . . **23**

Hao Zhang, Huijie Zhao, Qi Guo, Dong Xu and Wenjie Teng
Polarization-Multiplexed High-Throughput AOTF-Based Spectral Imaging System
Reprinted from: *Materials* **2023**, *16*, 4243, doi:10.3390/ma16124243 . . . . **32**

Zhongpeng Ji, Zhiping He, Yuhua Gui, Jinning Li, Yongjian Tan and Bing Wu et al.
Research and Application Validation of a Feature Wavelength Selection Method Based on Acousto-Optic Tunable Filter (AOTF) and Automatic Machine Learning (AutoML)
Reprinted from: *Materials* **2022**, *15*, 2826, doi:10.3390/ma15082826 . . . . **47**

Kai Yu and Huijie Zhao
Analysis on the Influence of Incident Light Angle on the Spatial Aberrations of Acousto-Optical Tunable Filter Imaging
Reprinted from: *Materials* **2022**, *15*, 4464, doi:10.3390/ma15134464 . . . . **59**

Hao Zhang, Huijie Zhao, Qi Guo and Yan Xuan
Calibration of Acousto-Optic Interaction Geometry Based on the Analysis of AOTF Angular Performance
Reprinted from: *Materials* **2023**, *16*, 3708, doi:10.3390/ma16103708 . . . . **72**

Kai Yu, Qi Guo, Huijie Zhao and Chi Cheng
The Calibration Methods of Geometric Parameters of Crystal for Mid-Infrared Acousto-Optic Tunable Filter-Based Imaging Systems Design
Reprinted from: *Materials* **2023**, *16*, 2341, doi:10.3390/ma16062341 . . . . **87**

Ildus Sh. Khasanov, Boris A. Knyazev, Sergey A. Lobastov, Alexander V. Anisimov, Pavel A. Nikitin and Oleg E. Kameshkov
Optical Characterization of Thin Films by Surface Plasmon Resonance Spectroscopy Using an Acousto-Optic Tunable Filter
Reprinted from: *Materials* **2023**, *16*, 1820, doi:10.3390/ma16051820 . . . . **102**

Pavel Alekseevich Nikitin, Vasily Valerievich Gerasimov and Ildus Shevketovich Khasanov
Acousto–Optic Modulation and Deflection of Terahertz Radiation
Reprinted from: *Materials* **2022**, *15*, 8836, doi:10.3390/ma15248836 . . . . **114**

Pavel A. Nikitin, Vasily V. Gerasimov and Ildus S. Khasanov
Temperature Effects in an Acousto-Optic Modulator of Terahertz Radiation Based on Liquefied SF_6 Gas
Reprinted from: *Materials* **2021**, *14*, 5519, doi:10.3390/ma14195519 . . . . **125**

Pavel Alekseevich Nikitin and Vasily Valerievich Gerasimov
Optimal Design of an Ultrasound Transducer for Efficient Acousto-Optic Modulation of Terahertz Radiation
Reprinted from: *Materials* **2022**, *15*, 1203, doi:10.3390/ma15031203 . . . . **136**

Vladimir Balakshy, Maxim Kupreychik, Sergey Mantsevich and Vladimir Molchanov
Acousto-Optic Cells with Phased-Array Transducers and Their Application in Systems of Optical Information Processing
Reprinted from: *Materials* **2021**, *14*, 451, doi:10.3390/ma14020451 . . . . **146**

Konstantin B. Yushkov, Natalya F. Naumenko and Vladimir Ya. Molchanov
Analysis of Acousto-Optic Figure of Merit in KGW and KYW Crystals
Reprinted from: *Materials* **2022**, *15*, 8183, doi:10.3390/ma15228183 . . . . **158**

Alexey Osipkov, Mstislav Makeev, Elizaveta Konopleva, Natalia Kudrina, Leonid Gorobinskiy and Pavel Mikhalev et al.
Optically Transparent and Highly Conductive Electrodes for Acousto-Optical Devices
Reprinted from: *Materials* **2021**, *14*, 7178, doi:10.3390/ma14237178 . . . . **170**

Mikhail V. Marunin and Nataliya V. Polikarpova
Polarization of Acoustic Waves in Two-Dimensional Phononic Crystals Based on Fused Silica
Reprinted from: *Materials* **2022**, *15*, 8315, doi:10.3390/ma15238315 . . . . **182**

Sergey A. Titov, Alexander S. Machikhin and Vitold Ed. Pozhar
Evaluation of Acoustic Waves in Acousto-Optical Devices by Ultrasonic Imaging
Reprinted from: *Materials* **2022**, *15*, 1792, doi:10.3390/ma15051792 . . . . **195**

Preface

It is our pleasure to present you this collection of papers.

Acousto-optic technologies have increasingly attracted the attention of scientists and engineers. Acousto-optic interactions have become the flexible basis of compact and robust devices for the analysis of the intensity, spectrum, polarization, and other properties of light. In this book, modern trends of acousto-optical spectral technologies, including the fundamentals and optimization of existing techniques, are highlighted and discussed. This reprint provides a forum for reports on technical developments that allow the spectral analysis of various objects to take place. The goal of this reprint is to provide readers with an overview of the hot topics and the state of the art regarding the applications of acousto-optic techniques, including new materials and structures, theoretical and experimental studies of light diffraction by ultrasonic waves, spectroscopy, and spectral imaging.

Alexander S. Machikhin and Vitold Pozhar
Editors

materials

Review

Applications of AOTF Spectrometers in In Situ Lunar Measurements

Jinning Li [1,2,†], Yuhua Gui [1,2,†], Rui Xu [1,2], Zehong Zhang [3], Wei Liu [3], Gang Lv [1,2], Meizhu Wang [1], Chunlai Li [1,2,*] and Zhiping He [1,2,*]

1 Key Laboratory of Space Active Opto-Electronics Technology, Shanghai Institute of Technical Physics of the Chinese Academy of Sciences, Shanghai 200083, China; lijinning@mail.sitp.ac.cn (J.L.); yhgui@mail.ustc.edu.cn (Y.G.); xurui@mail.sitp.ac.cn (R.X.); lvgang@mail.sitp.ac.cn (G.L.); wangmeizhu2013@163.com (M.W.)
2 University of Chinese Academy of Sciences, Beijing 100049, China
3 No.26 Research Institutes of China Electronics Technology Group Corporation, Chongqing 400060, China; 13637790568@163.com (Z.Z.); liuwei@163.com (W.L.)
* Correspondence: lichunlai@mail.sitp.ac.cn (C.L.); hzping@mail.sitp.ac.cn (Z.H.); Tel.: +021-2505-1072 (C.L.); +021-2505-1697 (Z.H.)
† These authors have contributed equally to this work.

Abstract: Spectrometers based on acousto-optic tunable filters (AOTFs) have several advantages, such as stable temperature adaptability, no moving parts, and wavelength selection through electrical modulation, compared with the traditional grating and Fourier transform spectrometers. Therefore, AOTF spectrometers can realize stable in situ measurement on the lunar surface under wide temperature ranges and low light environments. AOTF imaging spectrometers were first employed for in situ measurement of the lunar surface in the Chinese Chang'e project. The visible and near-infrared imaging spectrometer and the lunar mineralogical spectrometer have been successfully deployed on board the Chang'e-3/4 and Chang'e-5 missions. In this review, we investigate the performance indicators, structural design, selected AOTF performance parameters, data acquisition of the three lunar in situ spectral instruments used in the Chang'e missions. In addition, we also show the scientific achievement of lunar technology based on in situ spectral data.

Keywords: AOTF spectrometer; in situ measurement; lunar surface

check for updates

Citation: Li, J.; Gui, Y.; Xu, R.; Zhang, Z.; Liu, W.; Lv, G.; Wang, M.; Li, C.; He, Z. Applications of AOTF Spectrometers in In Situ Lunar Measurements. *Materials* **2021**, *14*, 3454. https://doi.org/10.3390/ma14133454

Academic Editors: Alexander S. Machikhin and Vitold Pozhar

Received: 25 April 2021
Accepted: 14 June 2021
Published: 22 June 2021

1. Introduction

Compared to remote sensing spectral measurement and post-sampling laboratory spectral measurement, in situ spectral measurement allows one to investigate targets at close range with millimeter resolution, without destroying the original state of the target. This leads to intriguing possibilities for studying surface topography and material composition [1]. However, in situ measurement is challenging for spectroscopic instruments. Especially, deep space exploration requires instruments to obtain high-quality and reliable spectral data under extremely complex environmental conditions [2].

The acousto-optic tunable filter (AOTF) is an electronically tunable dispersive optical device without any moving parts. It can change the wavelength of the output diffracted light by controlling the input radio frequency (RF) [3]. Harris and Wallace first proposed a collinear AOTF in 1969 [4]. In 1974, Chang et al. used TeO_2 to construct a noncollinear AOTF [5], which overcame the drawbacks of the collinear AOTF in terms of limited crystal availability and complicated design. In 1987, an AOTF spectrometer was used for ocean observations in the Soviet satellite "Ocean-O1-N2" [6]. In 2003, the SPICAM instrument onboard the European Space Agency's Mars Express mission realized the first deep space application of AOTF for Martian atmospheric analysis [7]. Since then, several AOTF instruments have been used in deep-space missions [8–12].

Spectrometers and imaging spectrometers using acousto-optic tunable filters have many advantages, such as small size, light weight, absence of moving parts, optional wavelength, high environmental adaptability and high spectral and spatial resolution. [13–15] Therefore, it is suitable for the in situ spectral and imaging measurement of extraterrestrial bodies. The Chinese Chang'e Project has implemented three AOTF spectrometers for in situ spectral measurement of the lunar surface: the visible and near-infrared imaging spectrometers (VNIS) onboard Chang'e-3 (2013) [16] and Chang'e-4 (2018) [17] unmanned lunar rovers and the lunar mineralogical spectrometer (LMS) onboard Chang'E-5 (2020) [18]. VNIS onboard Chang'e-3 was the first instrument to realize in situ imaging spectral measurement of the lunar surface. The MicrOmega onboard the Japanese Hayabusa-2 (2014) mission realized the in situ spectral measurement of asteroids [19,20]. The SuperCam mounted onboard NASA's MARS 2020 Perseverance rover is also equipped with an infrared in situ spectra submodule to realize in situ spectra of the Martian surface [21–23]. In addition, the ExoMars scheduled for launch in 2022 will also be equipped with an in situ spectral instrument, ISEM [24,25], to perform in situ spectral surveys of the Martian surface.

In this review, we report the design, performance specifications, and performance test results of the three AOTF spectrometers used for lunar in situ measurement in the Chang'e mission. We also report the material, tuning relationship and resolution of AOTFs used on these three AOTF in situ spectrometers. In addition, the effectiveness and results of the AOTF spectrometers are briefly described.

2. Lunar In Situ Spectrometers

The Yutu lunar rover in the Chang'e-3 mission carried the VNIS, which was the first AOTF hyperspectral imager to realize in situ measurement in deep space [16,26–29]. The main objectives of the mission are to perform visible and near-infrared spectral imaging (400–900 nm) and short-wave infrared spectral measurements (900–2400 nm) of the lunar surface targets. The VNIS can obtain spectral and geometric imaging data of objects on the lunar surface. The mission also focused on accomplishing in situ analyses of the mineral composition, content (abundance), and chemical composition of probe sites in the patrol area [30]. In 2018, Chang'e-4 achieved the first soft landing and roving survey on the far side of the Moon [31], and the Yutu-2 unmanned lunar rover also carried the VNIS instrument. Compared to the VNIS of the Chang'e 3 mission, the VNIS of the Chang'e 4 mission has optimized software and data acquisition logic with the addition of options for background light acquisition and subtraction [32]. The spectral resolution and the signal-to-noise ratio (SNR) were also improved. A specific performance comparison of the instruments is listed in Table 1.

Table 1. Main performance parameters of VNIS and LMS.

Parameters	VNIS/Chang'e-3		VNIS/Chang'e-4		LMS/Chang'e-5	
	VIS-NIR	SWIR	VIS-NIR	SWIR	VIS-NIR	IR
Spectral coverage/nm	449–950	900–2400	450–950	900–2400	480–1450	1400–3200
Spectral resolution/nm	2–7	3–12	2.4–6.5	3.6–9.5	2.4–9.4	7.6–24.9
FOV/deg	8.5 × 8.5	ø3.6	8.5 × 8.5	ø3.6	4.17 × 4.17	4.17 × 4.17
Effective pixels	256 × 256	1	256 × 256	1	256 × 256	1
Quantization/bits	10	16	10	16	10	16
SNR/dB	≥31@ albedo is 9% and solar incident angle is 45°	≥32@ albedo is 9% and solar incident angle is 75°	≥33@ albedo is 9% and solar incident angle is 45°	≥31@ albedo is 9% and solar incident angle is 75°	≥34@ albedo is 9% and solar incident angle is 45°	≥39@ albedo is 9% and solar incident angle is 45°
Sampling interval/nm	5		5		5	
Power/w	19.8		16.95		15.17	
Weight/kg	4.675/probe ~0.7/electronics		4.675/probe ~0.7/electronics		≤5.57	
Operating temperture	−20 °C~+55 °C		−20 °C~+55 °C		−25 °C~+65 °C	

In 2020, Chang'e-5 (China Lunar Exploration Program Phase III) realized China's first lunar sampling return mission [18]. LMS was one of the key scientific payloads for Chang'e-5. LMS is inherited from the previous generation of VNIS [27]. Compared with VNIS, the spectral range of LMS was extended to 3200 nm and the SNR was also improved, as shown in Table 1. Benefitting from the improvement of spectral range and SNR, LMS can not only analyze the mineral composition, but also explore OH/H_2O of lunar regolith.

2.1. VNIS on Board Chang'e-3 and Chang'e-4

VNIS is comprised of a measurement head and an electric cabinet, as shown in Figure 1, where the probe is installed outside the patrol cabin and the electric cabinet is installed inside the patrol cabin. The probe consists of a calibration diffuse reflector plate, an optical unit, and a probe electronics unit. The probe electronics unit includes the radio frequency (RF) driver unit, signal acquisition unit, and the main control circuit, comprised of several parts, such as the RF power amplifier, spectral measurement control, secondary power supply, and the load electric cabinet common unit.

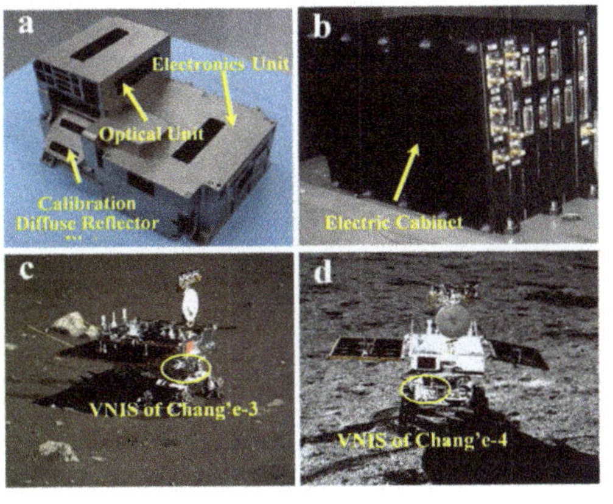

Figure 1. The probe composition of VNIS (**a**) and its electric cabinet (**b**); installation location of VNIS on the rover for the Chang'e-3 (**c**) and Chang'e-4 (**d**) missions.

VNIS contains two measurement channels, the visible and near-infrared (VIS-NIR 480–950 nm) and the shortwave infrared (SWIR 900–2400 nm). VIS-NIR is the imaging channel, and SWIR is the spectral acquisition channel. The VNIS on board Chang'e-3 has a spectral resolution of 2–7 nm in the VIS-NIR band and 3–12 nm in the SWIR band [29]. The VNIS on board Chang'e-4 has an improved spectral resolution of 2.4–6.5 nm in the VIS-NIR band and 3.6–9.5 nm in the SWIR band. The two channels have a field of view (FOV) of $8.5 \times 8.5°$ and ø3.6, the measurement distance is 0.7 to 1.3 m, and the instrument can observe targets within 0.2 m^2.

2.2. LMS on Board Chang'e-5

LMS is comprised of a mounting base, optical unit, electronics unit, 2D pointing mechanism, and the dust-proofing and calibration units fitted inside the Chang'e-5 lander's −Y + Z inclined side plate. The inclined side plate is perforated with a dust-proof device and a 2D pointing mechanism protruding from the perforations. The AOTF is encapsulated in an optical unit, which is isolated from the electronics unit for thermal control. The principle of operation is similar to that of VNIS, with the addition of a 2D pointing mechanism for selective measurement of the sampling area. The calibration plate was embedded in a dust-proof unit to realize in-flight calibration. An image and the installation location of LMS are shown in Figure 2.

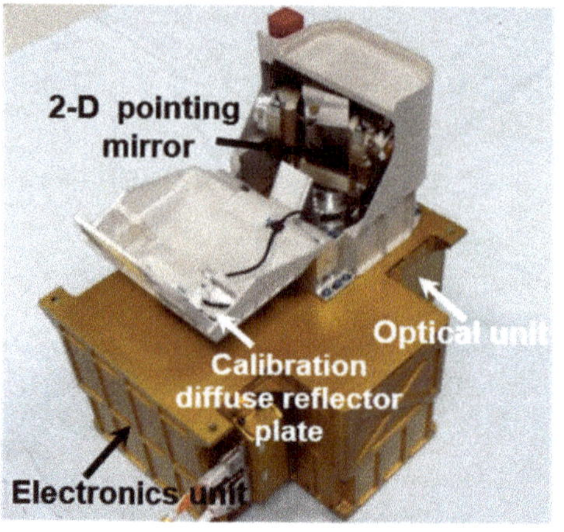

Figure 2. The composition of LMS.

LMS contains four measurement channels: the visible (VIS 480–950 nm), the near-infrared (NIR 900–1450 nm), the short-wave infrared (SWIR 1400–2300 nm), and the medium-wave infrared (MWIR 2200–3200 nm) [18]. Except for the VIS channel, which is imaged by complementary metal oxide semiconductor (CMOS), the other three channels (collectively referred to as IR) are the spectral acquisition channels. The infrared unit detectors were developed by the Shanghai Institute of Technical Physics, including InGaAs detectors for the NIR and SWIR channels, and the HgCdTe detectors for the MWIR channels. Both channels have a FOV of $4.17 \times 4.17°$, and can observe over a distance of 1.65 m. The VIS channel has a spectral resolution of 2.4–9.4 nm with an SNR of >34 dB, and the IR channel has a spectral resolution of 7.6–24.9 nm with an SNR of >39 dB. The pointing resolution of the 2D pointing mechanism is greater than 0.2°.

3. Main Characteristics of the AOTFs in VNIS and LMS

The performance of the AOTF significantly affects the quality of the spectrometer's spectral characteristics. Therefore, testing the performance indicators of AOTF and investigating its ability to adapt to complex conditions on the lunar surface is the key to ensure that spectrometers obtain high-quality data. Based on the spectroscopic principle of the AOTF device, the testing scheme shown in Figure 3 was designed to test the performance indicators of six AOTFs corresponding to the VNIS on Chang'e-3 and Chang'e-4 and the LMS on Chang'e-5.

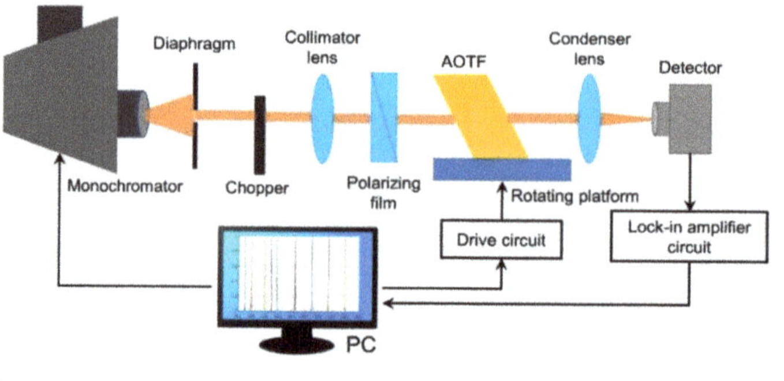

Figure 3. The AOTF testing system.

The monochromator produces narrow line-width light, which passes through a chopper to form an optical flux of variable intensity. The chopper also provides a reference signal to the lock-in amplifier circuit. A polarizing film is used in the optical path as a polarizing device to set the linear polarization direction of the light source incident on the crystal surface. By adjusting the size of the diaphragm aperture, collimated monochromatic light fills the effective aperture of the AOTF crystal to be measured. The crystal was placed on a rotating platform, and the RF drive signal was provided by a computer-controlled AOTF drive source. After the optical signal passes through the driven AOTF crystal, it is separated into the undiffracted 0-order light in the alternating state and the diffracted 1-order light in the corresponding polarization state. Both 0-order light and 1-order light can be used for measurement. This is because of the 0-order light measurement is easy to achieve optical alignment and improve measurement efficiency. The system measures the 0-order light. After passing through a condenser lens, the undiffracted 0-order light can be focused on the focal plane, and a photodetector is used to receive the undiffracted 0-order light. The detector converts the transformed variable-intensity optical flux into an alternating current (AC), which is amplified by the preamplifier, and then input to the lock-in amplifier. The amplified electrical signal is displayed on a computer. The main indicators include the tuning relationship between the diffraction wavelength and drive frequency, the relationship between the diffraction wavelength and diffraction efficiency, and the relationship between the diffraction wavelength and spectral resolution, as shown in Table 2.

Table 2. Main performance parameters of VNIS and LMS.

Parameters	VNIS/Chang'e-3		VNIS/Chang'e-4		LMS/Chang'e-5	
	VIS-NIR	**SWIR**	**VIS-NIR**	**SWIR**	**VIS-NIR**	**IR**
Material	TeO_2	TeO_2	TeO_2	TeO_2	TeO_2	TeO_2
Spectral coverage/nm	449–950	899–2402	450–950	900–2400	480–1450	1400–3200
FWHM/nm	2.3–6.3 @ < 630 nm 2.4–6.8 @ > 630 nm	3.1–8.9 @ < 1380 nm 4.4–11.6 @ > 1380 nm	2.0–5.8 @ < 630 nm 2.8–6.4 @ > 630 nm	3.75–8.4 @ < 1380 nm 4.2–9.6 @ > 1380 nm	2.6–9.4 @ < 780 nm 2.4–9.0 @ > 780 nm	7.6–20.8 @ 1400–2300 nm 11.6–24.9 @ 2200–3200 nm
RF/MHz	70.7–178.6	41.9–118.9	71.2–178.7	42.0–118.8	45.2–163.6	27.7–66.2
Angular aperture/°	>7	>8	>7	>8	>7	>3
Diffraction angle/°	>5.6	>7.5	>5.6	>7.5	>5.6	>7
Power/W	~2		~2		~2	

3.1. AOTFs on Chang'e-3/4 and Their Indicators

VNIS spectrometers onboard Chang'e-3 and Chang'e-4 have two spectral measurement channels: VIS-NIR and SWIR. Each channel uses one AOTF crystal, and the two instruments use a total of four AOTF crystals. Four AOTFs were made of TeO_2 material, and the performance indicators of four AOTFs were shown in Table 2, which were tested using the above testing system. Figure 4 shows AOTFs tuning relationship between the wavelength and the driving frequency and the relationship with the spectral resolution which used in VNIS spectrometers of Chang'e-3 and Chang'e-4.

AOTFs indicators used by the Chang'e-3 VNIS and the Chang'e-4 VNIS are similar, so here, we mainly discuss the AOTF used by the Chang'e-3 VNIS. The AOTF used for the VIS-NIR channel can achieve monochromatic light output from 449 to 950 nm with RF driving frequency varies from 70.7 to 178.6 MHz. The spectral resolution of the output monochromatic light varies with wavelength, and there is a boundary at 630 nm. Before 630 nm, the spectrometer resolution increased from 2.3 to 6.3 nm, and after 630 nm, the spectral resolution increased from 2.4 to 6.8 nm. The reason that the resolution is demarcated at 630 nm is due to the limited bandwidth of a single piezoelectric transducer. In order to improve the working bandwidth of AOTF, two different transducers of high frequency and low frequency are made to the switching of the two piezoelectric transducers, as shown in Figure 5. Due to the switching of the two piezoelectric transducers, the spectral resolution has a boundary at 630 nm.

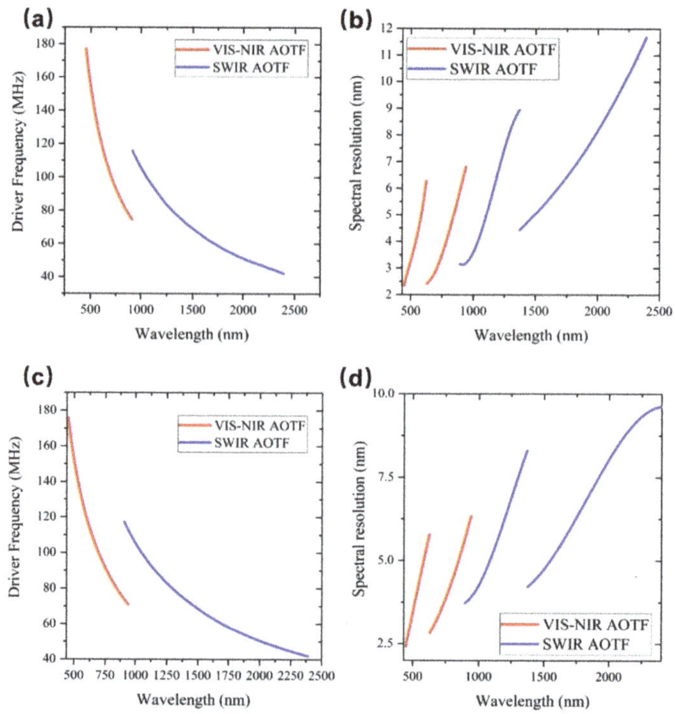

Figure 4. Driving frequency tuning curves of the acousto-optic tunable filters of the visible and near infrared imaging spectrometers onboard the Chang'e-3 (**a**) and Chang'e-4 (**c**); spectral resolution of the acousto-optic tunable filters vs wavelength for the Chang'e-3 (**b**) and Chang'e-4 (**d**) VNIS spectrometers.

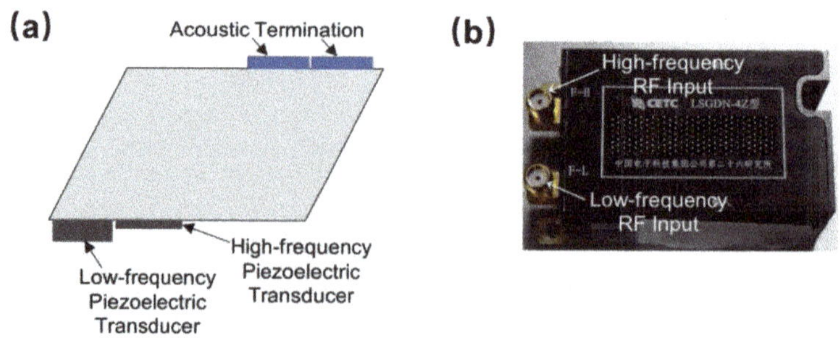

Figure 5. AOTF used on VNIS. (**a**) Main structure of the AOTF; (**b**) The physical object of AOTF.

For the AOTF used in the SWIR channel, the RF driving frequency changes from 41.9 to 118.9 MHz, and the spectral range varies from 899 to 2402 nm. The spectral resolution has a boundary at 1380 nm. When the spectral range is less than 1380 nm, the spectral resolution will vary from 3.1 to 8.9 nm. When the spectral range is greater than 1380 nm, the spectral resolution will vary from 4.4 to 11.6 nm.

3.2. AOTFs on Chang'e-5 and Their Indicators

Compared to the VNIS spectrometer, the LMS onboard Chang'e-5 has an extended measurement band of 3200 nm, with two channels in the VIS-NIR and SWIR-MWIR, and each channel uses one AOTF crystal. The wavelength-frequency tuning curves and spectral resolution distributions of the two AOTFs are shown in Figure 6. The AOTF used in the VIS-NIR channel can be driven by a RF of 45.2 to 163.6 MHz to obtain monochromatic light from 480 to 1450 nm. The spectral resolution is at 780 nm as the boundary, which ranges from 2.6 to 9.4 nm before 780 nm and from 2.4 to 9.0 nm after 780 nm. The RF of the SWIR-MWIR channel is 27.7 to 66.2 MHz, and the spectral range is from 1400 to 3200 nm. The spectral resolution ranges from 7.6 to 20.8 nm before 1400 nm and from 11.6 to 24.9 nm after 1400 nm.

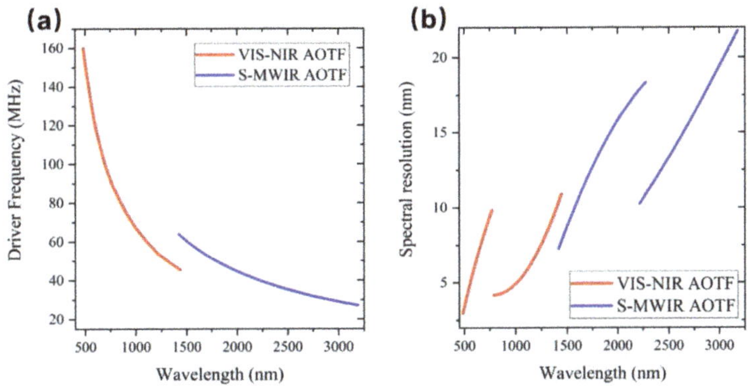

Figure 6. (**a**) The wavelength-frequency tuning curve of the AOTFs in the Chang'e-5 LMS; (**b**) Spectral resolution of the acousto-optic tunable filters vs. wavelength for the Chang'e-5 LMS.

4. Application Results of Lunar In situ Spectrometers

The VNIS onboard Chang'e-3 and 4 and the LMS onboard Chang'e-5 have successfully implemented in situ measurement applications on the lunar surface, and obtained in situ imaging and spectral data with millimeter-level resolution. The scientific data obtained present a unique perspective for research on the mineral composition of the lunar surface, weathering in space, and the origin and evolution of the moon. In this section, we focus on the data obtained by the three in situ spectrometers and the corresponding scientific research reports.

4.1. Application Results of VNIS on Board Chang'e-3

Chang'e-3 successfully landed on the northeastern Mare Imbrium on 14 December 2013. The VNIS was booted up for the first time on 23 December 2013, and images and spectral data of the lunar soil in different regions were collected [16]. The images and spectral data acquired by the VNIS on Chang'e-3 can be used to investigate the mineral composition and chemical content of the lunar surface. Various scientific results have been published by analyzing and processing these data [33–40]. These results not only confirm the results of previous remote sensing studies, but also conclusively establish that the mineral composition of olivine tends towards iron-rich mineral end members.

Lin et al. used the modified Gaussian model (MGM) method to extract the mineralogical modal composition of lunar soil from VNIS spectral data [34]. Investigations on the mineral and chemical composition have revealed that the lunar soil was flooded with lava from volcanic eruptions approximately 2.5 billion years ago.

4.2. Application Results of VNIS on Board Chang'e-4

Chang'e-4 landed in the Von Kármán crater on the far side of the Moon on 3 January 2019, and commenced its scientific patrol and exploration mission. On the next day, the VNIS onboard Chang'e-4 booted up and obtained high-quality spectral data successfully from multiple test points in the landing area. The Chang'e-4 and Yutu-2 lunar rovers ended the lunar night hibernation at 21:43 and 3:54 on 6 April 2021, and entered the 29th lunar day. Until 6 April, the Yutu-2 lunar rover traveled a cumulative distance of approximately 682.8 m.

Based on the unique in situ spectral data obtained by VNIS, we have gained more information about the origin and evolution of the moon and space weathering [1,31,41–53]. Li et al. further analyzed the REFF spectra collected by VNIS on the first lunar day. They demonstrated that a deep medium dominated by olivine and low-calcite pyroxene exists in the south pole-Aitken basin on the far side of the moon by analyzing the VNIS spectra [41]. This discovery provides direct evidence for explaining the composition of the lunar mantle, which has been unclear until now. These results will also improve the models on the formation and evolution of the moon. Lin et al. reported that the spectra of the observed area showed maximum absorption at 1 μm and 2 μm, which can more reliably determine the mineral composition. The analysis shows that the rock in the region is olivine–norite [42].

4.3. Application Effects of LMS on Board Chang'e-5

The main scientific mission of the Chang'e-5 mission is to perform lunar sampling return. Therefore, there is no rover similar to the Chang'e-3 and Chang'e-4 missions. The LMS is installed on the Chang'e-5 lander to detect the spectra of the sampling area. The detected spectral data can contribute to the survey of the lunar surface material composition and resources in the sampling area, and to provide scientific data for comparative research on sample laboratory measurements.

The LMS conducted lunar observations from 21:08 1 December 2020 (UTC) until 17:04 2 December 2020 (UTC), after the Chang'e-5 lander-ascender combination successful soft landing. First, a multi-spectral full-view mode observation was conducted to obtain a wide view of the lunar sampling area, and a six-band panoramic spectral image was acquired in the VIS channel and 14-band spectra in the IR channels (i.e., NIR, SWIR, and MWIR). The REFF image of the sampling area is shown in Figure 7. The spatial resolution of the image was 0.4–1 mm.

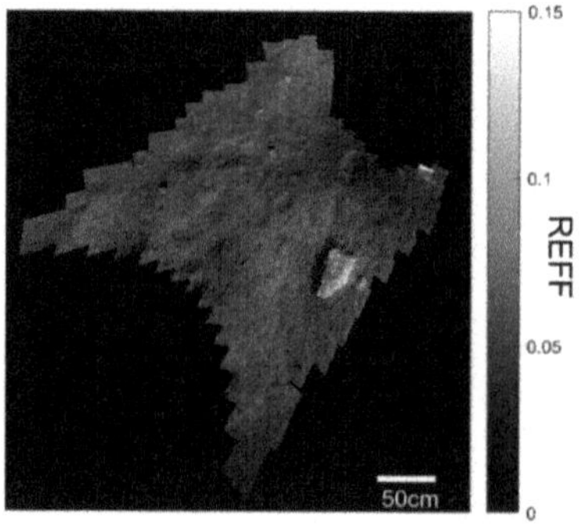

Figure 7. REFF image from the LMS at 900 nm using the full-view mode.

During the intervals of sampling transfer by the robotic arm, the LMS conducted full-band observations of the significant targets, sampling targets, and candidate points, and performed in-flight calibration.

5. Conclusions

In situ spectral analysis can investigate the surface material composition and weathering conditions of extraterrestrial objects without destroying the original environment. It also provides a unique perspective of data for the study of extraterrestrial objects. Payloads in the form of AOTF spectrometers have the advantages of small volume, low mass, and high spatial adaptability, which are very suitable for in situ spectral detection at close range on the lunar surface. Three AOTF spectrometers have been successfully applied to lunar in situ spectral measurement, including the VNIS on board the Chinese Chang'e-3/4 missions and the LMS on board the Chang'e-5 mission. These instruments have overcome harsh and low light environment and wide temperature variations on the lunar surface, and have successfully achieved spectral acquisition and in situ calibration of lunar objects. These results provide a unique perspective on scientific data for analyzing the mineral composition and evolutionary history of the Moon. This paper discusses three in situ spectrometers successfully applied to the lunar surface and their functionalities, performance parameters of the installed AOTF crystals, data acquisition, and scientific applications of the data. The results can provide reference for the design and application of in situ AOTF spectroscopic instruments applied to extraterrestrial bodies in the future.

Author Contributions: Conceptualization, J.L. and Z.H.; methodology, J.L.; software, Y.G.; validation, R.X., Z.H. and C.L.; formal analysis, J.L.; investigation, Y.G.; resources, R.X., G.L., Z.Z. and W.L.; data curation, J.L., G.L., Z.Z. and W.L.; writing—original draft preparation, J.L., Y.G. and M.W.; writing—review and editing, J.L. and M.W.; visualization, J.L.; supervision, Z.H. and C.L.; project administration, C.L. and Z.H.; funding acquisition, C.L. and Z.H. All authors have read and agreed to the published version of the manuscript.

Funding: This work was supported by the China Lunar Exploration Program (CLEP) and China National Space Administration (CNSA), and was also funded by the National Natural Science Foundation of China (NSFC) (Grant No. 21105109, No. 61605231, and No. 11941002), the Youth Innovation Promotion Association CAS (No. 2016018, and No. 2017286), the Program of Shanghai Aca-demic/Technology Research Leader (No. 19XD1424100), and the Shanghai Outstanding Academic Leaders Plan.

Institutional Review Board Statement: Not applicable.

Informed Consent Statement: Not applicable.

Data Availability Statement: Not applicable.

Conflicts of Interest: The authors declare no conflict of interest.

References

1. Gou, S.; Yue, Z.; Di, K.; Wan, W.; Liu, Z.; Liu, B.; Peng, M.; Wang, Y.; He, Z.; Xu, R. In situ spectral measurements of space weathering by Chang'e-4 rover. *Earth Planet. Sci. Lett.* **2020**, *535*, 116117. [CrossRef]
2. Hu, X.Y.; Ma, P.; Yang, Y.Z.; Zhu, M.H.; Jiang, T.; Lucey, P.G.; Sun, L.Z.; Zhang, H.; Li, C.L.; Xu, R.; et al. Mineral abundances inferred from In Situ reflectance measurements of Chang'E-4 landing site in South Pole-Aitken Basin. *Geophys. Res. Lett.* **2019**, *46*, 9439–9447. [CrossRef]
3. Chang, I.C. Acousto-Optic tunable filters. *Opt. Eng.* **1981**, *20*, 824–829. [CrossRef]
4. Harris, S.E.; Wallace, R.W. Acousto-Optic Tunable Filter. *J. Opt. Soc. Am.* **1969**, *59*, 744–747. [CrossRef]
5. Chang, I.C. Noncollinear acoustooptic filter with large angular aperture. *Appl. Phys. Lett.* **1974**, *25*, 370–372. [CrossRef]
6. Pustovoit, V.I.; Pozhar, V.E. Acousto-optic spectrometers for Earth remote sensing. *Earth Obs. Syst.* **1999**, *3750*, 243–249.
7. Korablev, O.; Bertaux, J.-L.; Fedorova, A.; Fonteyn, D.; Stepanov, A.; Kalinnikov, Y.; Kiselev, A.; Grigoriev, A.; Jegoulev, V.; Perrier, S.; et al. SPICAM IR acousto-optic spectrometer experiment on Mars Express. *J. Geophys. Res.* **2006**, *111*. [CrossRef]
8. Korablev, O.I.; Belyaev, D.A.; Dobrolenskiy, Y.S.; Trokhimovskiy, A.Y.; Kalinnikov, Y.K. Acousto-optic tunable filter spectrometers in space missions. *Appl. Opt.* **2018**, *57*, C103–C119. [CrossRef] [PubMed]
9. Korablev, O.; Fedorova, A.; Bertaux, J.-L.; Stepanov, A.V.; Kiselev, A.; Kalinnikov, Y.K.; Titov, A.Y.; Montmessin, F.; Dubois, J.P.; Villard, E.; et al. SPICAV IR acousto-optic spectrometer experiment on Venus Express. *Planet. Space Sci.* **2012**, *65*, 38–57. [CrossRef]
10. Nevejans, D.; Neefs, E.; Van Ransbeeck, E.; Berkenbosch, S.; Clairquin, R.; De Vos, L.; Moelans, W.; Glorieux, S.; Baeke, A.; Korablev, O.; et al. Compact high-resolution spaceborne echelle grating spectrometer with acousto-optical tunable filter based order sorting for the infrared domain from 2.2 to 4.3 mu m. *Appl. Opt.* **2006**, *45*, 5191–5206. [CrossRef] [PubMed]
11. Korablev, O.I.; Trokhimovskii, A.Y.; Vinogradov, I.I.; Fedorova, A.A.; Ivanov, A.Y.; Venkstern, A.A.; Barke, V.V.; Roste, O.Z.; Kalinnikov, Y.K.; Titov, A.Y.; et al. The RUSALKA device for measuring the carbon dioxide and methane concentration in the atmosphere from on board the International Space Station. *J. Opt. Technol.* **2011**, *78*, 317–327. [CrossRef]
12. Korablev, O.; Ivanov, A.; Fedorova, A.; Kalinnikov, Y.K.; Shapkin, A.; Mantsevich, S.; Viazovetsky, N.; Evdokimova, N.; Kiselev, A.V. Development of a mast or robotic arm-mounted infrared AOTF spectrometer for surface Moon and Mars probes. *Infrared Remote Sens. Instrum.* **2015**, *9608*, 960807.

13. Batshev, V.; Machikhin, A.; Martynov, G.; Pozhar, V.; Boritko, S.; Sharikova, M.; Lomonov, V.; Vinogradov, A. Polarizer-Free AOTF-Based SWIR hyperspectral imaging for biomedical applications. *Sensors* **2020**, *20*, 4439. [CrossRef]
14. Machikhin, A.; Batshev, V.; Pozhar, V. Aberration analysis of AOTF-based spectral imaging systems. *J. Opt. Soc. Am. A Opt. Image Sci. Vis.* **2017**, *34*, 1109–1113. [CrossRef]
15. Pozhar, V.; Machihin, A. Image aberrations caused by light diffraction via ultrasonic waves in uniaxial crystals. *Appl. Opt.* **2012**, *51*, 4513–4519. [CrossRef]
16. He, Z.P.; Wang, B.Y.; Lv, G.; Li, C.L.; Yuan, L.Y.; Xu, R.; Chen, K.; Wang, J.Y. Visible and near-infrared imaging spectrometer and its preliminary results from the Chang'E 3 project. *Rev. Sci. Instrum.* **2014**, *85*, 083104. [CrossRef]
17. Li, C.L.; Xu, R.; Lv, G.; Yuan, L.Y.; He, Z.P.; Wang, J.Y. Detection and calibration characteristics of the visible and near-infrared imaging spectrometer in the Chang'e-4. *Rev. Sci. Instrum.* **2019**, *90*, 103106. [CrossRef]
18. Xu, R.; Li, C.; Yuan, L.; Lv, G.; Xu, S.; Li, F.; Jin, J.; Wang, Z.; Pan, W.; Wang, R.; et al. Lunar Mineralogical Spectrometer on Chang'E-5 Mission. *Space Sci. Rev.*. (under review).
19. Bibring, J.P.; Hamm, V.; Langevin, Y.; Pilorget, C.; Arondel, A.; Bouzit, M.; Chaigneau, M.; Crane, B.; Darie, A.; Evesque, C.; et al. The MicrOmega investigation onboard Hayabusa2. *Space Sci. Rev.* **2017**, *208*, 401–412. [CrossRef]
20. Pilorget, C. *NIR Hyperspectral Microscopy for In Situ Analyses: The MicrOmega Experiment Onboard PHOBOS Grunt, Hayabusa-2 and ExoMars Missions*; Université Paris Sud—Paris XI: Orsay, France, 2012.
21. Sodnik, Z.; Cugny, B.; Karafolas, N.; Maurice, S.; Cais, P.; Montmessin, F.; Lapauw, L.; Bernardi, P.; Réess, J.M.; Wiens, R.C.; et al. The supercam infrared instrument on the NASA Mars2020 mission: Optical design and performance. In Proceedings of the International Conference on Space Optics—ICSO 2016, Biarritz, France, 18–21 October 2016. [CrossRef]
22. Reess, J.M.; Bonafous, M.; Lapauw, L.; Humeau, O.; Fouchet, T.; Bernardi, P.; Cais, P.; Deleuze, M.; Forni, O.; Maurice, S.; et al. The SuperCam infrared instrument on the NASA MARS2020 mission: Performance and qualification results. In Proceedings of the International Conference on Space Optics—ICSO 2018, Chania, Greece, 9–12 October 2018. [CrossRef]
23. Royer, C.; Poulet, F.; Reess, J.M.; Pilorget, C.; Hamm, V.; Fouchet, T.; Maurice, S.; Forni, O.; Bernardi, P.; Montmessin, F.; et al. Pre-launch radiometric calibration of the infrared spectrometer onboard SuperCam for the Mars2020 rover. *Rev. Sci. Instrum* **2020**, *91*, 063105. [CrossRef] [PubMed]
24. Dobrolenskiy, Y.; Mantsevich, S.; Evdokimova, N.; Korablev, O.; Fedorova, A.; Kalinnikov, Y.; Vyazovetskiy, N.; Titov, A.; Stepanov, A.; Sapgir, A.; et al. Acousto-Optic Spectrometer ISEM for ExoMars-2020 space mission: Ground measurements and calibrations. In *Fourteenth School on Acousto-Optics and Applications*; International Society for Optics and Photonics: Bellingham, WA, USA, 2019; Volume 11210, p. 112100F. [CrossRef]
25. Korablev, O.I.; Dobrolensky, Y.; Evdokimova, N.; Fedorova, A.A.; Kuzmin, R.O.; Mantsevich, S.N.; Cloutis, E.A.; Carter, J.; Poulet, F.; Flahaut, J.; et al. Infrared spectrometer for ExoMars: A mast-mounted instrument for the Rover. *Astrobiology* **2017**, *17*, 542–564. [CrossRef]
26. He, Z.P.; Li, J.N.; Li, C.L.; Xu, R. Measurement and correction model for temperature dependence of an Acousto-Optic Tunable Filter (AOTF) Infrared spectrometer for lunar surface detection. *Appl. Spectrosc.* **2020**, *74*, 81–87. [CrossRef] [PubMed]
27. He, Z.P.; Li, C.L.; Xu, R.; Lv, G.; Yuan, L.Y.; Wang, J.Y. Spectrometers based on acousto-optic tunable filters for in-situ lunar surface measurement. *J. Appl. Remote Sens.* **2019**, *13*. [CrossRef]
28. He, Z.P.; Xu, R.; Li, C.L.; Lv, G.; Yuan, L.Y.; Wang, B.Y.; Shu, R.; Wang, J.Y. Visible and Near-Infrared Imaging Spectrometer (VNIS) for In Situ lunar surface measurements. Spie-Int Soc Optical Engineering: Bellingham. *Sens. Syst. Next Gener. Satell.* **2015**, *9639*, 96391S.
29. He, Z.P.; Wang, B.Y.; Lu, G.; Li, C.L.; Yuan, L.Y.; Xu, R.; Liu, B.; Chen, K.; Wang, J.Y. Operating principles and detection characteristics of the Visible and Near-Infrared Imaging Spectrometer in the Chang'e-3. *Res. Astron. Astrophys.* **2014**, *14*, 1567–1577. [CrossRef]
30. Li, C.; Liu, J.; Ren, X.; Zuo, W.; Tan, X.; Wen, W.; Li, H.; Mu, L.; Su, Y.; Zhang, H.; et al. The Chang'e 3 mission overview. *Space Sci. Rev.* **2015**, *190*, 85–101. [CrossRef]
31. Li, C.; Zuo, W.; Wen, W.; Zeng, X.; Gao, X.; Liu, Y.; Fu, Q.; Zhang, Z.; Su, Y.; Ren, X.; et al. Overview of the Chang'e-4 mission: Opening the frontier of scientific exploration of the lunar far side. *Space Sci. Rev.* **2021**, *217*. [CrossRef]
32. Li, C.L.; Wang, Z.D.; Xu, R.; Lv, G.; Yuan, L.Y.; He, Z.P.; Wang, J.Y. The scientific information model of Chang'e-4 visible and Near-IR Imaging Spectrometer (VNIS) and In-Flight Verification. *Sensors* **2019**, *19*, 2806. [CrossRef]
33. Ling, Z.; Jolliff, B.L.; Wang, A.; Li, C.; Liu, J.; Zhang, J.; Li, B.; Sun, L.; Chen, J.; Xiao, L.; et al. Correlated compositional and mineralogical investigations at the Chang'e-3 landing site. *Nat. Commun.* **2015**, *6*, 8880. [CrossRef]
34. Zhang, J.; Yang, W.; Hu, S.; Lin, Y.; Fang, G.; Li, C.; Peng, W.; Zhu, S.; He, Z.; Zhou, B.; et al. Volcanic history of the Imbrium basin: A close-up view from the lunar rover Yutu. *Proc. Natl. Acad. Sci. USA* **2015**, *112*, 5342–5347. [CrossRef]
35. Wu, Y.; Wang, Z.; Cai, W.; Lu, Y. The absolute reflectance and new calibration site of the moon. *Astron. J.* **2018**, *155*, 213. [CrossRef]
36. Wu, Y.; Hapke, B. Spectroscopic observations of the moon at the lunar surface. *Earth Planet. Sci. Lett.* **2018**, *484*, 145–153. [CrossRef]
37. Wang, Z.C.; Wu, Y.Z.; Blewett, D.T.; Cloutis, E.A.; Zheng, Y.C.; Chen, J. Submicroscopic metallic iron in lunar soils estimated from the In Situ spectra of the Chang'E-3 mission. *Geophys. Res. Lett.* **2017**, *44*, 3485–3492. [CrossRef]
38. Lin, H.; Zhang, X.; Yang, Y.; Wu, X.; Guo, D. Mineral abundance and particle size distribution derived from In-Situ spectra measurements of yutu rover of Chang'e-3. *ISPRS* **2017**, *XLII-3/W1*, 85–89. [CrossRef]

39. Zhang, H.; Yang, Y.Z.; Yuan, Y.; Jin, W.D.; Lucey, P.G.; Zhu, M.H.; Kaydash, V.G.; Shkuratov, Y.G.; Di, K.C.; Wan, W.H.; et al. In Situ optical measurements of Chang'E-3 landing site in Mare Imbrium: 1. Mineral abundances inferred from spectral reflectance. *Geophys. Res. Lett.* **2015**, *42*, 6945–6950. [CrossRef]
40. Liu, B.; Li, C.L.; Zhang, G.L.; Xu, R.; Liu, J.J.; Ren, X.; Tan, X.; Zhang, X.X.; Zuo, W.; Wen, W.B. Data processing and preliminary results of the Chang'e-3 VIS/NIR Imaging Spectrometer In-Situ analysis. *Res. Astron. Astrophys.* **2014**, *14*, 1578–1594. [CrossRef]
41. Li, C.; Liu, D.; Liu, B.; Ren, X.; Liu, J.; He, Z.; Zuo, W.; Zeng, X.; Xu, R.; Tan, X.; et al. Chang'E-4 initial spectroscopic identification of lunar far-side mantle-derived materials. *Nature* **2019**, *569*, 378–382. [CrossRef]
42. Lin, H.L.; He, Z.P.; Yang, W.; Lin, Y.T.; Xu, R.; Zhang, C.; Zhu, M.H.; Chang, R.; Zhang, J.H.; Li, C.L.; et al. Olivine-norite rock detected by the lunar rover Yutu-2 likely crystallized from the SPA-impact melt pool. *Natl. Sci. Rev.* **2020**, *7*, 913–920. [CrossRef]
43. Wu, Y.; Kührt, E.; Grott, M.; Jin, Q.; Xu, T.; Helbert, J.; Hamm, M.; Qin, N.; Ma, J.; Vincent, J.B.; et al. Chang'E-4 rover spectra revealing micro-scale surface thermophysical properties of the moon. *Geophys. Res. Lett.* **2021**, *48*. [CrossRef]
44. Lin, H.; Li, S.; Lin, Y.; Liu, Y.; Wei, Y.; Yang, W.; Yang, Y.; Hu, S.; Wu, X.; Xu, R.; et al. Thermal modeling of the lunar regolith at the Chang'E-4 landing site. *Geophys. Res. Lett.* **2021**, *48*. [CrossRef]
45. Yang, Y.; Lin, H.; Liu, Y.; Lin, Y.; Wei, Y.; Hu, S.; Yang, W.; Xu, R.; He, Z.; Zou, Y. The effects of viewing geometry on the Spectral analysis of lunar regolith as inferred by In Situ spectrophotometric measurements of Chang'E-4. *Geophys. Res. Lett.* **2020**, *47*, e2020GL087080. [CrossRef]
46. Lin, H.L.; Yang, Y.Z.; Lin, Y.T.; Liu, Y.; Wei, Y.; Li, S.; Hu, S.; Yang, W.; Wan, W.H.; Xu, R.; et al. Photometric properties of lunar regolith revealed by the Yutu-2 rover. *Astron. Astrophys.* **2020**, *638*, 6. [CrossRef]
47. Lin, H.L.; Lin, Y.T.; Yang, W.; He, Z.P.; Hu, S.; Wei, Y.; Xu, R.; Zhang, J.H.; Liu, X.H.; Yang, J.F.; et al. New insight into lunar regolith-forming processes by the lunar rover Yutu-2. *Geophys. Res. Lett.* **2020**, *47*, 9. [CrossRef]
48. Huang, J.; Xiao, Z.; Xiao, L.; Horgan, B.; Hu, X.; Lucey, P.; Xiao, X.; Zhao, S.; Qian, Y.; Zhang, H.; et al. Diverse rock types detected in the lunar South Pole–Aitken Basin by the Chang'E-4 lunar mission. *Geology* **2020**, *48*, 723–727. [CrossRef]
49. Gou, S.; Yue, Z.; Di, K.; Wang, J.; Wan, W.; Liu, Z.; Liu, B.; Peng, M.; Wang, Y.; He, Z.; et al. Impact melt breccia and surrounding regolith measured by Chang'e-4 rover. *Earth Planet. Sci. Lett.* **2020**, *544*, 116378. [CrossRef]
50. Gou, S.; Di, K.; Yue, Z.; Liu, Z.; He, Z.; Xu, R.; Liu, B.; Peng, M.; Wan, W.; Wang, Y.; et al. Forsteritic olivine and magnesium-rich orthopyroxene materials measured by Chang'e-4 rover. *Icarus* **2020**, *345*, 113776. [CrossRef]
51. Chen, J.; Ling, Z.; Qiao, L.; He, Z.; Xu, R.; Sun, L.; Zhang, J.; Li, B.; Fu, X.; Liu, C.; et al. Mineralogy of Chang'e-4 landing site: Preliminary results of visible and near-infrared imaging spectrometer. *Sci. China Inf. Sci.* **2020**, *63*. [CrossRef]
52. Gou, S.; Di, K.C.; Yue, Z.Y.; Liu, Z.Q.; He, Z.P.; Xu, R.; Lin, H.L.; Liu, B.; Peng, M.; Wan, W.H.; et al. Lunar deep materials observed by Chang'e-4 rover. *Earth Planet. Sci. Lett.* **2019**, *528*, 9. [CrossRef]
53. Lin, H.; Xu, R.; Yang, W.; Lin, Y.; Wei, Y.; Hu, S.; He, Z.; Qiao, L.; Wan, W. In Situ photometric experiment of lunar regolith with visible and near-infrared imaging spectrometer on board the Yutu-2 lunar rover. *J. Geophys. Res. Planets* **2020**, *125*. [CrossRef]

materials

MDPI

Article

Color Reproduction by Multi-Wavelength Bragg Diffraction of White Light

Alexander Machikhin [1,*], Alina Beliaeva [1,2,3,*], Galina Romanova [2] and Egor Ershov [4]

1 Acousto-Optic Spectroscopy Laboratory, Scientific and Technological Center of Unique Instrumentation of the Russian Academy of Sciences, 15 Butlerova, 117342 Moscow, Russia
2 Higher Engineering and Technical School, ITMO University, 49A Kronverksky Pr., 197101 Saint Petersburg, Russia; romanova_g_e@mail.ru
3 Department of Design and Technologies of Electronic and Laser Means, Saint Petersburg State University of Aerospace Instrumentation, 67A Bolshaya Morskaya, 190000 Saint Petersburg, Russia
4 Sector of Color Reproduction and Synthesis, Institute for Information Transmission Problems of the Russian Academy of Sciences (Kharkevich Institute), 19 Bolshoy Karetny, 127051 Moscow, Russia; ershov@iitp.ru
* Correspondence: machikhin@ntcup.ru (A.M.); belyaevalina@inbox.ru (A.B.)

Abstract: Accurate color reproduction is highly important in multiple industrial, biomedical and scientific applications. Versatile and tunable light sources with high color-rendering quality are very much in demand. In this study, we demonstrate the feasibility of multi-wavelength Bragg diffraction of light for this task. Tuning the frequencies and amplitudes of bulk acoustic waves in the birefringent crystal demonstrates high precision in setting the number, wavelengths and intensities of the monochromatic components necessary to reproduce a specific color assigned according to its coordinates in the CIE XYZ 1931 space. We assembled a setup based on multi-bandpass acousto-optic (AO) filtration of white light and verified the reproduced color balance in multiple experiments. The proposed approach delivers almost full coverage of the CIE XYZ 1931 space and facilitates building compact color reproduction systems (CRSs) for various purposes.

Keywords: color reproduction; spectral power distribution; acousto-optic interaction; Bragg diffraction; multi-wavelength light filtration; image processing

Citation: Machikhin, A.; Beliaeva, A.; Romanova, G.; Ershov, E. Color Reproduction by Multi-Wavelength Bragg Diffraction of White Light. *Materials* **2023**, *16*, 4382. https://doi.org/10.3390/ma16124382

Academic Editor: Dirk Poelman

Received: 15 May 2023
Revised: 9 June 2023
Accepted: 12 June 2023
Published: 14 June 2023

1. Introduction

Light sources with the ability to accurately display colors play an important role in solving various tasks including physiological studies of visual perception [1], colorimetry [2] and multi-spectral imaging [3]. These devices are a viable alternative to printed color charts and checkers that degrade and change their properties over time. Such tunable illuminants combine light beams from a few sources with different emission spectra, e.g., lasers [4,5] or LEDs [6–12]. By optical coupling and adjusting the relative intensities of these beams, one can reproduce a specific color with additive color mixing. Because of metamerism, the color may appear the same even when the spectral power distributions (SPDs) of the illuminants differ.

To measure color appearance (from a human observer point of view), the Commission international de l'eclairage (CIE) introduced a special standard CIE XYZ in 1931 [13]. Obviously, it is possible to introduce such a standard because the majority of people with normal color vision have almost the same color perception mechanism. The CIE XYZ 1931 standard introduced the shape of human eye spectral sensitivity functions calculated from the results of color-matching experiments [14] on 17 middle-aged British people with normal color vision. Later similar experiments have been conducted many times to verify the obtained results and investigate the influence of different factors, such as increased field of view for color stimuli [15]. Nowadays, this standard is one of the most important foundations in color reproduction science, because it facilitates the assignment of a numerical color value to the observed spectrum. CIE XYZ 1931 paves the way to separate the brightness (Y) of a color (determined by the total energy of radiation) and the chromaticity by introducing x, y coordinates. In this paper, we use the CIE XYZ 1931 standard to demonstrate CIE x, y coverage using different CRSs and estimate the color reproduction quality.

Figure 1 illustrates the areas in this space occupied by typical CRSs based on four lasers [4] and three [11] and four [8] LEDs (Figure 1). We may see that a significant number of colors is not available for reproduction using such devices. To expand their scope, more light sources with different SPDs are necessary, which means a complication of the CRS's optical design, adjustment and management.

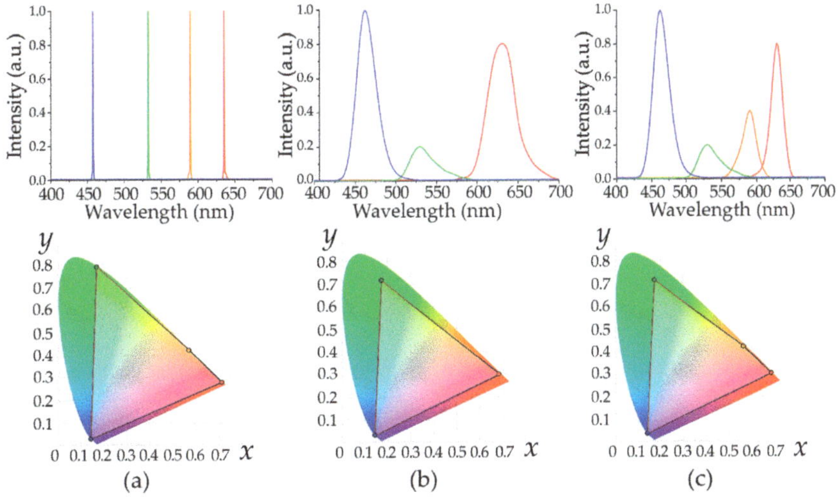

Figure 1. Typical SPDs (upper row) of CRSs based on four lasers (**a**), three (**b**) and four LEDs (**c**) and color areas that they cover in CIE XYZ 1931 chromaticity space (lower row).

An alternative approach to full-color reproduction might be related to tunable multi-bandpass spectral filtration of white light. Simultaneous selection and mixing of a few monochromatic or narrow-band components of the required wavelengths and intensities could potentially reproduce any color. To implement this approach, one needs a spectral component that is able to transmit light in a few selectable wavelength bands simultaneously with a tunable transmission for each of them. None of the conventional spectral elements (prisms, diffraction gratings, liquid crystal or Fabry–Perot tunable filters, etc.) delivers such a collection of features.

AO interaction, i.e., diffraction of light by ultrasonic waves, is a unique technique that allows tunable multi-bandpass filtration. It became a flexible and repeatable method for selecting the required spectrum of light [16]. That is why we believe that AO interaction may become an effective tool for selecting a specific color from wide-band light and propose applying it for color reproduction tasks. This feasibility study aims to experimentally validate this idea and comprehensively evaluate its advantages and limitations.

AO spectral filtration of light as well as other AO effects are based on photoelasticity, i.e., the change in the refractive index of a medium due to the presence of sound waves that create a volume diffraction grating. By changing the parameters of ultrasound, one may vary the structure of this grating and the diffraction pattern. In Bragg mode, when $2\pi\lambda L >> \Lambda^2$ (λ and Λ are wavelengths of light and sound waves, respectively, and L is a length of AO interaction), the diffraction pattern consists of two maxima (the zeroth and the first orders). The first diffraction order summarizes selective reflections of light from the wave fronts of ultrasonic waves and thus is informative for the analysis of the light spectrum.

Being compact, solid-state and PC-controlled, AO tunable filters (AOTFs) are a robust and versatile tool to manage the spectrum of transmitted light [17,18]. Arbitrary spectral access, multi-wavelength selection and transmission function apodization are key features of AO interaction that deliver a good compromise in terms of spectral bandwidth, tuning range, aperture and switching time.

Thus, Bragg diffraction of white light using multiple ultrasound waves seems an attractive basis for building reliable and flexible CRSs. In this study, we experimentally demonstrate the feasibility of this technique for accurate color reproduction.

2. Proposed Approach

The concept of the proposed AOTF-based CRS is shown in Figure 2. It includes an AO crystal and a piezoelectric transducer bonded to it. When electrical signals are applied to the transducer, it generates ultrasonic waves propagating through the crystal. These waves modulate the refractive index and thereby form a three-dimensional diffraction grating. The simplest grating is achieved when applying a single acoustic frequency f_0. In this mode with the Bragg phase-matching conditions fulfilled, there is only the first diffraction order, and the AOTF transmission function is described with a squared sinc function: $T(\lambda) \sim \text{sinc}^2((\lambda - \lambda_0)/\delta\lambda)$, where $\lambda_0 \sim 1/f_0$ and $\delta\lambda \sim {\lambda_0}^2$ [16]. Diffraction light intensity is proportional to the acoustic power P radiated from the transducer. If N frequencies f_j ($j = 1, 2 \ldots N$) are launched simultaneously, then $T(\lambda)$ is the sum of transmission functions $T(\lambda - \lambda_j)$ defined by each of them. By varying the values of N, f_j and P_j, one can select the required number N of the spectral bands as well as precisely tune their positions λ_j and intensities I_j.

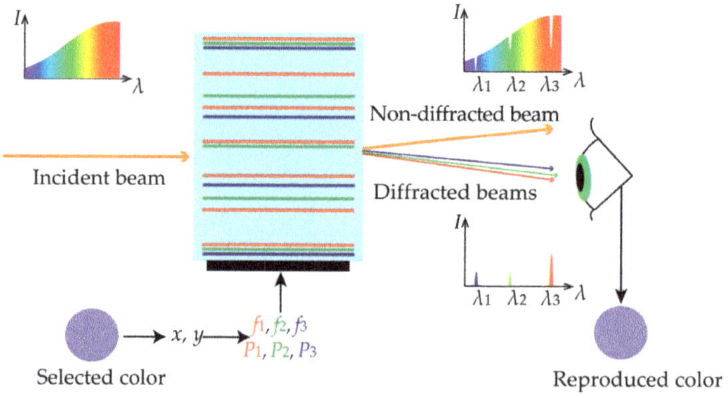

Figure 2. Proposed CRS concept.

The number of selected bands N determines the palette of available colors. If the AOTF operates in dual-frequency mode, then reproducible colors lay on the line segment between the points defined by λ_1 and λ_2 (Figure 3a). Changing the frequencies f_1 and f_2 varies the inclination angle of this line on the CIE XYZ 1931 chromaticity plane and thus achieves segments with different colors. The proper power ratio P_1/P_2, i.e., intensities I_1/I_2, enables accurate selection of a particular color within the segment. If the AOTF selects 3, 4 or more spectral bands simultaneously, then the available gamut is described using a triangle (Figure 3b), quadrangle (Figure 3c) or polygon with more vertices defined by the selected wavelengths λ_j, i.e., the acoustic frequencies f_j launched in the crystal. The relative intensities I_j of the light beams in this case have to be assigned in accordance with the center of gravity law [19] and, therefore, the acoustic power P_j has to be adjusted accordingly.

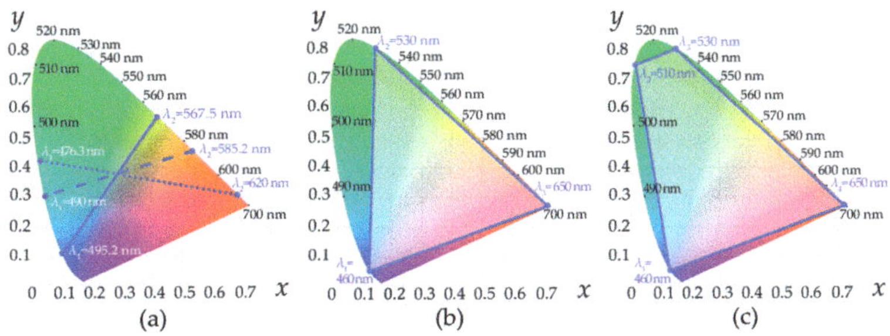

Figure 3. Color areas covered in CIE XYZ 1931 chromaticity space with AOTF-based CRS operating in dual- (**a**), three- (**b**) and four-frequency (c) mode.

Thus, to reproduce the color x, y with an AOTF-based CRS, it is enough in principle to choose the number N and positions λ_j of the basic wavelengths, calculate the relative intensities I_j of the diffracted beams and properly adjust the values of acoustic power P_j.

3. Experimental Setup

Birefringent crystals are the most efficient AO medium in terms of diffraction efficiency, availability and ease of handling. They enable anisotropic AO interaction, which changes the polarization (e→o or o→e) of the incident optical beam and enables easy separation of the first diffraction order from the zeroth one using a linear polarizer [16]. Tellurium dioxide (TeO_2) is one of the best AO crystals that is widely used because of its high figure of merit $M_2 = n_o^3\, n_e^3 p^2/\rho V^3$ (n_o and n_e are refractive indices for ordinary and extraordinary polarized light, p is effective photoelastic constant, ρ is density and V is ultrasound wave velocity), which is 1200×10^{-15} s^3kg^{-1} at the wavelength $\lambda = 632.8$ nm. A high value of M_2 ensures the low driving power $P = (\lambda^2 h)/(2M_2 l)$ (h and l are the height and length of the acoustic beam, respectively) that is required for the high efficiency of Bragg diffraction and eases the fabrication of AO devices.

For our experiments, we produced a typical TeO_2 AOTF with a cut angle of 7° and a wide-aperture geometry of AO interaction. The AOTF had an optical aperture of 10 × 10 mm and angular aperture of 4° × 4°. The front facet of the crystal was perpendicular to the incident light beam. For easier mixing of the diffracted light beams, the back facet angle of the crystal corresponds to their minimal chromatic shift in the range of 430–780 nm [20]. Because of the poor transparency of tellurium dioxide below 430 nm, the violet part 380–430 nm of the visible spectrum is unavailable. With some other AO crystals, e.g., quartz (SiO_2), it is possible to cover the entire visible range of 380–780 nm.

To generate bulk ultrasonic waves in the TeO_2 crystal, we bonded a thin-plate piezoelectric transducer to it via cold indium welding. The transducer had 2 3.5 × 8 mm sections with a 0.5 mm gap between them. The sections were electrically connected in series to reduce the effective capacitance. The thickness of the transducer corresponds to the acoustic frequency 85 MHz which is the central frequency of the operating range. The transducer was made of X-cut lithium niobate ($LiNbO_3$) because of its large piezoelectric and electromechanical coupling coefficients, stable functioning at high frequencies and well-established manufacturing process. To place ground and drive electrodes, thin films of chromium and gold were evaporated and deposited on the transducer. To ensure a good acoustic impedance matching of the piezoelectric transducer and AO crystal, a specific acoustic bond consisting of chromium, gold and indium was evaporated. After welding, electrical matching networks were installed to match the transducer and the electrical impedance of the driving radio frequency generator. When a radio frequency signal was applied to the piezoelectric transducer, it vibrated and generated bulk shear ultrasonic waves that propagated in the TeO_2 crystal and were absorbed by the acoustic absorber placed at its opposite facet.

By varying the acoustic frequency f from 56 to 137 MHz, the wavelength λ of filtered light may be precisely tuned in the range 430–850 nm according to the tuning curve $\lambda(f)$ obtained after AOTF calibration using a certified monochromator. The total driving acoustic power $P = \Sigma P_j$ is variable from 0 to 2.5 W and defaults to 1 W at all wavelengths. To enable multi-frequency mode, we used a four-channel driver based on a direct digital synthesizer (Analog Devices AD9959) and two homemade two-channel amplifiers, which enabled precise assignment of the position and intensity of each selected spectral band.

Figure 4 illustrates the experimental setup. To ensure anisotropic o→e diffraction and cut off non-diffracted light, we placed the AO cell between crossed polarizers P1 and P2. Lens L1 with a focal distance of 35 mm collimated wide-band light from a fiber-coupled 150 W halogen lamp LS and sent it to the AOTF. After AO diffraction, lens L2 with a focal distance of 50 mm focused the filtered light onto the sensor of color camera CAM (TheImagingSource DFM 42BUC03). CAM acquired images of the diffracted beams and thus facilitated the comparison of the reproduced colors with the reference ones. Its quantum efficiency is spectrally dependent and must be taken into account during color rendering experiments. That is why, between L2 and CAM, there was a 50/50 beam splitter BS that directed the reflected light to spectrometer SP (Ocean Insight Flame-T-UV-VIS).

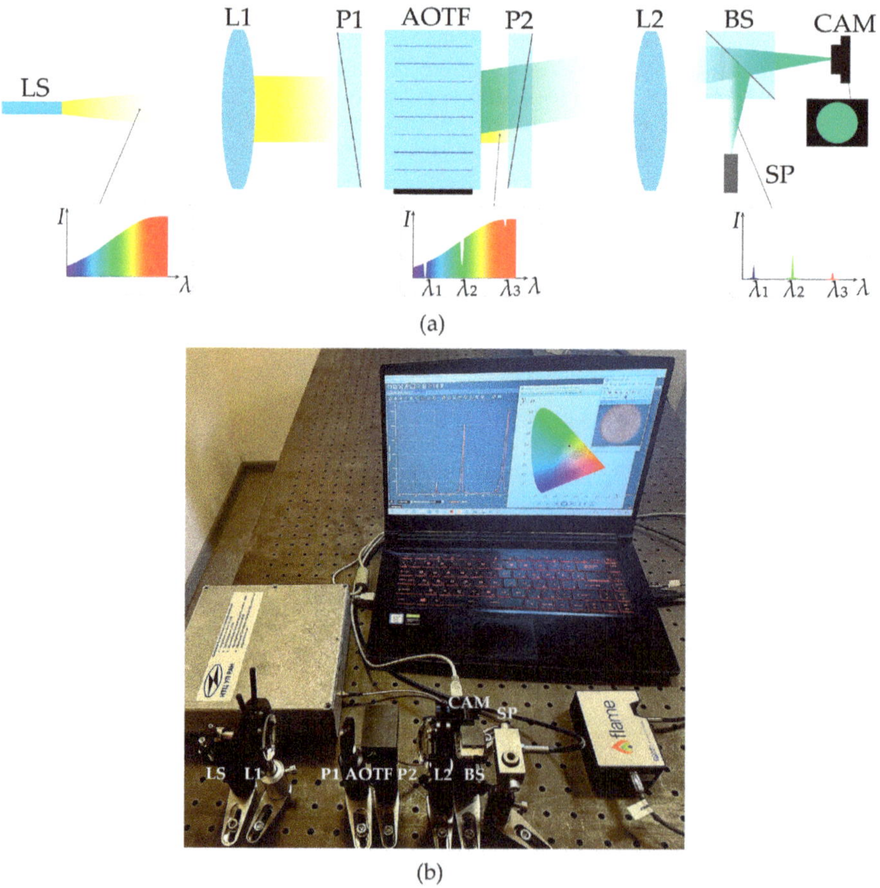

Figure 4. Scheme (**a**) and appearance (**b**) of the experimental setup for color reproduction. LS—light source, L1 and L2—lenses, P1 and P2—polarizers, AOTF—acousto-optical tunable filter, BS—beam splitter, CAM—color camera, SP—spectrometer.

SP is necessary for the interactive measurement and correction of the light intensities I_j in the selected bands and managing the acoustic power P_j in AOTF channels with respect to the quantum efficiency of the CAM sensor in order to ensure that the ratio of the intensities corresponds to the assigned color.

All the experiments presented below were carried out in typical laboratory conditions (temperature 23 °C and relative humidity 50%). In fact, all optical (lenses, polarizers, AOTF, etc.) and electronic (LS, AOTF driver, CAM) components of the setup can operate at a wider range of ambient temperatures (10–40 °C) and humidity (30–70%). To validate the repeatability of the proposed CRS in various tolerable combinations of temperature and humidity, further study is necessary.

4. Experimental Results

We demonstrated the proposed system's capabilities by conducting multiple color reproduction experiments. Figure 5 shows images of the reference and reproduced colors obtained with an AOTF operating in single- (points A, E and I), dual- (points B, D and G) and three-frequency (points C, F and H) modes. Here, the amount and positions of the colors for color rendering as well as the spectral bands for their reproduction were selected just to illustrate the flexibility of the AOTF-based CRS. Depending on the application, these parameters may be optimized for the stable and repeatable reproduction of colors located in a certain area of the CIE XYZ 1931 chromaticity plane without spectral tuning. The images formed by diffracted light beams are presented after correction, i.e., averaging over the cross-section of the diffracted beams' profiles. Otherwise, they look non-uniform because of vignetting, diffraction-caused distortions and optical aberrations (see below in Figure 6d). The exposure time was equal to 20 ms in all experiments, as shown in Figure 6. The relative intensities presented in Figure 6 were calculated from the measured values with respect to the quantum efficiency of the camera sensor.

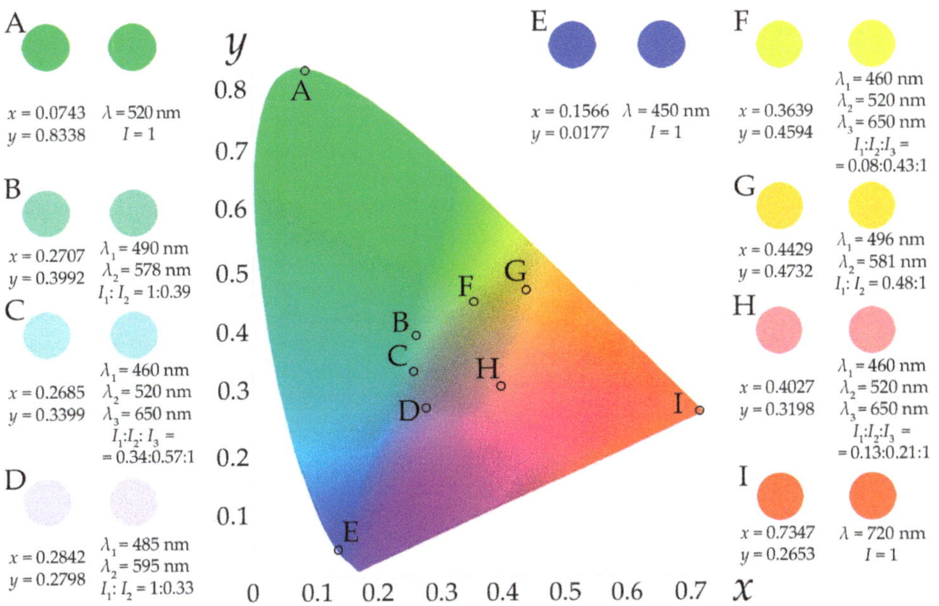

Figure 5. Reference (left circles) and reproduced (right circles) colors plotted in CIE XYZ 1931 space. Reference colors are assigned according to their x, y coordinates. Reproduced colors are presented according to the wavelengths λ_j and relative intensities I_j of the selected monochromatic components.

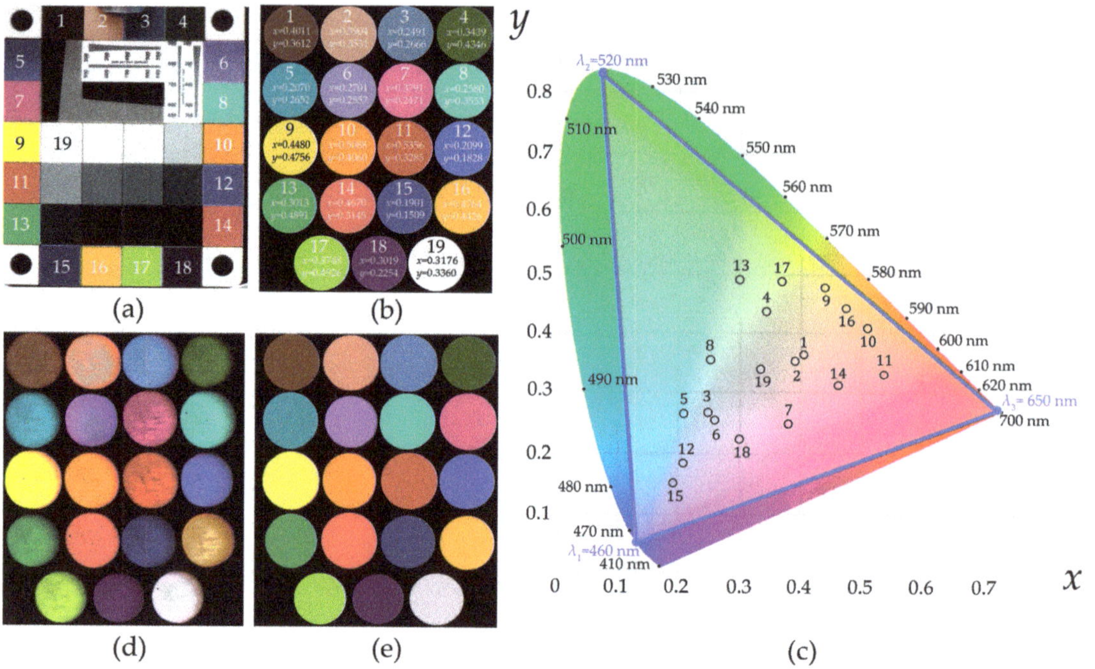

Figure 6. Color checker (**a**), its selected colors presented as reference images (**b**) and plotted in CIE XYZ 1931 space (**c**) and acquired raw (**d**) and averaged (**e**) images.

Figure 5 demonstrates the high visual similarity of the assigned colors and the colors in the acquired images. To obtain a more objective merit of color rendition, we took a certified color checker (Edmund Optics Rez Checker) widely used for testing true color balance and comparing with reference patterns. Figure 6a shows the color checker's image acquired with the same lens L2 and the same color camera CAM operating in the same mode as in the color reproduction experiments [18] (Figure 6b). We chose the wavelengths λ_1 = 460 nm, λ_2 = 520 nm and λ_3 = 650 nm as the basic ones to form a wide enough triangle in CIE XYZ 1931 and cover all 19 selected colors (Figure 6c). In this experiment, we again calculated the necessary ratio of the intensities I_j and applied the acoustic power P_j accordingly. The total acoustic power ΣP_j applied to the AOTF was adjusted in order to achieve the same intensity of the maximal component in all acquired images as in the reference ones shown in Figure 6b. Figure 6d,e illustrates the raw and corrected (averaged) color images formed by the diffracted light. Visually, the reproduced colors (Figure 6e) seem almost indistinguishable from the ones in the color checker (Figure 6a). Correspondingly, the average and median reproduction quality according to ΔE_{2000} are about 1.25 and 1.12, respectively, which is close to the just noticeable difference (ΔE_{2000} = 1) threshold, while those for the angular errors between the reference and generated tristimulus are about 0.68 and 0.61, respectively.

Figure 7 illustrates the spectra $I(\lambda)$ selected using AOTF and measured with spectrometer SP for all 19 colors in the color checker. Coordinates x, y of each color as well as the calculated relative intensities I_1:I_2:I_3 of the respective monochromatic components 460 nm, 520 nm and 650 nm necessary for its reproduction are shown inside the circles.

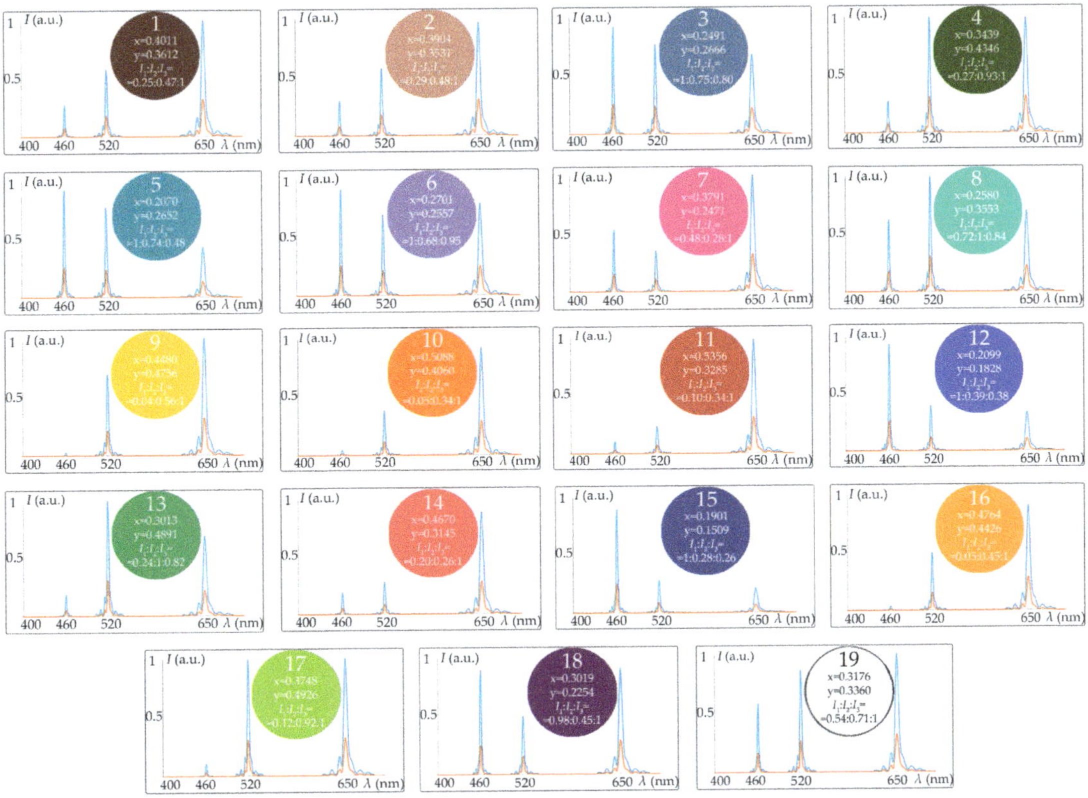

Figure 7. Spectra required to reproduce 19 colors in color checker for visual perception (red) and digital registration (blue), i.e., regardless of and with regard to the spectral quantum efficiency of the camera sensor, respectively.

5. Discussion

Though the CRS implementation demonstrated in this study has multiple advantages including almost full CIE XYZ space coverage, high speed and repeatability, it still has the potential for higher efficacy.

First, accurate spatio-temporal and radiometric calibration of the setup will allow obtaining the dependence of image intensity I on acoustic power P in each image pixel with respect to the transmittance of optical components and the quantum efficiency of the sensor and thus will simplify the system by excluding the spectrometer. Regular calibration and refinement of correction factors will compensate for long-term temporal variations in CRS performance.

Second, accurate optical CRS design will lead to uniform and non-distorted light beams acceptable for human visual color perception study in the entire color triangle. To achieve this, the simulation and optical design of the setup must take into account multiple effects related to light transmittance through fiber-optic cable and AOTFs: material and waveguide dispersion, scattering and absorption losses, wave-front distortions caused by AO diffraction, etc.

Third, the adjustable width $\delta\lambda$ and shape $T(\lambda)$ of the transmission function, which is an important advantage of AO diffraction, was not used in this feasibility study but is quite helpful in many rendering tasks and may widen the variety of AOTF-based CRS applications. A common approach to interactive management of the transmission function is dispersive synthesis, i.e., varying the ultrasonic signal with a complex-valued spectrum [17]. With this feature, the proposed CRS may become even more flexible and versatile and may deliver multiple scenarios of color reproduction. Interactive and broad-range change in the number N, shapes $T_j(\lambda)$ and intensities $I_j(\lambda)$ of the selected spectral channels is an ultimate solution for multiple color rendering tasks but requires advanced hardware for programmable radio frequency waveform synthesis.

Fourth, one of the main AOTF drawbacks critical for some color reproduction applications is its low optical throughput due to its small angular aperture, narrow spectral band, linear polarization selection under anisotropic diffraction and other factors. Depending on the application, this problem may be solved in a few ways: using a light source with higher power, optimizing the optical coupling between the AOTF and other components, frequency modulation of ultrasonic waves [21], complicating the optical scheme to ensure simultaneous effective AO diffraction of both polarization components of a randomly polarized light, etc.

Fifth, the metrological characteristics of the proposed AOTF-based illuminants must be further investigated. Evaluation of their accuracy and repeatability is a challenging task and requires detailed theoretical and experimental analysis of the contribution of multiple systematic and random factors to the color reproduction error. Spatial, temporal, spectral and intensity variations are inevitable in systems of this type and must be calibrated over the entire range of multiple physical parameters with respect to the operating mode and conditions.

With these issues taken into account, the proposed AOTF-based approach to color rendering may have multiple potential applications in biomedicine, industry and metrology. One of them deserves more thorough consideration. Now, there are still no color-appearance models that ensure high-quality color representation and prediction within the entire color triangle of a standard observer. One of the main reasons for this is the lack of tools and techniques for studying the color perception of an arbitrary reference color located close to the border of the color triangle. The most popular methods are based on wide-coverage displays and do not allow such a capability. Colorimetric setups require installation and changing multiple certified color filters, which makes large-scale measurements almost impossible. The new AOTF-based technique proposed in this study may help overcome these limitations and bring new opportunities to vision science. It demonstrates an efficient method of building laboratory setups for human vision system research including presenting stimuli, measuring thresholds and quantitative characteristics and analyzing the color perception properties. Such systems will be a pioneer in understanding human visual perception over the entire area of the color triangle of the standard CIE XYZ 1931 observer.

6. Conclusions

AO interaction is a flexible tool for the spectral filtration of light by varying the parameters of the acoustic waves. AO interaction enables the stable selection of monochromatic components with the required wavelengths and amplitudes. Launching a few ultrasound frequencies with tunable power in AO crystal and mixing the diffracted light beams can enable the creation of an almost infinite number of colors. Together with comprehensive radiometric calibration of the optical setup and digital image processing, this technique delivers unique color reproduction capabilities with just one white light source. This type of tunable illuminant provides nearly full coverage of the CIE XYZ 1931 chromaticity space and demonstrates high color-rendering quality, high luminous efficiency, fast tuning and easy adjustment. We believe that an AOTF-based CRS may have multiple applications in industry and biomedicine and can be helpful for developing new up-to-date color reproduction standards in colorimetry that are extremely important for the worldwide scientific community.

7. Patents

Machikhin A., Beliaeva A., Romanova G.: color reproduction method based on polychromatic acousto-optic filtering of broadband radiation 17 March 2022 (RU 2786365 C1) (In Russian).

Author Contributions: Conceptualization, A.M.; methodology, A.M. and E.E.; experiment, A.B.; validation, G.R. and E.E.; formal analysis, A.M., G.R. and E.E.; investigation, A.B.; data curation, A.B. and A.M.; writing—original draft preparation, A.M.; writing—review and editing, A.B., G.R. and E.E.; supervision, A.M.; project administration, A.M. and G.R.; funding acquisition, A.M. All authors have read and agreed to the published version of the manuscript.

Funding: This research was funded by Federal State Task Program of Scientific and Technological Center of Unique Instrumentation of the Russian Academy of Sciences (project #FFNS-2022-0010).

Institutional Review Board Statement: Not applicable.

Informed Consent Statement: Not applicable.

Data Availability Statement: The data that support the findings of this study are available from the corresponding author upon reasonable request.

Acknowledgments: The experiments were carried out on the base of the Center for Collective Use of the Scientific and Technological Center of Unique Instrumentation of the Russian Academy of Sciences.

Conflicts of Interest: The authors declare no conflict of interest.

References

1. Land, E.H. Experiments in color vision. *Sci. Am.* **1951**, *200*, 84–99. [CrossRef] [PubMed]
2. Ohta, N.; Robertson, A. *Colorimetry: Fundamentals and Applications*; John Wiley & Sons: Hoboken, NJ, USA, 2006.
3. Park, J.I.; Lee, M.H.; Grossberg, M.D.; Nayar, S.K. Multispectral imaging using multiplexed illumination. In Proceedings of the IEEE 11th International Conference on Computer Vision, Rio de Janeiro, Brazil, 14–20 October 2007; pp. 1–8.
4. Neumann, A.; Wierer, J.J.; Davis, W.; Ohno, Y.; Brueck, S.R.J.; Tsao, J.Y. Four-color laser white illuminant demonstrating high color-rendering quality. *Opt. Exp.* **2001**, *19*, 982–990. [CrossRef] [PubMed]
5. Chang, S.W.; Liao, W.C.; Liao, Y.M.; Lin, H.I.; Lin, H.Y.; Lin, W.J.; Lin, S.Y.; Perumal, P.; Haider, G.; Tai, C.T.; et al. A white random laser. *Sci. Rep.* **2008**, *8*, 2720. [CrossRef] [PubMed]
6. Salters, B.A.; Krijn, M.P.C.M. Color reproduction for LED-based general lighting. In Proceedings of the SPIE Optics + Photonics: Nonimaging Optics and Efficient Illumination Systems III 6338, San Diego, CA, USA, 13–17 August 2006; Volume 6338, pp. 106–116.
7. Tanaka, M.; Horiuchi, T.; Tominaga, S. Color control of a lighting system using RGBW LEDs. In Proceedings of the IS&T/SPIE Electronic Imaging—Color Imaging XVI: Displaying, Processing, Hardcopy, and Applications, San Francisco, CA, USA, 24–27 January 2011; Volume 7866.
8. Oh, J.H.; Yang, S.J.; Sung, Y.G.; Do, Y.R. Excellent color rendering indexes of multi-package white LEDs. *Opt. Exp.* **2012**, *20*, 20276–20285. [CrossRef] [PubMed]
9. Lei, J.; Xin, G.; Liu, M. Spectral assemblage using light emitting diodes to obtain specified lighting characteristics. *App. Opt.* **2014**, *53*, 8151–8156. [CrossRef] [PubMed]
10. Yoon, H.C.; Kang, H.; Lee, S.; Oh, J.H.; Yang, H.; Do, Y.R. Study of perovskite QD down-converted LEDs and six-color white LEDs for future displays with excellent color performance. *ACS Appl. Mater Interfaces* **2016**, *8*, 18189–18200. [CrossRef] [PubMed]
11. Murdoch, M.J. Dynamic color control in multiprimary tunable LED lighting systems. *J. Soc. Inf. Disp.* **2019**, *27*, 570–580. [CrossRef]
12. Choi, H.; Yeom, J.Y.; Ryu, J.M. Development of a multiwavelength visible-range-supported opto–ultrasound instrument using a light-emitting diode and ultrasound transducer. *J. Sens.* **2018**, *18*, 3324. [CrossRef] [PubMed]
13. Smith, T.; Guild, J. The CIE colorimetric standards and their use. *Trans. Opt. Soc.* **1931**, *33*, 73. [CrossRef]
14. Judd, D.B.; Wyszecki, G. *Color in Science, Business and Industry*; John Wiley and Sons: New York, NY, USA, 1975.
15. Westland, S. Review of the CIE system of colorimetry and its use in dentistry. *J. Esthet. Restor. Dent.* **2003**, *15*, S5–S12. [CrossRef] [PubMed]
16. Goutzoulis, A.P. *Design and Fabrication of Acousto-Optic Devices*; CRC Press: Boca Raton, FL, USA, 2021.
17. Yushkov, K.B.; Makarov, O.Y.; Molchanov, V.Y. Novel protocol of hyperspectral data acquisition by means of an acousto-optical tunable filter with synthesized transmission function. *Opt. Lett.* **2019**, *44*, 1500–1503. [CrossRef] [PubMed]
18. Machikhin, A.S.; Khokhlov, D.D.; Pozhar, V.E.; Kozlov, A.B.; Batshev, V.I.; Gorevoy, A.V. Acousto-optical tunable filter for a swept light source with variable transmission function. In Proceedings of the SPIE/COS Photonics Asia: Optical Design and Testing VIII, Beijing, China, 11–13 October 2018; Volume 10815, pp. 129–134.

19. Hunt, R.W.G. *The Reproduction of Colour*; John Wiley & Sons: Hoboken, NJ, USA, 2005.
20. Machikhin, A.S.; Gorevoy, A.V.; Batshev, V.I.; Pozhar, V.E. Modes of wide-aperture acousto-optic diffraction in a uniaxial birefringent crystal. *J. Opt.* **2021**, *23*, 125607. [CrossRef]
21. Mazur, M.M.; Mazur, L.I.; Suddenok, Y.A.; Shorin, V.N. Increase of an Output Optical Signal of an Acousto-Optic Monochromator upon Frequency Modulation of a Control Signal. *Opt. Spectrosc.* **2018**, *125*, 594–598. [CrossRef]

materials

Article

Spectral Imaging Experiments with Various Optical Schemes Based on the Same AOTF

Vladislav Batshev [1,2], Alexander Machikhin [1,3,*], Alexey Gorevoy [1], Grigoriy Martynov [1], Demid Khokhlov [1,3], Sergey Boritko [1,4], Vitold Pozhar [1,2] and Vladimir Lomonov [5]

1 Scientific and Technological Center of Unique Instrumentation Russian Academy of Sciences, Butlerova Str. 15, 117342 Moscow, Russia; batshev@bmstu.ru (V.B.); gorevoy.av@ntcup.ru (A.G.); martynov.gn@ntcup.ru (G.M.); khokhlov.dd@ntcup.ru (D.K.); boritko@ntcup.ru (S.B.); vitold@ntcup.ru (V.P.)
2 Laser and Optical-Electronic Systems Department, Bauman Moscow State Technical University (National Research University), 2-nd Baumanskaya Str. 5, 105005 Moscow, Russia
3 Institute of Information Technologies and Computer Science, Moscow Power Engineering University (National Research University), Krasnokazarmennaya 14, 111250 Moscow, Russia
4 Moscow Institute of Physics and Technology, National Research University, 9 Institutskiy Per., Dolgoprudny, 141701 Moscow, Russia
5 Federal Scientific Research Center "Crystallography and Photonics", Russian Academy of Sciences, 119333 Moscow, Russia; yupisarev@yandex.ru
* Correspondence: machikhin@ntcup.ru

Abstract: Spectral image filtration by means of acousto-optical tunable filters (AOTFs) has multiple applications. For its implementation, a few different optical schemes are in use. They differ in image quality, number of coupling components, dimensions and alignment complexity. To choose the optical system of AOTF-based spectral imager properly, many factors have to be considered. Though various schemes of acousto-optic (AO) filtration have been tested and discussed, their comparative analysis has not been reported up to now. In this study, we assembled the four most popular schemes (confocal, collimating, tandem and double-path) using the same AO cells and experimentally compared their main features. Depending on the application, each scheme may be the basis of compact cost-effective spectral imaging devices.

Keywords: spectral imaging; acousto-optic interaction; image quality; confocal scheme; collimating scheme

Citation: Batshev, V.; Machikhin, A.; Gorevoy, A.; Martynov, G.; Khokhlov, D.; Boritko, S.; Pozhar, V.; Lomonov, V. Spectral Imaging Experiments with Various Optical Schemes Based on the Same AOTF. *Materials* **2021**, *14*, 2984. https://doi.org/10.3390/ma14112984

Academic Editor: Alain Moissette

Received: 30 April 2021
Accepted: 29 May 2021
Published: 31 May 2021

1. Introduction

AOTFs have become a popular tool for various hyperspectral imaging applications in biomedicine [1,2], agriculture [3,4], aerospace [5,6] and other fields. Based on anisotropic Bragg diffraction of wide-band light by ultrasound in crystalline media, these spectral elements provide a good combination of optical (high spectral and spatial resolution, wide tuning range, etc.) and technical (compactness, absence of moving parts, etc.) features [7].

The imaging capabilities of AOTF-based systems are defined by multiple factors. Most of them, e.g., geometry of AO interaction, shape of AO cell (AOC) and structure of ultrasound beam [8–10], have been examined and discussed in detail. Besides these AOTF characteristics, the optical scheme of AO spectral filtration has a strong influence on the features of AOTF-based images [11]. In practice, a few different schemes are used.

A typical AOTF configuration includes a single AOC located between two polarizers crossed with respect to each other [12]. More effective in terms of spectral resolution and residual image distortion are double AOTFs that may consist of tandem AOCs [13] or utilize a double light pass through a single AOC by means of optical feedback [14] and back reflection [15].

Moreover, there are three image formation methods common for all AOTFs: collimating, confocal and convergent [10]. Collimating (telescopic) scheme is the most common and the simplest one and provides high image quality and spectral resolution. However, the non-uniformity of the central wavelength of filtered light across the field of view results in a specific image spatio-spectral distortion [16]. Confocal telecentric scheme, which forms the image inside AOC, is free from this drawback but leads to visualization of the inhomogeneous ultrasound structure in the filtered image and to spectral resolution degradation [17]. Specific AO aberrations (distortion and chromatic image drift) are present in both schemes. Converging scheme is used quite rarely [18].

Tandem AOTF consisting of two identical AOCs provides higher spatial and spectral resolution, but at the same time, lower light transmission and larger dimensions. Normally, tandem AOTF utilizes a collimating scheme to avoid additional couplers necessary for the confocal one. To reduce the dimensions and price of a tandem AOTF, a single AOC operating in double-pass collimating scheme can be used [19].

All these optical schemes differ in image quality, number of coupling components, dimensions, alignment complexity and other features. For each specific task and for each AOC, it is necessary to optimize the optical system for AO image filtration. Though the schemes mentioned above have been already tested and discussed, their comparative analyses have not been reported up to now. We need to extract the characteristic features of each scheme by separating them from intrinsic features of AOC itself.

In this study, we assemble, test and analyze four various schemes (Figure 1). Two of them are commonly used schemes: collimating (coll) and confocal (conf) ones. Two others imply two-stage light diffraction: by two AOC in series (tandem configuration) and by the same AOC (double-pass configuration). All of them are described in Section 2. The comparison procedure and results are presented in Section 3, while the general conclusions are presented in the last section.

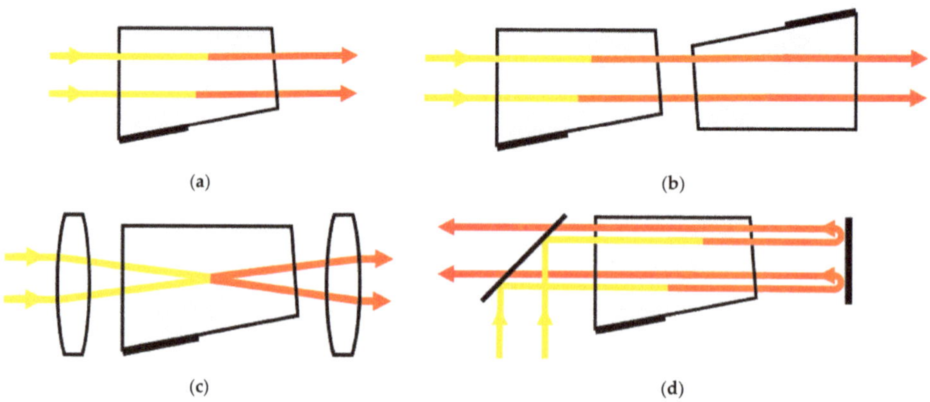

Figure 1. Typical schemes of AO filtration image: (**a**) single-crystal single-pass collimating scheme (S-coll); (**b**) double-crystal single-pass (tandem) collimating scheme (T-coll); (**c**) single-crystal single-pass confocal scheme (S-conf); (**d**) single-crystal double-pass collimating scheme (D-coll).

2. Materials and Methods

In the experiments, we used a typical AOC made from paratellurite (α-TeO_2), which is the most widely used AO uniaxial crystal. The geometry of this AOC is optimized for minimization of the chromatic image drift in the confocal optical scheme. It provides good spectral and spatial resolution and may be used in a wide spectral range [20]. Residual angular dispersion is 0.4°. The AOC has a cut angle $\gamma = 7°$ (Figure 2). The wide-aperture diffraction geometry of e-polarized light [8,12] is realized for incident angle $\theta = 73.85°$. Basic parameters: the piezotransducer length L = 12 mm, clear input aperture is 10×12 mm^2, the angular aperture is $4° \times 4°$. An incident light beam must be directed normally to the front facet. The back-facet inclination angle is with respect to incident facet $\beta = 2.3°$. By varying the acoustic frequency within intervals of 60–110 MHz, one can tune the wavelength of filtered light in the range 450–850 nm. Driving acoustic power interval is 2.3–2.5 W. It is adjusted to equalize the AOTF's light transmission in this range. In tandem AOTF two identical AOCs of this type was used.

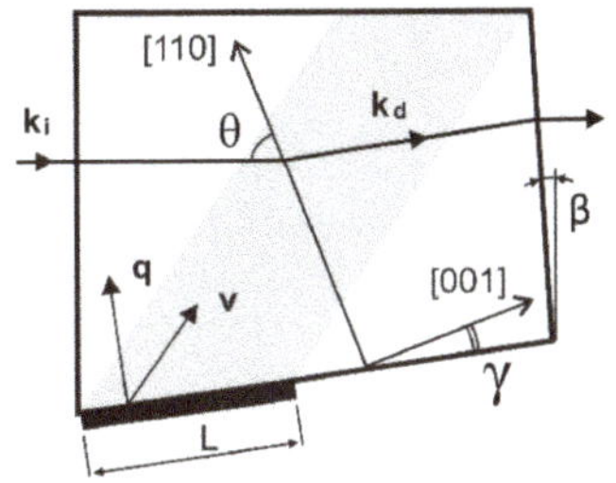

Figure 2. AO cell configuration. Group velocity **v** deviates from the ultrasound wave vector q due to walk-off effect; deflection of diffracted light wave vector k_d from incident k_i depends on θ and γ, while direction of output light beam is governed by back-facet inclination angle β. All the variants of experimental schemes have the same structure: object → input optics → AO spectral unit → output optics → photodetector.

In all the experiments presented below (Figures 3–6), we used both standard and custom optical components. For image acquisition, the camera TheImagingSource DMK 37BUX178 with 1/1.8″ 6.3 MP CMOS sensor was installed. Depending on the inspected object, two different lenses L0 were utilized: an infinity-corrected microscope objective Carl Zeiss Planachromat 10/0.25 and photography lens Minolta MAXXUM 135/2.8. Afocal system L1, L2 consists of two identical standard lenses: 25 mm board lenses TheImagingSource TBL 25 or Thorlabs AC127-025-AB achromatic doublets. Focusing lens L3 is 75 mm Kowa LM75HC lens or Thorlabs AC127-075-AB achromatic doublet. Focusing lens in confocal scheme is 35 mm board lens TheImagingSource TBL 35. Coupling lenses L4 and L5 are 36 mm doublets specially designed and manufactured to compensate for the chromatic focal shift introduced by the AO cell in the confocal scheme. Lens L7 in the last setup is a 50 mm achromatic doublet.

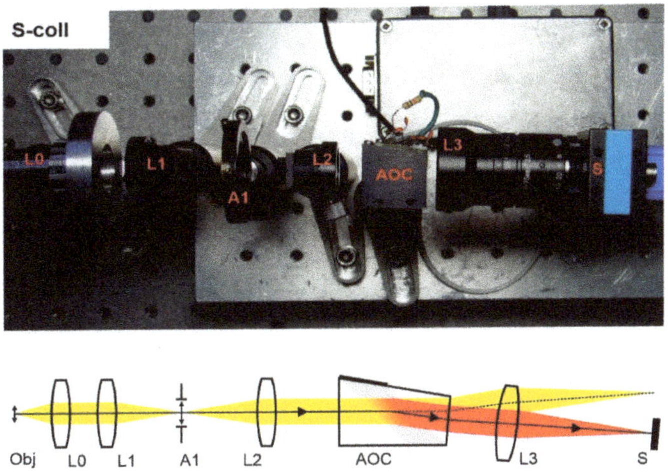

Figure 3. S-coll scheme. Breadboard model and optical beam-pass diagram.

A single-crystal single-pass collimating scheme (S-coll) is comprised of the following input optical part elements: input lens L0, field stop A1 inside the afocal system L1–L2, which define the collimated beam divergence preventing sensor irradiation by non-diffracted (unfiltered) light (Figure 3). An output lens L3 forms the spectral image on the sensor S.

Single-crystal single-pass confocal scheme (S-conf) comprises two similar optical systems in the input and output: a pair of lenses with intermediate diaphragm (Figure 4). The input system forms an intermediate image inside the AOC, while the output projects its filtered component onto the sensor S. The aperture stop A2 defines the convergence angle and output stop A3 blocks the non-diffracted (unfiltered) beam. Coupling lenses L4 and L5 are 36 mm custom-designed to compensate for the chromatic focal shift introduced by the AO cell. Other components are standard machine vision lenses.

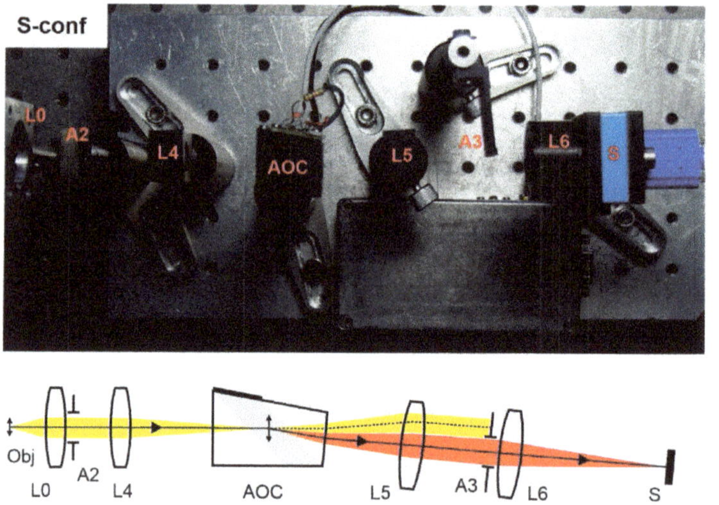

Figure 4. S-conf scheme. Breadboard model and optical beam-pass diagram.

Double-crystal single-pass (tandem) collimating scheme (T-coll) needs an afocal system L2–L3 in the input and output focusing lens L3 to transfer the object image into the sensor S (Figure 5). Tandem AOTF consists of two identical AO cells and three crossed Glan–Taylor calcite polarizers P1, P2, and P3. The second AO cell is rotated by 180° with respect to the first one. The polarizers are necessary to block the non-diffracted light beam leaving the filter at the same direction as the twice-diffracted one, whereas the lateral separation is insufficient due to the rather small deflecting angle. As the extinction ratio of polarizers is not always enough to eliminate non-diffracted light completely, we additionally utilized a field stop A1.

Since AO cells are identical and opposite turned, the same wide-aperture diffraction mode is realized in both AO cells, and, thus, image aberrations of distortion and chromatic drift are compensated. The drawback of the scheme is significantly increased energy losses. The advantages are higher spectral and spatial resolution.

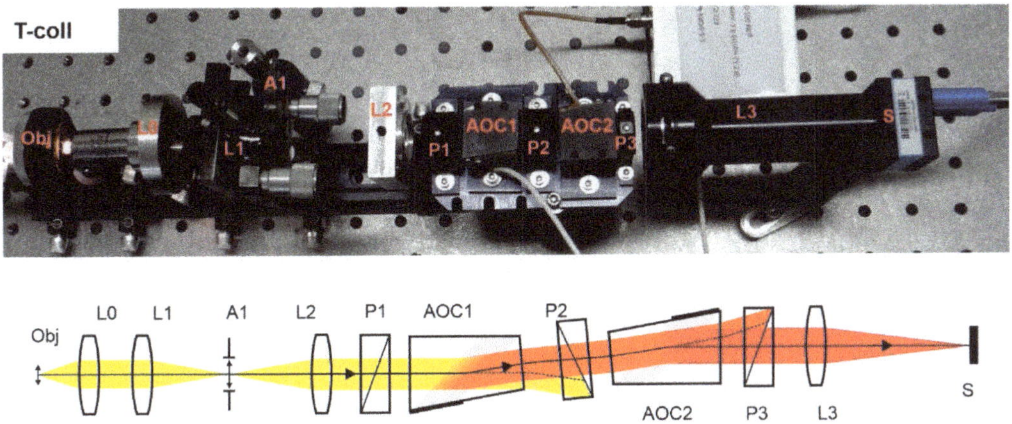

Figure 5. T-coll scheme. Breadboard model and optical beam-pass diagram.

Double-pass single-crystal collimating scheme (D-coll) contains beamsplitter BS, which divides it into three parts (Figure 6). The input optical system consists of a lens L0 and the afocal system L1-L2 with the field stop A1 inside. The spectral unit additionally comprises an AO cell, a retroreflector (lens L6 with the mirror M), which directs back the diffracted light emerging from AO cell, while the rest of the light does not return. The reflected spectrally selected light diffracts in the AO cell and is filtered one more time. In terms of ray tracing and spectral selection, the double-pass scheme is very similar to the tandem one with two identical AO cells.

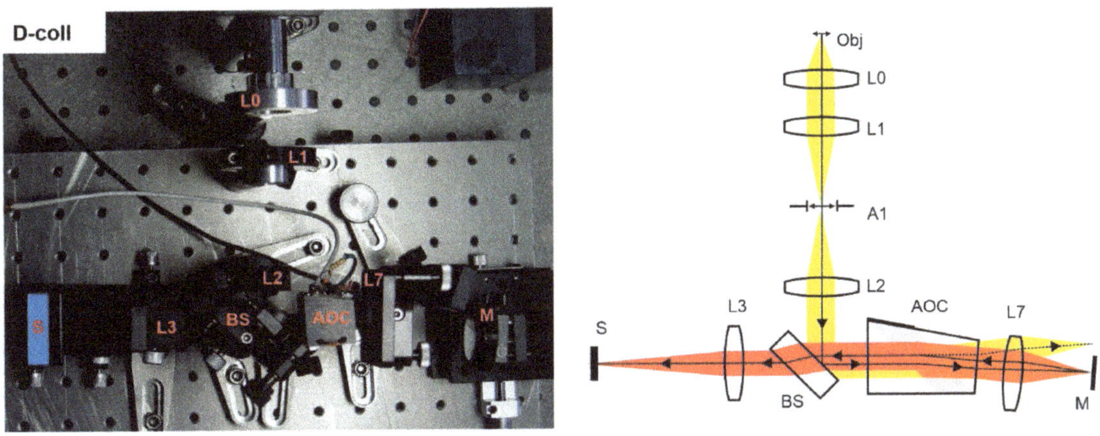

Figure 6. D-coll scheme. Breadboard model and optical beam-pass diagram.

In this scheme, we used a Thorlabs BSF10-A beamsplitter. A plate beamsplitter is preferable, as reflections from the edges of that usually degrade the image contrast. In general, the normal incidence on any flat surface located after BS should be avoided to protect the sensor from stray light that would reduce the image contrast. Even AR coating does not fully solve this problem, since the filtered light has a narrow spectrum and therefore its intensity is much less than the intensity of unfiltered light. To minimize stray light, the AOC is slightly tilted in the sagittal plane and its faces are coated with the antireflection coating. These measures, however, do not ensure solving the problem, because usually the tilt angle is small (otherwise it will cause optical aberrations, in particular astigmatism) and so reflections emerging in a wide spectral range make the selection of narrowband filtered light difficult. For this reason, polarizers are not suitable in this scheme. To solve the problem, we placed a field stop A1, like in the S-coll scheme (Figure 3), and thus have managed to adjust the scheme.

Generally, in this scheme, the transmission is significantly reduced by the beamsplitter, but there is an additional optical channel which can be used for white-light imaging or auxiliary purposes. Therefore, the double-pass scheme is promising for multimodal operation. One can also use an optional beamsplitter and other considered schemes to provide a similar multimodal operation. However, this will lead to additional losses in all other schemes.

3. Experimental Results

Imaging capabilities of the described schemes were estimated with use of the test targets (Figure 7a) and specific color print (Figure 7b), for characterization of the spatial resolution, image distortion and spectral imaging contrast features.

(**a**) (**b**)

Figure 7. Spatial resolution test target (**a**) and color painting (**b**) used in the experiments.

For adequate comparison, one needs to equalize the image settings. To make the fields of view of all four systems roughly equal, we chose appropriate parameters of optical components L0–L6. The image longitudinal chromatic aberrations in the confocal scheme caused by the dispersion of the α-TeO_2 crystal were compensated for by custom designed lenses L4 and L5. In collimating schemes, defocusing is negligible, and other optical elements were standard achromatic lenses.

To equalize the image brightness, we optimized the intensity of the halogen light source as well as the exposure time and gain of the camera. For example, images at wavelength 600 nm (Figure 8) were recorded with the same lighting conditions, but different exposure times: 0.01 s (S-call and S-conf), 0.25 s (T-coll) and 0.5 s (D-coll).

The spectral resolution was measured in the central part of the field of view by the diffraction grating spectrometer Ocean Insight FLAME.

As can be seen from Figures 8 and 9, classical single-crystal single-pass schemes (S-coll and S-conf) demonstrate rather similar features. The main differences are higher spatial resolution in the confocal scheme, with lower spectral resolution [17,18]. Double-stage schemes (T-coll and D-coll) exhibit significantly higher spectral and spatial resolution, in exchange for higher light losses [19]. The double-pass scheme (D-coll) combines the advantages of single-crystal and tandem AOTFs (high resolution at lower cost), but requires laborious adjustment. Generally, all the schemes demonstrate their usability so a given AO cell can be used in different schemes.

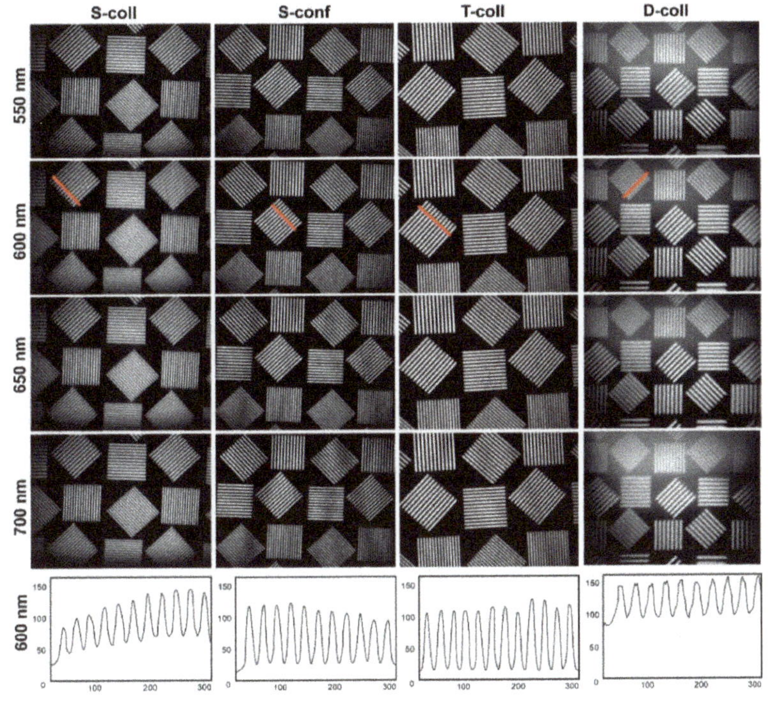

Figure 8. Spectral images of the test target and intensity cross sections along the red lines.

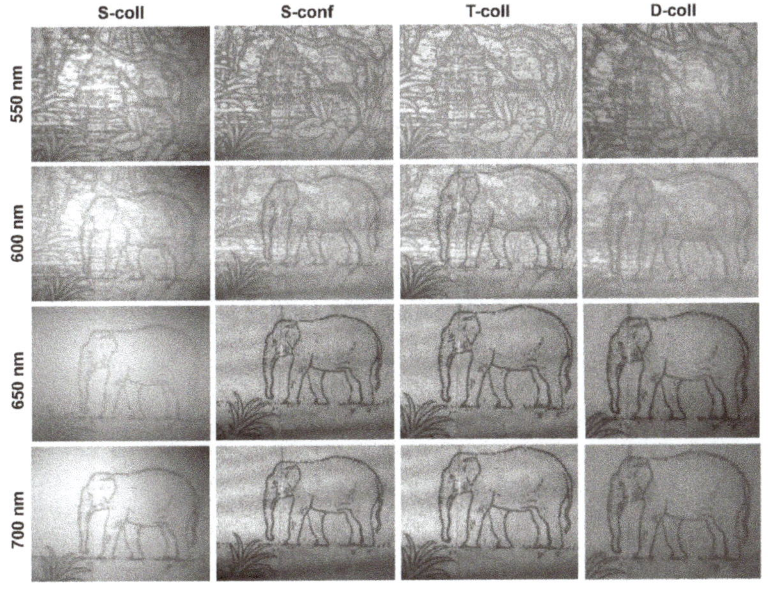

Figure 9. Spectral images of test color print recorded in four different schemes.

4. Discussion and Conclusions

In this experimental study, we assembled four popular optical schemes for AOTF-based spectral imaging (Figures 3–6). The same typical AOCs operating in a conventional wide-aperture mode were exploited. In this way, we eliminated the influence of the AOC's features and highlighted the differences caused by the optical scheme. Though all schemes are capable of providing acceptable image quality, there are considerable differences in spectral and spatial resolution and other imaging characteristics and, therefore, adequate choice of the optical scheme permits optimization of particular features important for solving the problem.

A detailed analysis of the obtained images (Figures 8 and 9) is presented in Table 1. These results should be interpreted primarily as a comparative assessment of the studied schemes and not as a merit of the best performance achievable in each scheme.

Table 1. Experimental results.

Scheme	Spectral Resolution (at 600 nm), nm	Contrast at 15 mm^{-1}	Chromatic (450–850 nm) Drift, %	Exposure Time (at 600 nm), s	Rel. Cost
S-coll	4.8	0.50	5%	0.01	1
S-conf	5.5	0.60	0.5%	0.01	1.1
T-coll	3	0.65	1%	0.25	2.5
D-coll	3.8	0.20	0.5%	0.5	1.5

The spectral resolution, as expected, is higher in schemes with double AO filtering (D-coll and T-coll). The best image contrast at the selected spatial frequency (~15 mm^{-1}) is achieved in a scheme with two AOCs (T-coll) and in a confocal scheme. The worst contrast is demonstrated by the double-pass scheme (D-coll), in our opinion, for two reasons. First is stray light. Second, light energy losses make it necessary to work with long exposure times, and therefore, along with the signal, the background light is also exposed for longer.

In the experiments, we used AOCs with the geometry optimized for the confocal scheme. In particular, it compensates for the chromatic drift by the optimal output facet angle. Therefore, the chromatic drift is almost absent in S-conf, but it is noticeable in S-coll. In double-pass schemes, chromatic drift is compensated. Its residual non-zero values can be explained by inaccurate alignment. Exposure time in double-pass schemes is significantly higher due to double absorption in the AOC, a narrower spectral bandwidth, and the presence of a beamsplitter (in the D-coll scheme). In all schemes, the AOCs are the most expensive components, followed by polarizers. Thus, T-coll scheme seems to be the most cost ineffective.

This study paves the way to understanding the main issues related to the optical design of AOTF-based spectral imagers. A good result is always a compromise between the AOTF configuration and the complexity of the optical coupling and adjustment. Depending on the application, each scheme may be effective. The study results are helpful for the development of new AOTF-based imagers. Each scheme is worth detailed studying and discussing in a separate article.

Author Contributions: Conceptualization and original draft preparation, V.B. and A.M.; methodology, V.B.; formal analysis and supervision, V.P. and S.B.; investigation, G.M. and D.K.; resources, V.L.; review and editing, G.M., A.G. and V.P.; data processing, A.G.; funding acquisition, S.B. All authors have read and agreed to the published version of the manuscript.

Funding: This research was funded by the Russian Foundation for Basic Research, grant number 18-29-20095. The production of TeO_2 crystals for AOCs was supported by Federal State Task Program of the Federal Scientific Research Center "Crystallography and Photonics".

Institutional Review Board Statement: Not applicable.

Informed Consent Statement: Not applicable.

Data Availability Statement: The data presented in this study are available on request from the corresponding author.

Acknowledgments: This work was performed using the equipment of the Center for Collective Use of the Scientific and Technological Center of Unique Instrumentation of the Russian Academy of Sciences [21].

Conflicts of Interest: The funders had no role in the design of the study; in the collection, analyses, or interpretation of data; in the writing of the manuscript, or in the decision to publish the results.

References

1. Lu, G.; Fei, B. Medical Hyperspectral Imaging: A Review. *J. Biomed. Opt.* **2014**, *19*, 010901. [CrossRef] [PubMed]
2. Li, Q.; He, X.; Wang, Y.; Liu, H.; Xu, D.; Guo, F. Review of spectral imaging technology in biomedical engineering: Achievements and challenges. *J. Biomed. Opt.* **2013**, *18*, 100901. [CrossRef] [PubMed]
3. Inoue, Y.; Peñuelas, J. An AOTF-Based Hyperspectral Imaging System for Field Use in Ecophysiological and Agricultural Applications. Int. *J. Remote Sens.* **2001**, *22*, 3883–3888. [CrossRef]
4. Hagen, N.; Kudenov, M.W. Review of Snapshot Spectral Imaging Technologies. *Opt. Eng.* **2013**, *52*, 090901. [CrossRef]
5. Chanover, N.J.; Voelz, D.G.; Glenar, D.A.; Young, E.F. AOTF-Based Spectral Imaging for Balloon-Borne Platforms. *J. Astron. Instrum.* **2014**, *3*, 1440005. [CrossRef]
6. Korablev, O.I.; Belyaev, D.A.; Dobrolenskiy, Y.S.; Trokhimovskiy, A.Y.; Kalinnikov, Y.K. Acousto-optic tunable filter spectrometers in space missions. *Appl. Opt.* **2018**, *57*, C103–C119. [CrossRef] [PubMed]
7. Abdlaty, R.; Sahli, S.; Hayward, J.; Fang, Q. Hyperspectral imaging: Comparison of acousto-optic and liquid crystal tunable filters. *Proc. SPIE* **2018**, 105732P. [CrossRef]
8. Pozhar, V.; Machihin, A. Image Aberrations Caused by Light Diffraction via Ultrasonic Waves in Uniaxial Crystals. *Appl. Opt.* **2012**, *51*, 4513. [CrossRef] [PubMed]
9. Voloshinov, V.B.; Yushkov, K.B.; Linde, B.B.J. Improvement in Performance of a TeO2 Acousto-Optic Imaging Spectrometer. *J. Opt. A Pure Appl. Opt.* **2007**, *9*, 341–347. [CrossRef]
10. Balakshy, V.I. Application of Acousto-Optic Interaction for Holographic Conversion of Light Fields. *Opt. Laser Technol.* **1996**, *28*, 109–117. [CrossRef]
11. Machikhin, A.; Batshev, V.; Pozhar, V. Aberration Analysis of AOTF-Based Spectral Imaging Systems. *J. Opt. Soc. Am. A* **2017**, *34*, 1109. [CrossRef] [PubMed]
12. Chang, I.C. Noncollinear Acousto-optic Filter with Large Angular Aperture. *Appl. Phys. Lett.* **1974**, *25*, 370–372. [CrossRef]
13. Machikhin, A.; Pozhar, V.; Batshev, V. Double-AOTF-Based Aberration-Free Spectral Imaging Endoscopic System for Bio-medical Applications. *J. Innov. Opt. Health Sci.* **2015**, *08*, 1541009. [CrossRef]
14. Zhang, C.; Wang, H.; Zhang, Z.; Yuan, J.; Shi, L.; Sheng, Z.; Zhang, X. Narrowband Double-Filtering Hyperspectral Imaging Based on a Single AOTF. *Opt. Lett.* **2018**, *43*, 2126. [CrossRef] [PubMed]
15. Mazur, M.M.; Pustovoit, V.I. Non-Collinear Acousto-Optical Filter. Russian Patent RU 2388030 C1, 27 April 2010.
16. Gorevoy, A.V.; Machikhin, A.S.; Martynov, G.N.; Pozhar, V.E. Spatiospectral Transformation of Noncollimated Light Beams Diffracted by Ultrasound in Birefringent Crystals. *Photon. Res. PRJ* **2021**, *9*, 687–693. [CrossRef]
17. Suhre, D.R.; Denes, L.J.; Gupta, N. Telecentric Confocal Optics for Aberration Correction of Acousto-Optic Tunable Filters. *Appl. Opt.* **2004**, *43*, 1255. [CrossRef] [PubMed]
18. Machikhin, A.S.; Pozhar, V.E. Spatial and Spectral Image Distortions Caused by Diffraction of an Ordinary Polarised Light Beam by an Ultrasonic Wave. *Quantum Electron.* **2015**, *45*, 161–165. [CrossRef]
19. Pustovoit, V.I.; Pozhar, V.E.; Mazur, M.M.; Shorin, V.N.; Kutuza, I.B.; Perchik, A.V. Double-AOTF Spectral Imaging System. *Proc. SPIE* **2005**, *5953*, 200–203. [CrossRef]
20. Batshev, V.I.; Machikhin, A.S.; Kozlov, A.B.; Boritko, S.V.; Sharikova, M.O.; Karandin, A.V.; Pozhar, V.E.; Lomonov, V.A. Tunable Acousto-Optic Filter for the 450–900 and 900–1700 nm Spectral Range. *J. Commun. Technol. Electron.* **2020**, *65*, 800–805. [CrossRef]
21. Center for Collective Use of STC UI RAS. Available online: http://ckp.ntcup.ru/ (accessed on 20 May 2021).

materials MDPI

Article

Polarization-Multiplexed High-Throughput AOTF-Based Spectral Imaging System

Hao Zhang [1,2], Huijie Zhao [1,2,3,*], Qi Guo [1,2], Dong Xu [1,2] and Wenjie Teng [1,2]

1 School of Instrumentation Science and Opto-Electronics Engineering, Beihang University, Beijing 100191, China; hao2017@buaa.edu.cn (H.Z.)
2 Aerospace Optical-Microwave Integrated Precision Intelligent Sensing, Key Laboratory of Ministry of Industry and Information Technology, Beihang University, Beijing 100191, China
3 Institute of Artificial Intelligence, Beihang University, Beijing 100191, China
* Correspondence: hjzhao@buaa.edu.cn

Abstract: Spectral imaging detection using acousto-optical tunable filters (AOTFs) faces a significant challenge of low throughput due to the traditional design that only receives a single polarization light. To overcome this issue, we propose a novel polarization multiplexing design and eliminate the need for crossed polarizers in the system. Our design allows for simultaneous collection of ±1 order light from the AOTF device, resulting in a more than two-fold increase in system throughput. Our analysis and experimental results validate the effectiveness of our design in improving system throughput and enhancing the imaging signal-to-noise ratio (SNR) by approximately 8 dB. In addition, AOTF devices used in polarization multiplexing applications require optimized crystal geometry parameter design that does not follow the parallel tangent principle. This paper proposes an optimization strategy for arbitrary AOTF devices which can achieve similar spectral effects. The implications of this work are significant for target detection applications.

Keywords: AOTF; spectral imaging system; polarization multiplexed; high throughput

check for updates

Citation: Zhang, H.; Zhao, H.; Guo, Q.; Xu, D.; Teng, W. Polarization-Multiplexed High-Throughput AOTF-Based Spectral Imaging System. *Materials* **2023**, *16*, 4243. https://doi.org/10.3390/ma16124243

Academic Editors: Alexander S. Machikhin and Vitold Pozhar

Received: 10 May 2023
Revised: 4 June 2023
Accepted: 4 June 2023
Published: 8 June 2023

1. Introduction

The acousto-optic tunable filter (AOTF) is a narrow-band spectral filter that can be electrically tuned. It operates based on the principle of acousto-optic interaction, where the refractive index of the acousto-optic medium undergoes periodic changes in response to an input acoustic wave [1]. This phenomenon resembles the behavior of a diffraction grating [2]. Figure 1 illustrates the AOTF devices, featuring an acoustic wave generated by the transducer and absorbed by an absorber. By switching the radio frequency (RF) signals applied to the transducer, the AOTF device can scan the spectral regions of interest [3]. The AOTF device offers fast response time, high spectral resolution, and wavelength configurability, making it an ideal choice for various applications, including remote sensing and medical imaging [4–8]. However, the low throughput limits the performance of the AOTF spectral imaging system [9,10]. This limitation stems from the optical aperture constraints of the AOTF devices and the extinction design of the system. The typical optical aperture limitations for the visible AOTF devices with TeO_2 crystal are a through-aperture of up to 25×25 mm^2 [11] and a real aperture angle of 9° [12]. These constraints are primarily influenced by the properties of the crystal material, and ongoing research explores alternative materials [13,14]. Meanwhile, the extinction design of the system is mainly attributed to the traditional optical design, which only receives a single polarization of light using crossed polarizers. This design choice leads to a significant throughput loss of over 50% [15].

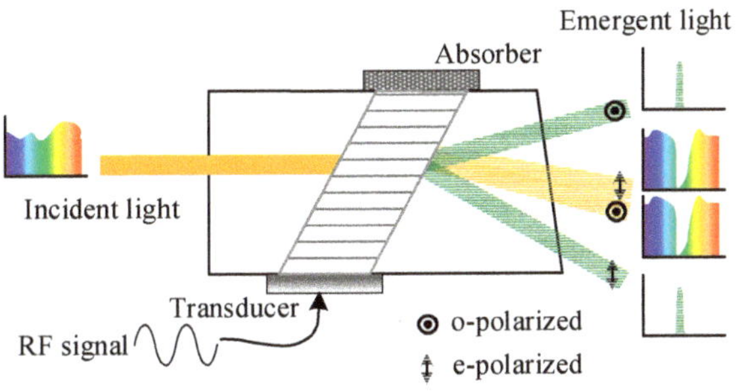

Figure 1. Principle diagram of the AOTF device.

To address the low-throughput issue of AOTF devices, two solutions are commonly employed: non-polarization design and ultrasonic frequency modulation [16,17]. Non-polarization design aims to enhance throughput by eliminating the polarizer from the system. This approach offers distinct advantages over the ultrasonic frequency modulation method, including the avoidance of multi-frequency crosstalk and limitations on acoustic power, as well as the elimination of specific polarization requirements for incident light. In 1999, Voloshinov introduced a design in which the incident light with different polarization states enters the AOTF device at specific angles to ensure that the ± 1 order light is imaged at the same position on the detector. This design utilized a monochromatic laser with beam expansion and the incident light passed through a polarization beam splitter (PBS), separating the ordinary (o) light from the extraordinary (e) light. The o light was reflected into the AOTF device at a predetermined angle [18]. However, this design exhibited optical inconsistency and was only suitable for near-axis fields of view. Subsequently, Fang et al. proposed a dual-arm design, with simultaneous collection of ± 1 order light using two mirrors behind the AOTF device [19]. For two-dimensional (2D) imaging of the sample, the design was combined with a motorized scanning stage [20]. Nonetheless, this design resulted in low-resolution images or was limited to single-point detection applications. Subsequently, they developed a design that can be used for spectral imaging applications, involving the modulation of the polarization state of both beams entering the AOTF device through the use of a half-wave plate [21]. However, this design significantly compresses the real aperture angle of the AOTF device, which relies on the principle of separating diffracted (± 1 order) and transmitted (0 order) light [2]. The suppression of panchromatic transmitted light in spectral imaging systems is crucial, as its intensity is generally much higher than the effective spectral intensity. Additionally, stray light resulting from the phase-delay variation across the application spectrum range of the half-wave plate also affects image quality. Moreover, the front design of the system is based on collimation optics, which does not facilitate the simultaneous detection of ± 1 order light (see Section 2.3 for a detailed analysis). Furthermore, Beliaeva et al. analyzed the feasibility of highly efficient tunable light source using a similar principle [22]. In conclusion, the previous AOTF system designs exhibit limitations in spectral imaging applications, including inconsistent spectral response, compressed real aperture angle, transmission stray light interference, and complex structure.

In this paper, we present a high-throughput AOTF system with a polarization multiplexing design for spectral imaging applications. The front telecentric confocal optics with the optimized AOTF device in the system demonstrates its unique suitability for polarization multiplexing applications. To achieve stray light suppression in spectral imaging systems, the rear blocking optics effectively suppress transmitted light while allowing diffracted light to pass unaffected. Compared to the traditional single polarization detection systems, our proposed design enables simultaneous collection of the ± 1 order light of the AOTF device and eliminates the need for crossed polarizers, resulting in an over two-fold throughput improvement. Furthermore, the AOTF devices used in polarization multiplexing applications need optimization of crystal geometry parameters, which does not follow the parallel tangent principle. To address this issue, we propose an optimization strategy for arbitrary AOTF devices that can achieve similar goals, avoiding the cost and long manufacturing periods associated with traditional methods. The proposed design and optimization strategy are introduced and discussed in detail in Section 2, with experimental results from desktop systems provided in Section 3 to verify their effectiveness in achieving high throughput.

2. Methods and Analysis

2.1. Basic Principles of the Noncollinear AOTF Device

There are two fundamental configurations of AOTFs: collinear and noncollinear. In the collinear configuration, the interacting optical and acoustic waves propagate in the same direction, whereas in the noncollinear configuration, the directions of the optical and acoustic waves are different [23]. As shown in Figure 2, a typical noncollinear AOTF device, which usually has three crystal geometry parameters, including the front facet angle (θ_i^*), the acoustic cut-angle (θ_α), and the back facet angle (θ_β), which greatly affect its performance. In particular, most noncollinear AOTF devices are designed based on the parallel tangent principle to achieve high diffraction efficiency within the angular aperture [24]. For arbitrarily polarized incidence, the o and e light must be analyzed separately due to their distinct refractive indices. In addition, this paper establishes two coordinate systems for analysis: the optical axis coordinate system (X_0OY_0) and the crystal axis coordinate system (X_1OY_1).

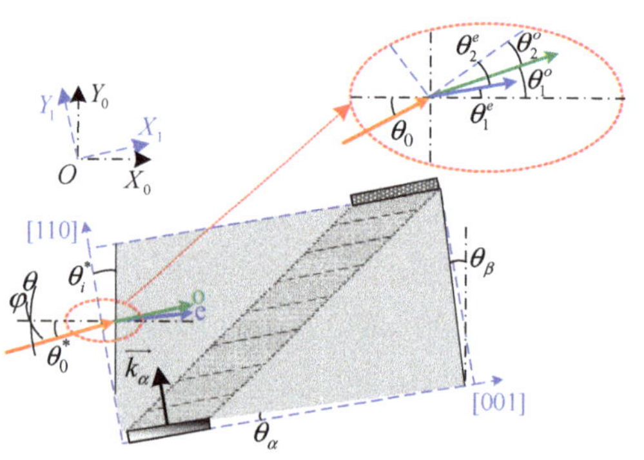

Figure 2. Structure diagram of the noncollinear AOTF device (top view).

2.2. Structure of the AOTF Spectral Imaging System

The proposed system consists mainly of the front optics, an AOTF device, the blocking optics, and an imager as shown in Figure 3. The AOTF device, which is the central component of the system, has certain constraints on the system design such as the working spectral range, the through aperture (D_{AOTF}), and the real aperture angle (θ_{AOTF}), as shown in Figure 4a. These constraints affect the design of the front optics. Typically, the front optics have two structures: collimation and confocal [25]. However, in our proposed design, we use telecentric confocal optics, where L_{input} is equivalent to the focal length (f_1) of the front lens1 [26,27]. For a $2\omega_0$ field-of-view application, the front optics must meet certain requirements as below:

$$\begin{cases} 2f_1 \tan\omega_0 = 2H_{\text{in}} \le D_{\text{AOTF}} \\ 2\arctan\left(D_{\text{input}} \Big/ 2f_1\right) = \theta_{\text{in}} \le \theta_{\text{AOTF}} \end{cases} \tag{1}$$

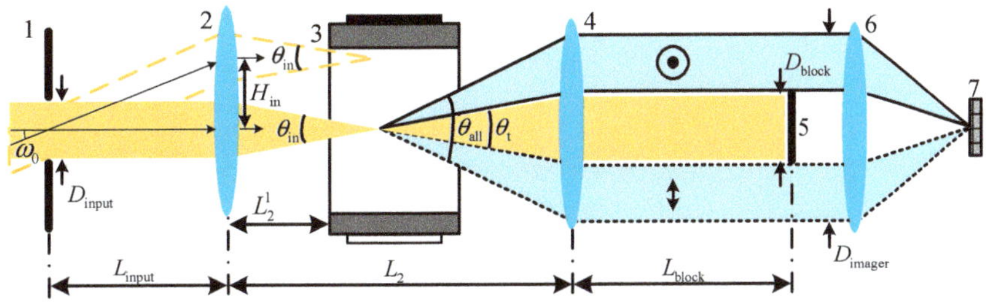

Figure 3. Structure diagram of the proposed AOTF spectral imaging system. 1—input aperture, 2—lens1, 3—AOTF device, 4—lens2, 5—light barrier, 6—lens3, and 7—detector.

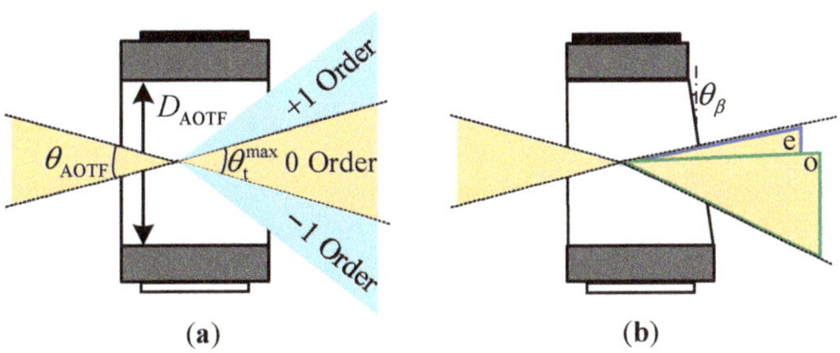

Figure 4. Diagram of AOTF devices. (**a**) AOTF device without a wedge angle for monochromatic light application, and (**b**) the extension of 0 order light caused by the wedge angle in the AOTF device.

In spectral imaging systems, the 0 order light represents the panchromatic spectrum and typically exhibits significantly higher intensity than the effective spectral intensity. Therefore, it is crucial to employ blocking optics to suppress the 0 order light. In this study, the blocking optics are composed of a lens set and a light barrier located at the exit pupil. The width of the light barrier is meticulously designed to fulfill specific criteria:

$$2f_2 \tan\left(\theta_t^{max} \Big/ 2\right) \geq D_{block} \geq 2f_2 \tan\left(\theta_t \Big/ 2\right) \tag{2}$$

where θ_t is the cone angle of the 0 order (panchromatic transmitted) light and f_2 is the focal length of lens2. Additionally, θ_t^{max} represents the maximum cone angle of 0 order light, corresponding to the angle between the ±1 order light. This design effectively blocks the 0 order light, while allowing the ±1 order light to pass unaffected. The real aperture angle can be estimated by the separation angle of the normal incident light. It is worth noting that some AOTF devices have a wedge compensation that extends the 0 order light, as depicted in Figure 4b, which should be considered when calculating the real aperture angle.

The imager is used to capture spectral images and consists of an objective lens and a focal plane array detector. To ensure efficient light collection, the objective lens requires a large entrance pupil diameter, represented as:

$$D_{imager} \geq 2f_2 \tan\left(\theta_{all} \Big/ 2\right) \tag{3}$$

where θ_{all} is the cone angle of all emitted light as Figure 3. Our proposed system satisfies the "$4f$" design criteria, where $L_{input} = f_1$, $L_2 = f_1 + f_2$, and $L_{block} = f_2$. During operation, the incident light forms the first image point inside the AOTF device, and all rear optics perform secondary imaging on this point. We find that the relative position parameter (L_2^1)

of the AOTF device is a key parameter, and optimizing this parameter ensures a better image quality of the ±1 order light with polarization multiplexing design. As shown in Figure 5a, the image quality is described by the half-size of the dispersed spot (δ_{oe}) in the diffraction direction of the AOTF device in this paper. The position of the first image point of the o and e light is obtained by ray-tracing the main light and the edge light based on geometric optics theory as Figure 5a, and the ray tracing process inside an AOTF device can be found in Ref. [28]. Then, the traversing method is used to find the optimized L_2^1 so that the lateral difference (Δy_{oe}) between the two image points is 0 as Figure 5b,c. Finally, the optimal first image point is searched in the interval of o and e light image points to minimize the comprehensive δ_{oe} of the two beams.

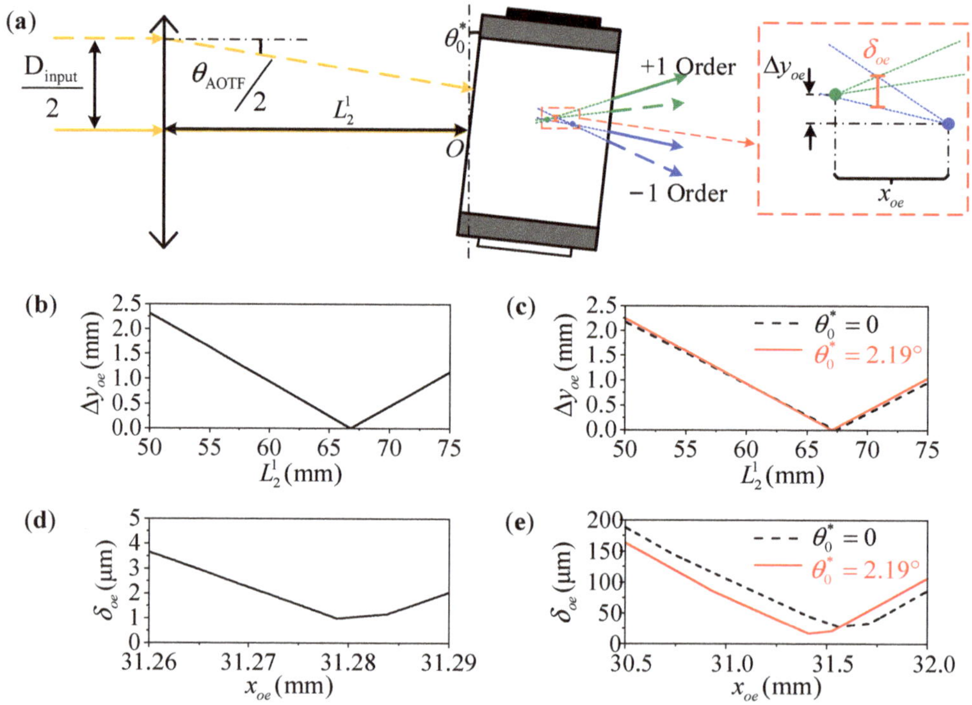

Figure 5. The parameter L_2^1 will affect the relative position of the first image points of the o and e light. (**a**) Schematic diagram of light path. Additionally, the relative position difference distribution is shown with the crystal geometry parameters of (**b**,**d**) $\theta_i^* = 14.14°$, $\theta_\alpha = 6.50°$, and $\theta_\beta = 0°$, (**c**,**e**) $\theta_i^* = 15.07°$, $\theta_\alpha = 6.49°$, and $\theta_\beta = -4.64°$. The red solid lines are the results of the 2nd AOTF device by the optimization strategy. x_{oe} in (**d**,**e**) is the relative distance to the incident point O.

In this paper, two AOTF devices are presented for comparison. The first device (the 1st AOTF) has optimized crystal geometry parameters of $\theta_i^* = 14.14°$, $\theta_\alpha = 6.50°$, and $\theta_\beta = 0$ (the 1st AOTF) for non-polarization applications [29,30]. The second device (the 2nd AOTF) is used in the experiments and closes to the parallel tangent principle for e-light input application, with crystal geometry parameters of $\theta_i^* = 15.07°$, $\theta_\alpha = 6.49°$, and $\theta_\beta = -4.64°$ [31]. As shown in Figure 5b,d, the optimized L_2^1 of the 1st AOTF device is 66.81 mm, and the minimum δ_{oe} is less than 1.0 μm. From Figure 5c,e, we can find that the optimized L_2^1 of the 2nd AOTF device is 67.36 mm, and the minimum δ_{oe} is 28.4 μm. As the optimization strategy is applied to the 2nd AOTF device, resulting in improved aberration values of $L_2^1 = 67.06$ mm and $\delta_{oe} = 16.7$ μm. Additionally, the optimization strategy is described in detail in Section 2.3. It is observed that using AOTF devices with unoptimized crystal geometry parameters results in large aberrations, which is not desirable for polarization multiplexing design. Additionally, the aberration will be improved with applying the optimization strategy. However, it is pointed out that the optimized AOTF device is a better choice and we are working on this for the future.

2.3. Optimization Strategy for Polarization Multiplexing Applications

Traditional AOTF devices, which follow the parallel tangent principle, are typically designed for a single polarization state (o or e light). However, when the 2nd AOTF device is directly used for polarization multiplexing applications, there can be a substantial spectral difference in the ± 1 order light, as depicted in Figure 6(a,c4). This issue can be mitigated by optimizing the crystal geometric parameters of the AOTF devices [29,30]. By optimizing these parameters, the normal incident o and e light of the same wavelength can satisfy the momentum matching principle at the same tuning frequency, resulting in a significant improvement in the matched wavelength distribution of o and e light within the real aperture angle, as demonstrated in Figure 6b,e. However, developing such a new AOTF device can be costly and time-consuming. Alternatively, we propose an optimization strategy to rotate the traditional AOTF device at an off-axis angle (θ_0^*) in the polar plane to achieve the same goal, as illustrated in Figure 2. This also enables the o and e light of the same wavelength to satisfy the momentum matching principle at the same tuning frequency, as depicted in Figure 6f. In this case, the matched wavelength distribution of o and e light within the real aperture angle is shown in Figure 6d. The difference between the optimization strategy and the optimized crystal geometric parameters is negligible, with differences of less than 0.01 nm. Moreover, AOTF devices are sensitive to the polarization state and incident angle of the light. It can be observed from Figure 6b,d that the optimized AOTF device can only ensure consistent matched wavelengths of o and e light at the given incident angle. As shown in Figure 6(c1), the optimized AOTF device has the same matched wavelengths of o and e light under normal incidence. Although the matched wavelengths at the edge have been improved, the difference is still about 9.7 nm, as shown in Figure 6(c2). Previous studies have reported the beneficial effects of the telecentric confocal design in suppressing optical sidelobe effects [27]. However, we have found that this design also has a great adaptability for polarization multiplexing applications. It enables the AOTF device to filter the arbitrary views equally, and the spectral response function is equal to the integral within the cone angle of the light entering the AOTF device as [32]:

$$\text{Response} \propto \int\int_{-\frac{\theta_{in}}{2}}^{\frac{\theta_{in}}{2}} T_0^o(\theta,\varphi,\lambda) I(\eta_o, L_o, \Delta k_o) d\theta d\varphi + \int\int_{-\frac{\theta_{in}}{2}}^{\frac{\theta_{in}}{2}} T_0^e(\theta,\varphi,\lambda) I(\eta_e, L_e, \Delta k_e) d\theta d\varphi \tag{4}$$

where θ is polar angle of the incident light and φ is the azimuth angle as shown in Figure 2. $T_0^o(\theta,\varphi,\lambda)$ and $T_0^e(\theta,\varphi,\lambda)$ are the intensities of o and e light entering the AOTF device, respectively. Additionally, $I(\eta, L, \Delta k)$ is the diffraction efficiency equation, given as:

$$I = \eta \frac{\sin^2\left[\eta + (\Delta k L/2)^2\right]^{1/2}}{\eta + (\Delta k L/2)^2} \tag{5}$$

where L is the acousto-optic interaction length that is affected by the length of the transducer. Additionally, the actual transducer length of the 2nd AOTF device used in the experiments is about 10 mm. The parameter Δk describes the momentum mismatch and η represents the peak diffraction efficiency [33]. By employing telecentric confocal optics, the spectral response curves of o and e light are closer to each other as Figure 6(c3). In contrast, adopting the collimation design would cause the occurrence of "double peaks" in the spectral response curves at some image points as shown in Figure 6(c2), which is not conducive to spectral detection.

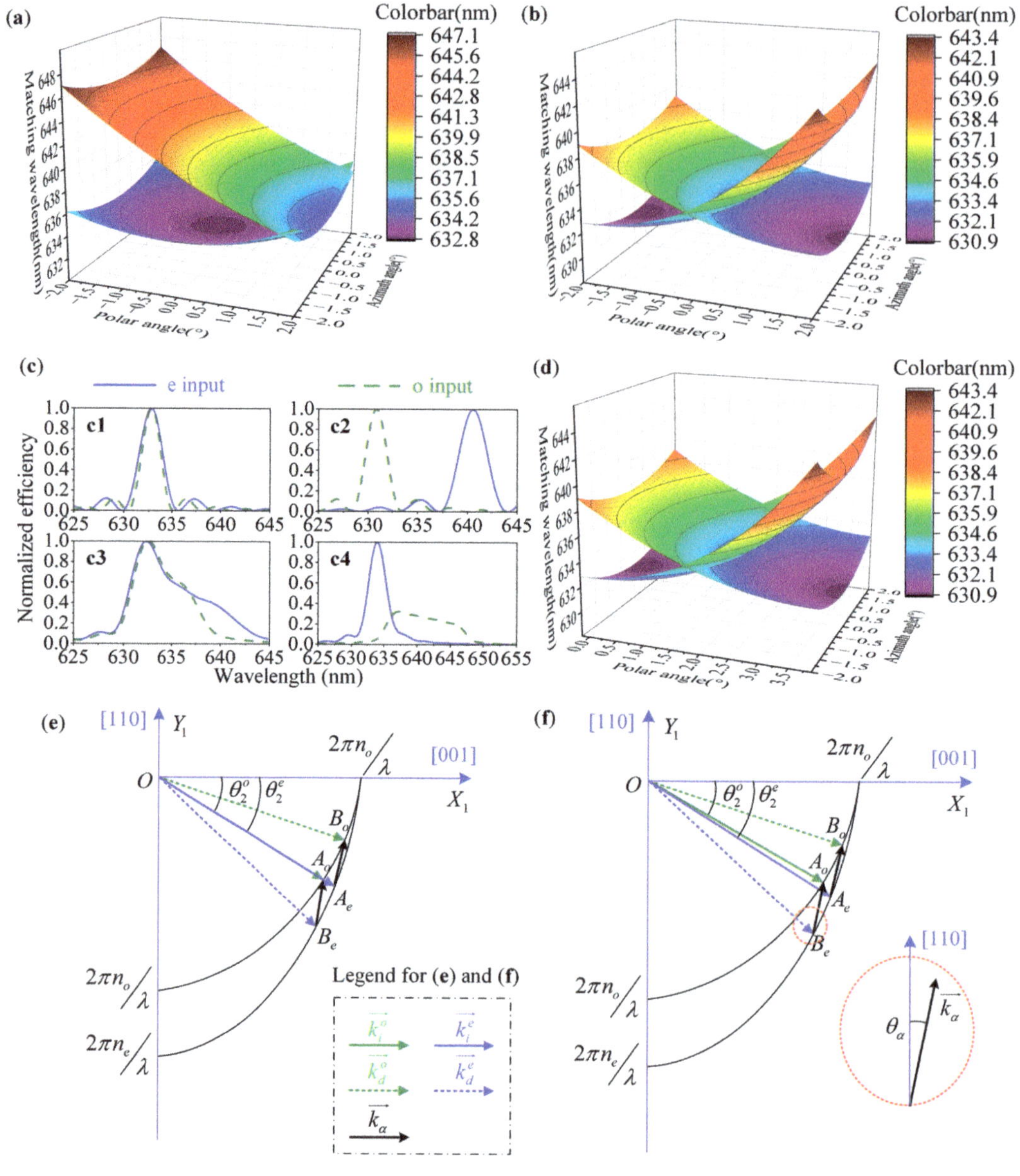

Figure 6. Matched wavelength distributions within the real aperture angle at 632.8 nm of (**a**) the 2nd AOTF device, (**b**) the 1st AOTF device, and (**d**) the 2nd AOTF device by the optimization strategy. (**c**) Spectral curves with the collimation design in which (**c1**) describes the normal incidence of the 1st AOTF device, and (**c2**) corresponds to the edge field. Spectral curves with telecentric confocal design that (**c3**) describes the integral response of the 1st AOTF device, and (**c4**) corresponds to the 2nd AOTF device. (**e**) The wave vector diagram of the 1st AOTF device, and (**f**) corresponds to the 2nd AOTF device by the optimization strategy.

The torsional strategy is introduced in detail below. The light with an off-axis angle θ_0 obeys the Snell's law as:

$$n_0 \sin\theta_0 = n_1 \sin\theta_1 \tag{6}$$

where θ_1 is the refraction angle in the crystal, and n_0 and n_1 are the refractive indices in the air and crystal, respectively. The n_0 is generally 1.0 in the air, and n_1 of the o and e light can be solved by [34]:

$$\begin{cases} n_1^o = n_o \\ n_1^e = \frac{n_o n_e}{\sqrt{{n_o}^2 \sin^2\theta_2^e + {n_e}^2 \cos^2\theta_2^e}} \end{cases} \tag{7}$$

with markers o and e to distinguish o and e light. n_o and n_e are the principal refractive indices, respectively. In the crystal coordinate system, the refraction angle θ_2 is:

$$\theta_2 = \theta_1 - \theta_i^* \tag{8}$$

Using Equations (6)–(8), we can easily obtain θ_2^o by:

$$\theta_2^o = \arcsin\left(\frac{\sin\theta_0}{n_o}\right) - \theta_i^* \tag{9}$$

while θ_2^e can be obtained by solving a quadratic equation:

$$\left(\cos^2\theta_i^* - \frac{\sin^2\theta_0}{{n_e}^2}\right)\tan^2\theta_2^e + 2\sin\theta_i^*\cos\theta_i^*\tan\theta_2^e + \left(\sin^2\theta_i^* - \frac{\sin^2\theta_0}{{n_o}^2}\right) = 0 \tag{10}$$

Now we can obtain θ_2^o and θ_2^e in the wave vector diagram (Figure 6f) at the incident angle of θ_0, and n_1^o and n_1^e can also be obtained from Equation (7). The incident optical wave vectors can be solved by:

$$\begin{cases} \left|\vec{k_i^o}\right| = \frac{2\pi n_1^o}{\lambda} \\ \left|\vec{k_i^e}\right| = \frac{2\pi n_1^e}{\lambda} \end{cases} \tag{11}$$

and the coordinates of the intersection A are:

$$\begin{cases} A_o = \frac{2\pi n_1^o}{\lambda}(\cos\theta_2^o, \sin\theta_2^o) \\ A_e = \frac{2\pi n_1^e}{\lambda}(\cos\theta_2^e, \sin\theta_2^e) \end{cases} \tag{12}$$

The linear equations of A_oB_e and A_eB_o are similar:

$$y - y_A = \frac{1}{\tan\theta_\alpha}(x - x_A) \tag{13}$$

The wave vector elliptic equations of o and e light are:

$$\begin{cases} \frac{x^2}{{n_o}^2} + \frac{y^2}{{n_o}^2} = \left(\frac{2\pi}{\lambda}\right)^2 \\ \frac{x^2}{{n_o}^2} + \frac{y^2}{{n_e}^2} = \left(\frac{2\pi}{\lambda}\right)^2 \end{cases} \tag{14}$$

Then, using Equations (12)–(14), the coordinate of the point $B_e(x_B^e, y_B^e)$ can be obtained by solving a quadratic equation:

$$\begin{aligned} \left(\frac{\tan^2\theta_\alpha}{{n_o}^2} + \frac{1}{{n_e}^2}\right){y_B^e}^2 & + \frac{4\pi\tan\theta_\alpha}{\lambda n_o}(\cos\theta_2^o - \sin\theta_2^o\tan\theta_\alpha)y_B^e \\ & + \frac{4\pi^2}{\lambda^2}\left(\cos^2\theta_2^o + \sin^2\theta_2^o\tan^2\theta_\alpha - 2\sin\theta_2^o\cos\theta_2^o\tan\theta_\alpha - 1\right) = 0 \end{aligned} \tag{15}$$

and

$$x_B^e = \tan\theta_\alpha(y_B^e - y_A^o) + x_A^o \tag{16}$$

Similarly, we can obtain the coordinates of point $B_o(x_B^o, y_B^o)$ by:

$$\begin{aligned}&(\tan^2\theta_\alpha+1){y_B^o}^2+\frac{4\pi n_1^e\tan\theta_\alpha}{\lambda}(\cos\theta_2^e-\sin\theta_2^e\tan\theta_\alpha)y_B^o\\&+\frac{4\pi^2}{\lambda^2}\left[(n_1^e)^2\cos^2\theta_2^e+(n_1^e)^2\sin^2\theta_2^e\tan^2\theta_\alpha-2(n_1^e)^2\sin\theta_2^e\cos\theta_2^e\tan\theta_\alpha-{n_o}^2\right]=0\end{aligned} \quad (17)$$

and

$$x_B^o=\tan\theta_\alpha(y_B^o-y_A^e)+x_A^e \quad (18)$$

Finally, the fitting torsion angle θ_0^* can be obtained by traversing method with $\left|\overrightarrow{A_oB_e}\right|=\left|\overrightarrow{A_eB_o}\right|$. When using a normal AOTF device with e light input design, θ_0^* is greater than 0. While for an AOTF device with o light input design, θ_0^* is less than 0. In this study, the 2nd AOTF device used had a fitting θ_0^* of 2.19°. The optimization strategy employed also allows for the same AOTF device to be used in o light input mode, which can be achieved through the calculation of the Bragg angle of the o light input application under the parallel tangent principle. Specifically, the Bragg angle of the o light input application should be 13.18° at $\theta_\alpha = 6.49°$, and the torsion angle required is 4.29°, switching to the o light input mode. We can now set the example system parameters as shown in Table 1.

Table 1. Example system parameters of the proposed design.

Parameters	The 1st AOTF	The 2nd AOTF (Optimization Strategy)
$2\omega_0$ (°)	±7.5	±7.5
D_{input} (mm)	5.2	5.2
f_1 (mm)	75	75
L_2^1 (mm)	66.8	67.1
D_{AOTF} (mm)	20	20
θ_{AOTF} (°)	4.0	4.0
f_2 (mm)	75	75
D_{block} (mm)	5.2	5.3
D_{imager} (mm)	15.9	16.1

We have further calculated the theoretical throughputs of the two AOTF devices using Equation (4) with the parameters listed in Table 1. The results of these calculations are presented in Figure 7 and Table 2. In our calculations, we accounted for both polarization states and utilized an angular sampling step of 0.1°. As a result, we obtained an input intensity of 3362 (41 × 41 × 2) at each wavelength within the real aperture angle of 4.0°. Compared to the traditional single polarization detection, the polarization multiplexing design proposed in this paper can improve the throughput by about two-fold. Additionally, we observed a slight increase of 0.7% in the peak diffraction efficiency.

Table 2. Theoretical throughput data and comparison results.

AOTF Device	The 1st AOTF			The 2nd AOTF		
Detection Mode	**Polarization Multiplexing**	**Single e Light**	**Single o Light**	**Polarization Multiplexing**	**Single e Light**	**Single o Light**
Theoretical throughput within 620–650 nm	78,090	38,458	39,119	77,680	37,979	38,402
Maximum diffraction efficiency	33.3%	32.4%	32.6%	33.1%	31.8%	32.4%
Ratio of throughput improvement		2.03	2.00		2.05	2.02

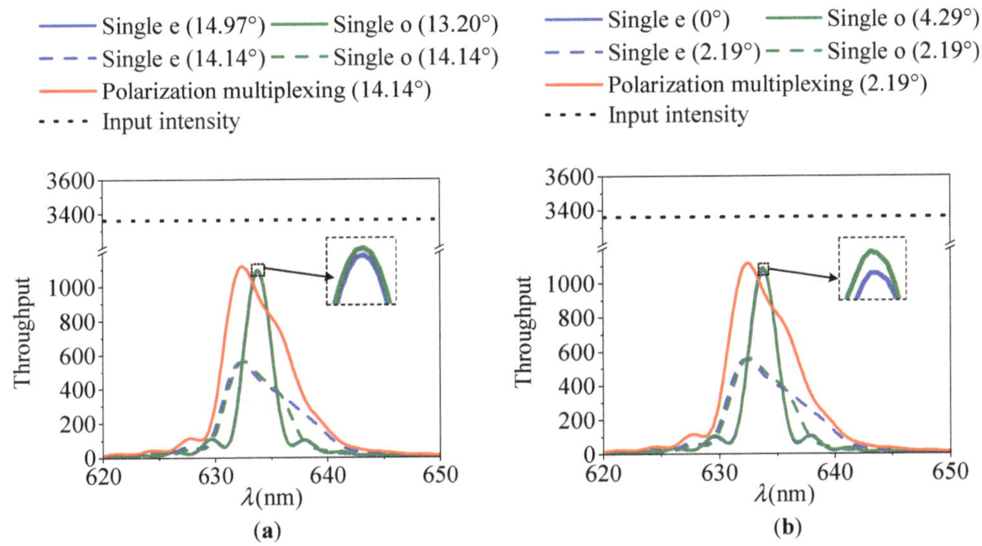

Figure 7. Spectral response curves. (**a**) The 1st AOTF device with optimization of crystal geometric parameters. $\theta_i^* = 14.97°$ and $\theta_i^* = 13.20°$ satisfy the parallel tangent principle of e and o light, respectively. (**b**) The 2nd AOTF device with the optimization strategy.

3. Results and Discussion

3.1. Test and Analysis of Optimization Strategy

A spectral test optics setup, as shown in Figure 8, was constructed. The light source was a wide-spectrum light source composed of an integrating sphere (Halogen lamp) and a collimator. The input aperture used is a circular aperture of 5.2 mm diameter. The AOTF device selected for the experiment was the SGL100-400/850-20LG-K, which was sourced from China. The commercial spectrometer used is the Ocean Optics USB2000+. Spectral sampling of o and e light was performed behind the light barrier (exit pupil). The results of spectral tests are shown in Figure 9. Additionally, the results show good agreement with the theoretical calculations. This demonstrates that spectral modulation of the AOTF device can be realized using the optimization strategy, making it suitable for polarization multiplexing applications. In addition, our previous work has revealed the consistent response of the incident light at a torsion angle of 2.19° [31].

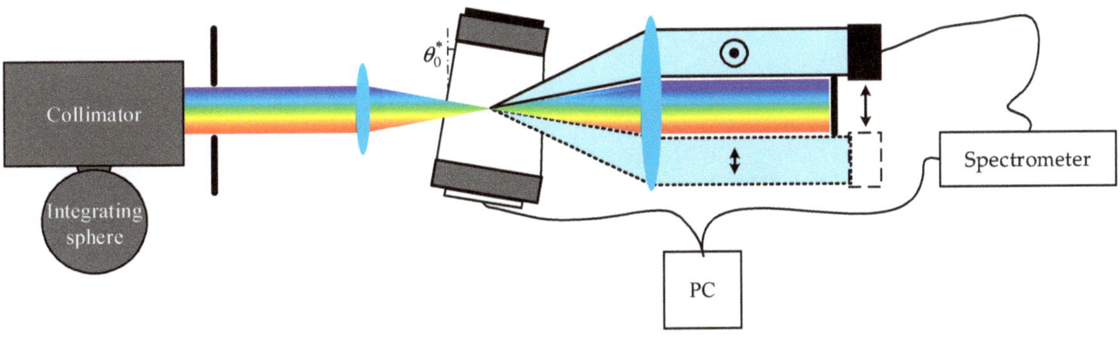

Figure 8. Structure diagram of the desktop system for testing the spectral curves.

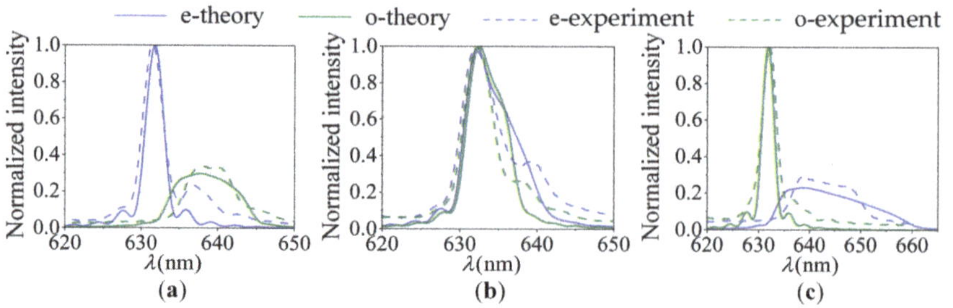

Figure 9. The tested and theoretical spectral response curves at (**a**) $\theta_0^* = 0$, (**b**) $\theta_0^* = 2.19°$, and (**c**) $\theta_0^* = 4.29°$.

3.2. Test and Analysis of Throughput

Then, the actual throughput of the proposed polarization multiplexing design and the traditional single polarization light design are compared in this paper. The evaluation factor used is the average Digital Number (DN) index of the detector within the effective area (green box area of the images), while the image signal-to-noise ratio (SNR) is evaluated using the same size of the red box area of the image. In the tests, len1 and lens2 temporarily use a single lens here, which can be optically optimized in the future. Additionally, further details of the system parameters can be found in Table 1. In the traditional design, cross polarizers are used before and after the AOTF device as shown in Figure 10, and the baffle used is wider to eliminate interference from the other polarization light. The polarizer used in the system has a transmittance of approximately 86% (GCL-050003). The imager uses a camera (Basler acA640-120gm) with 50 mm focal length lens (PENTAX B5014A). The light source is also the wide-spectrum light source composed of an integrating sphere and a collimator. The target is a hollow square target. Throughout the experiment, the camera maintained a consistent exposure time of 20 ms (milliseconds) and a fixed gain of 300. The captured images are listed in Figure 11, which show that the proposed polarization multiplexing design can effectively improve the system throughput by over 2.6-fold, as indicated in Table 3. By eliminating the losses incurred by the cross polarizers, the resulting increase in throughput is nearly twice as much as predicted by the theoretical calculations in Section 2.3. Simultaneously, the efficiency of conventional designs in separately collecting o light and e light is comparable.

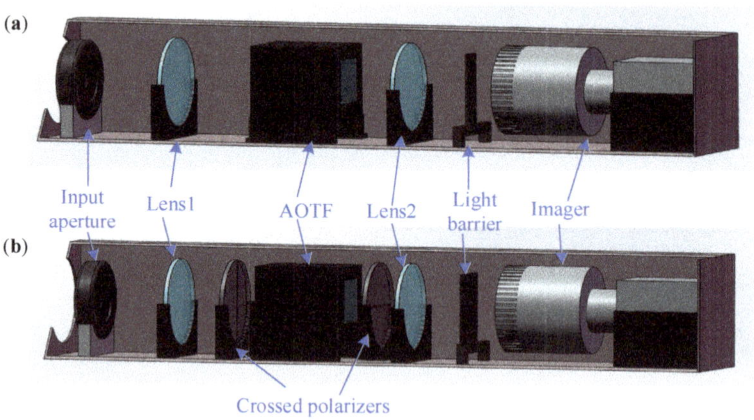

Figure 10. Structure diagram of the AOTF spectral imaging systems for testing the throughput. (**a**) Polarization multiplexing design. (**b**) Single polarization detection design.

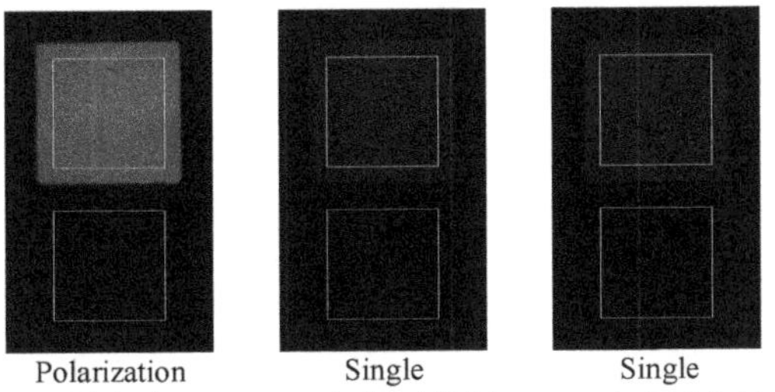

Figure 11. The captured images of the square target at 632.8 nm.

Table 3. The response DN values of the captured images.

Parameter	Test 1	Test 2	Test 3	Average
Signal of polarization multiplexing $\left(DN_s^{pm}\right)$	109.7	109.5	109.5	109.5
Signal of single e − polarized light (DN_s^e)	38.1	38.6	38.3	38.3
Signal of single o − polarized light (DN_s^o)	41.5	42.0	41.9	41.8
DN_s^{pm}/DN_s^e (Eliminating loss of polarizers)	2.88 (2.13)	2.84 (2.10)	2.86 (2.11)	2.85 (2.11)
DN_s^{pm}/DN_s^o (Eliminating loss of polarizers)	2.64 (1.96)	2.61 (1.93)	2.61 (1.93)	2.62 (1.94)
Noise of polarization multiplexing $\left(DN_n^{pm}\right)$	4.9	4.7	4.7	4.8
Noise of single e − polarized light (DN_n^e)	4.2	4.6	4.4	4.4
Noise of single o − polarized light (DN_n^o)	4.4	4.9	4.8	4.7
R^{pm}/dB	27.0	27.4	27.3	27.2
R^e/dB	19.1	18.5	18.8	18.8
R^o/dB	19.5	18.7	18.8	19.0

This paper also includes a comparison of the SNR of images captured by various systems. The SNR can be computed using the following formula:

$$R = 20\lg\left(DN_s \Big/ DN_n\right) \tag{19}$$

where DN_s and DN_n are the DN values of the signal and noise, respectively. The variable R represents the magnitude of the image SNR. The results demonstrate that the proposed polarization multiplexing design improves the image SNR by approximately 8 dB, as illustrated in Table 3.

3.3. Test and Analysis of Spatial Resolution

In addition, spatial resolution represents a crucial characteristic of the imaging system. To assess the image quality in the polarization multiplexing system, we conducted a comprehensive evaluation employing the USAF-1951 resolution target as the reference. Figure 12a displayed the captured image, revealing a notable improvement in quality compared to previously published works [18,35]. For quantitative assessment, we utilized the contrast transfer function (CTF) defined as follows:

$$F = \frac{DN_{\max} - DN_{\min}}{DN_{\max} + DN_{\min}} \tag{20}$$

where $DN_{\max}$ and $DN_{\min}$ are the maximum and minimum DN values along the orthogonal line direction as shown in Figure 12a, respectively. The variable F denotes the CTF value. Furthermore, we obtained the CTF analysis results in both horizontal and vertical directions, displayed in Figure 12b. The horizontal direction corresponds to the AOTF diffraction direction. The spatial resolution of the USAF-1951 target can be determined as follows:

$$\text{Resolution} = 2^{Group + \frac{(Element-1)}{6}} \tag{21}$$

where $Group$ is the group number and $Element$ is the element number as Figure 12a. In the red dot wireframe ($Group = 0$, $Element = 1$), the corresponding single linewidth is 0.5 mm, indicating a resolution of 1 Lp/mm. The CTF analysis results show that the horizontal resolution is slightly higher than the vertical resolution as shown in Figure 12b. This discrepancy is primarily influenced by polarization multiplexing, resulting from the distinction between o and e light [36,37]. In future research, we aim to enhance this aspect by optimizing the geometric parameters of the AOTF device (as detailed in Section 2.2) and optimizing the lens group of the overall optical system.

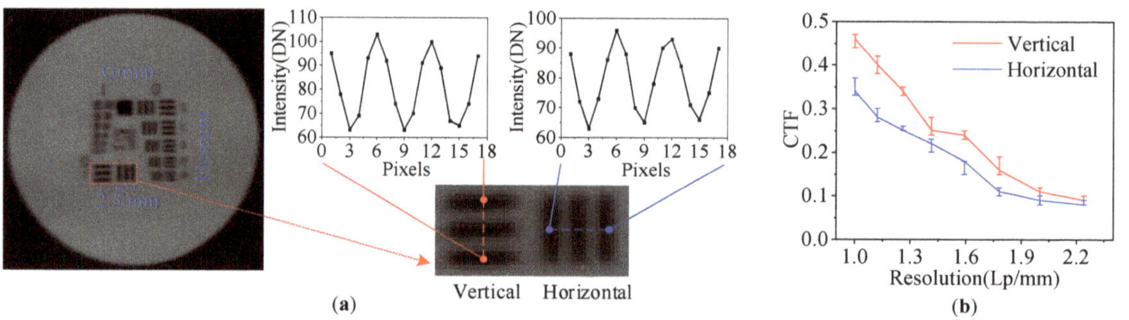

Figure 12. Resolution test and analysis results. (**a**) Captured image of USAF-1951 resolution target with polarization multiplexing application. (**b**) CTF analysis results.

4. Conclusions

In summary, we propose a novel design to address the low throughput of traditional AOTF spectral imaging systems. Our design utilizes polarization multiplexing to simultaneously capture ±1 order light of the AOTF device, resulting in a more than two-fold increase in system throughput. The experimental results validate the effectiveness of our design in improving system throughput and enhancing imaging SNR by approximately 8 dB. Moreover, our proposed design offers simplicity in structure compared to previously published works. Additionally, it is optimized for concerns such as inconsistent spectral response, real aperture angle compression, and transmission stray light interference. This research offers a valuable contribution to the field of AOTF spectral imaging systems in target detection applications. However, it should be noted that the polarization multiplexing design may lead to a reduction in spatial resolution, especially in the AOTF diffraction direction, which necessitates further efforts to overcome in future studies.

In addition, AOTF devices used in polarization multiplexing applications require a special design that does not follow the parallel tangent principle. This paper introduces an optimization strategy for arbitrary AOTF devices, which can achieve similar spectral effects without incurring high costs or long manufacturing periods. The test results demonstrate that the optimization strategy effectively corrects the spectral inconsistency of ±1 order light from the AOTF device at the same tuning frequency. This work is not only applicable to the proposed polarization multiplexing system, but also beneficial to AOTF polarization spectroscopy detection systems [38] and other AOTF detection systems that collect ±1 order light simultaneously.

Author Contributions: Conceptualization, H.Z. (Hao Zhang) and H.Z. (Huijie Zhao); Methodology, H.Z. (Hao Zhang); Investigation, H.Z. (Hao Zhang) and D.X.; Validation, H.Z. (Hao Zhang) and W.T.; Formal analysis, H.Z. (Hao Zhang); Resources, H.Z. (Huijie Zhao) and Q.G.; Writing—original draft preparation, H.Z. (Hao Zhang); Writing—review and editing, H.Z. (Huijie Zhao); Supervision, H.Z. (Huijie Zhao) and Q.G.; Project administration, H.Z. (Huijie Zhao); Funding acquisition, H.Z. (Huijie Zhao) and Q.G. All authors have read and agreed to the published version of the manuscript.

Funding: This research was funded by the National Natural Science Foundation of China (61875004, 62075202).

Institutional Review Board Statement: Not applicable.

Informed Consent Statement: Not applicable.

Data Availability Statement: Not applicable.

Acknowledgments: We would like to thank Xu Zefu for some discussions and Beihang Nano for generously providing access to their darkroom and some experimental equipment.

Conflicts of Interest: The authors declare no conflict of interest.

References

1. Dixon, R. Acoustic diffraction of light in anisotropic media. *IEEE J. Quantum Electron.* **1967**, *3*, 85–93. [CrossRef]
2. Voloshinov, V.B.; Mosquera, J.C. Wide-aperture acousto-optic interaction in birefringent crystals. *Opt. Spectrosc.* **2006**, *101*, 635–641. [CrossRef]
3. Georgiev, G.; Georgieva, E.; Konstantinov, L. Angular and power characteristics of noncollinear acousto-optic tunable filters. *Opt. Lasers Eng.* **1999**, *31*, 1–12. [CrossRef]
4. Del Bosque, D.; Mahoney, C.; Cheng, L.-J.; Hamilton, M.; Reyes, G.; Gundersen, H.; LaBaw, C. Acousto-optic tunable filter for hyperspectral imagery and dual-use applications. In Proceedings of the AIAA, Space Programs and Technologies Conference and Exhibit, Huntsville, AL, USA, 21–23 September 1993.
5. Korablev, O.; Fedorova, A.; Villard, E.; Joly, L.; Kiselev, A.; Belyaev, D.; Bertaux, J.L. Characterization of the stray light in a space borne atmospheric AOTF spectrometer. *Opt. Express* **2013**, *21*, 18354–18360. [CrossRef]
6. Yushkov, K.B.; Molchanov, V.Y. Hyperspectral imaging acousto-optic system with spatial filtering for optical phase visualization. *J. Biomed. Opt.* **2017**, *22*, 66017. [CrossRef]
7. Jiangwei, Y.; Chunguang, Z.; Hao, W.; Lei, S. Rapid Microscopic Spectral Imaging of Lung Cancer Tissue Based on Acousto-Optic Tunable Filter. *ChJL* **2018**, *45*, 0407003. [CrossRef]
8. Korablev, O.I.; Belyaev, D.A.; Dobrolenskiy, Y.S.; Trokhimovskiy, A.Y.; Kalinnikov, Y.K. Acousto-optic tunable filter spectrometers in space missions [Invited]. *Appl. Opt.* **2018**, *57*, C103–C119. [CrossRef]
9. Wachman, E.S.; Niu, W.H.; Farkas, D.L. Imaging acousto-optic tunable filter with 0.35-micrometer spatial resolution. *Appl. Opt.* **1996**, *35*, 5220–5226. [CrossRef]
10. Vila-Francés, J. Improving the performance of acousto-optic tunable filters in imaging applications. *J. Electron. Imaging* **2010**, *19*, 043022. [CrossRef]
11. Pannell, C. Recent advances in acousto-optic tunable filters for hyper-spectral imaging with real-time spectral unmixing. In Proceedings of the CLEO: Science and Innovations 2013, San Jose, CA, USA, 9–14 June 2013; pp. 1–3.
12. Voloshinov, V.B.; Molchanov, V.Y.; Mosquera, J.C. Spectral and polarization analysis of optical images by means of acousto-optics. *Opt. Laser Technol.* **1996**, *28*, 119–127. [CrossRef]
13. Gupta, N. Materials for imaging acousto-optic tunable filters. In Proceedings of the Image Sensing Technologies: Materials, Devices, Systems, and Applications, Baltimore, MD, USA, 21 May 2014; pp. 41–50.
14. Xu, Z.; Dai, S.; Lin, C.; Wu, Z. Research progress of acousto-optic crystals, glass materials and modulators. *Laser Optoelectron.* **2021**, *58*, 1516007.
15. Voloshinov, V.B.; Molchanov, V.Y. Acousto-optical modulation of radiation with arbitrary polarization direction. *Opt. Laser Technol.* **1995**, *27*, 307–313. [CrossRef]
16. Liu, J.; Rong, S.; Ma, Y.; Wang, J. A hyperspectral imager with adjustable spectral selectivity based on AOTF. *Proc. SPIE–Int. Soc. Opt. Eng.* **2010**, *7857*, 1–11. [CrossRef]
17. Mazur, M.M.; Mazur, L.I.; Suddenok, Y.A.; Shorin, V.N. Increase of an Output Optical Signal of an Acousto-Optic Monochromator upon Frequency Modulation of a Control Signal. *Opt. Spectrosc.* **2018**, *125*, 594–598. [CrossRef]
18. Voloshinov, V.B.; Molchanov, V.Y.; Babkina, T.M. Acousto-optic filter of nonpolarized electromagnetic radiation. *Tech. Phys* **2000**, *45*, 1186–1191. [CrossRef]
19. Yuan, Y.; Hwang, J.Y.; Krishnamoorthy, M.; Ye, K.; Zhang, Y.; Ning, J.; Wang, R.C.; Deen, M.J.; Fang, Q. High-throughput acousto-optic-tunable-filter-based time-resolved fluorescence spectrometer for optical biopsy. *Opt. Lett.* **2009**, *34*, 1132–1134. [CrossRef]

20. Nie, Z.; An, R.; Hayward, J.E.; Farrell, T.J.; Fang, Q. Hyperspectral fluorescence lifetime imaging for optical biopsy. *J. Biomed. Opt.* **2013**, *18*, 096001. [CrossRef]
21. Abdlaty, R.; Orepoulos, J.; Sinclair, P.; Berman, R.; Fang, Q.Y. High Throughput AOTF Hyperspectral Imager for Randomly Polarized Light. *Photonics* **2018**, *5*, 3. [CrossRef]
22. Romanova, G.; Beliaeva, A.; Ryvkina, Y.; Ryabov, D. Design Features of a Tunable Source Based on an Acousto-Optical Tunable Filter. In Proceedings of the 2022 Wave Electronics and its Application in Information and Telecommunication Systems (WECONF), St. Petersburg, Russia, 30 May–3 June 2022; pp. 1–4.
23. Chang, I.C. Tunable Acousto-Optic Filters: An Overview. *Opt. Eng.* **1977**, *16*, 455–460. [CrossRef]
24. Chang, I.C. Noncollinear acousto-optic filter with large angular aperture. *Appl. Phys. Lett.* **1974**, *25*, 370–372. [CrossRef]
25. Batshev, V.; Machikhin, A.; Gorevoy, A.; Martynov, G.; Khokhlov, D.; Boritko, S.; Pozhar, V.; Lomonov, V. Spectral Imaging Experiments with Various Optical Schemes Based on the Same AOTF. *Materials* **2021**, *14*, 2984. [CrossRef] [PubMed]
26. Romier, J.; Selves, J.; Gastellu-Etchegorry, J. Imaging spectrometer based on an acousto-optic tunable filter. *Rev. Sci. Instrum.* **1998**, *69*, 2859–2867. [CrossRef]
27. Suhre, D.R.; Denes, L.J.; Gupta, N. Telecentric confocal optics for aberration correction of acousto-optic tunable filters. *Appl. Opt.* **2004**, *43*, 1255–1260. [CrossRef] [PubMed]
28. Batshev, V.; Gorevoy, A.; Pozhar, V.; Machikhin, A. Aberration analysis of AOTF-based stereoscopic spectral imager using optical design software. *J. Physics: Conf. Ser.* **2021**, *2127*, 012035. [CrossRef]
29. Glenar, D.A.; Hillman, J.J.; Saif, B.; Bergstralh, J. Acousto-optic imaging spectropolarimetry for remote sensing. *Appl. Opt.* **1994**, *33*, 7412–7424. [CrossRef]
30. Georgiev, G.; Glenar, D.A.; Hillman, J.J. Spectral characterization of acousto-optic filters used in imaging spectroscopy. *Appl. Opt.* **2002**, *41*, 209–217. [CrossRef]
31. Zhang, H.; Zhao, H.; Guo, Q.; Xuan, Y. Calibration of Acousto-Optic Interaction Geometry Based on the Analysis of AOTF Angular Performance. *Materials* **2023**, *16*, 3708. [CrossRef]
32. Xu, Z.F.; Zhao, H.J.; Jia, G.R.; Sun, S.J.; Wang, X.Y. Optical schemes of super-angular AOTF-based imagers and system response analysis. *Opt. Commun.* **2021**, *498*, 127204. [CrossRef]
33. Chang, I.C.I. Acoustooptic Devices and Applications. *IEEE Trans. Sonics Ultrason.* **1976**, *23*, 2–21. [CrossRef]
34. Goutzoulis, A.P.; Pape, D.R. *Design and Fabrication of Acousto-Optic Devices*; Marcel Dekker Inc.: New York, NY, USA, 1994.
35. Yushkov, K.B.; Dupont, S.; Kastelik, J.C.; Voloshinov, V.B. Polarization-independent imaging with an acousto-optic tandem system. *Opt. Lett.* **2010**, *35*, 1416–1418. [CrossRef]
36. Gass, P.A.; Sambles, J.R. Accurate design of a noncollinear acousto-optic tunable filter. *Opt. Lett.* **1991**, *16*, 429–431. [CrossRef] [PubMed]
37. Machikhin, A.; Gorevoy, A.; Batshev, V.; Pozhar, V. Modes of wide-aperture acousto-optic diffraction in a uniaxial birefringent crystal. *JOpt* **2021**, *23*, 125607. [CrossRef]
38. Anchutkin, V.S.; Bel'skii, A.B.; Voloshinov, V.B.; Yushkov, K.B. Acoustooptical method of spectral-polarization image analysis. *J. Opt. Technol.* **2009**, *76*, 473–477. [CrossRef]

materials

MDPI

Article

Research and Application Validation of a Feature Wavelength Selection Method Based on Acousto-Optic Tunable Filter (AOTF) and Automatic Machine Learning (AutoML)

Zhongpeng Ji [1,2], Zhiping He [1,2,*], Yuhua Gui [1,2], Jinning Li [1,2], Yongjian Tan [1,2], Bing Wu [1,2], Rui Xu [1,2] and Jianyu Wang [1,2,*]

1 Key Laboratory of Space Active Opto-Electronics Technology, Shanghai Institute of Technical Physics of the Chinese Academy of Sciences, Shanghai 200083, China; jizhongpeng@aliyun.com (Z.J.); yhgui@mail.ustc.edu.cn (Y.G.); lijinning@mail.sitp.ac.cn (J.L.); tanyongjian@mail.sitp.ac.cn (Y.T.); wubing@mail.sitp.ac.cn (B.W.); xurui@mail.sitp.ac.cn (R.X.)

2 University of Chinese Academy of Sciences, Beijing 100049, China

* Correspondence: hzping@mail.sitp.ac.cn (Z.H.); jywang@mail.sitp.ac.cn (J.W.); Tel.: +86-021-2505-1697 (Z.H.); +86-139-1661-4280 (J.W.)

Abstract: Near-infrared spectroscopy has been widely applied in various fields such as food analysis and agricultural testing. However, the conventional method of scanning the full spectrum of the sample and then invoking the model to analyze and predict results has a large amount of collected data, redundant information, slow acquisition speed, and high model complexity. This paper proposes a feature wavelength selection approach based on acousto-optical tunable filter (AOTF) spectroscopy and automatic machine learning (AutoML). Based on the programmable selection of sub nm center wavelengths achieved by the AOTF, it is capable of rapid acquisition of combinations of feature wavelengths of samples selected using AutoML algorithms, enabling the rapid output of target substance detection results in the field. The experimental setup was designed and application validation experiments were carried out to verify that the method could significantly reduce the number of NIR sampling points, increase the sampling speed, and improve the accuracy and predictability of NIR data models while simplifying the modelling process and broadening the application scenarios.

Keywords: AOTF; AutoML; feature wavelength selection; near infrared detection system

Citation: Ji, Z.; He, Z.; Gui, Y.; Li, J.; Tan, Y.; Wu, B.; Xu, R.; Wang, J. Research and Application Validation of a Feature Wavelength Selection Method Based on Acousto-Optic Tunable Filter (AOTF) and Automatic Machine Learning (AutoML). *Materials* **2022**, *15*, 2826. https://doi.org/10.3390/ma15082826

Academic Editors: Alexander S. Machikhin and Vitold Pozhar

Received: 4 March 2022
Accepted: 8 April 2022
Published: 12 April 2022

1. Introduction

NIR spectroscopy has many advantages, such as being nondestructive and accurate, and has been widely applied in areas such as food safety [1,2], drug analysis [3], agricultural testing [4], and basic chemistry [5]. The current common approach to NIR spectroscopy is to obtain the full continuous spectral data of the sample in the spectral range and to use the corresponding algorithms to model the correlation between the sample and the spectral data. However, the NIR spectral data of a sample has a relatively high dimensionality and also suffers from inter-spectral overlap, covariance, and noise, which negatively affects the performance of the NIR spectral model [6]. The selection of the effective feature wavelengths of the sample is extremely important at this point. Feature wave-length selection extracts spectrally valid variables and removes useless or interfering wavelength data, improving the accuracy and predictiveness of the data model. Only spectral data from a specific band or specific wavelength points are required to build a well-performing detection model, requiring significantly fewer wavelength sampling points.

Numerous algorithms exist to select the characteristic wavelengths of the collected NIR spectra of the samples to build reliable models [7], such as the adaptive reweighted sampling (CARS) [8], the random frog hopping algorithm (RF) [9], the PLS–genetic algorithm (PLS-GA) [10], etc. These algorithms have different performance for different data and problems, and in practice modelling usually involves human experience in selecting the most suitable model, which increases the complexity of the modelling process. In contrast, the recently proposed automatic machine learning (AutoML) [11] allows for automated model selection without human intervention. The model automatically generates the network structure that is most efficient for the task at hand, and the model automatically searches for the best sequence of combinations of operations under the different structures produced by itself. This approach can effectively reduce the complexity of modelling and also ensures the robustness of the model.

After modelling the characteristic wavelengths of the sample, the current instruments still need to collect the full spectrum first and then select the characteristic spectra to input into the model to obtain the final results [12,13]. However, this full-spectrum acquisition method results in slow data acquisition and processing. Although common grating-based spectroscopy instruments can acquire data quickly, the cost is significantly higher due to the use of array detectors, and problems such as non-uniformity between detectors can also affect the signal-to-noise ratio of the acquisition. In some applications where fast real-time processing is required, such as industrial on-line analysis, faster spectral acquisition and higher data quality are often required. Therefore, an NIR spectrometer that can be coupled with a feature wavelength filtering algorithm to achieve variable wavelength acquisition is a solution that can guarantee both speed and data quality and model robustness. An AOTF spectrometer can change the diffraction wavelength by changing the frequency of the RF power signal added to it, which can achieve sub-nm-level central wavelength picking in the full spectrum and use a unit detector to obtain data, which can effectively improve the signal-to-noise ratio and is an ideal device that can be used with the feature wavelength selection algorithm.

AOTF-based NIR spectroscopy has been extensively studied in food inspection and agricultural applications. Several studies have been conducted to implement commercial AOTF-NIR spectrometers for nondestructive detection of dried apple and olive fruit [14–17]. Diffuse reflectance spectra were acquired in the wavelength range 1100–2300 nm at 2 nm intervals using the Luminar 5030 AOTF-NIR Miniature 'Hand-held' Analyzer (Brimrose Corporation, Baltimore, MD, USA). The feasibility of using AOTF-NIR spectroscopy in an intelligent drying system for nondestructive detection and monitoring of physicochemical changes in organic apple wedges during the drying process was investigated. Partial least squares (PLS) regression models were developed to monitor changes in water activity, moisture content, soluble solids content, and chroma during drying. The classification models were computed using K-means and Partial Least Squares Discriminant Analysis (PLS-DA) algorithms in sequence [14]. AOTF-NIR was also satisfactorily applied to predict phenolic compounds [15] and monitor the ripening of olives [16]. Water content, fat content, and free acidity in olive fruit were predicted by online NIR spectroscopy combined with chemometrics techniques [17]. In addition to the PLS regression algorithm, a sensor software based on an artificial neural network (SS-ANN) was designed by Allouche et al. [18] for monitoring olive malaxation. In hyperspectral imaging, a prototype on-line AOTF-based hyperspectral image acquisition system (450–900 nm) has also been developed for tenderness assessment of beef carcasses [19]. However, in these previous studies, it is usually necessary to collect all spectral data before making predictions, and for AOTF spectroscopy more research is needed to optimize the sampling strategy according to specific applications.

Therefore, this paper utilizes the features of AOTF programming to acquire specific wavelengths and combines an AOTF-based spectrometer with an automatic machine learning–based feature wavelength selection algorithm to achieve rapid output of target substance detection results in the field. This form of application will effectively improve the efficiency of online NIR spectroscopy systems and bring broader prospects for industry applications, as it can significantly reduce the amount of data to be collected and reduce the time required for the cyclic data collection process during online inspection.

2. AutoML-Based Feature Wavelength Selection

The efficiency of current machine learning algorithms often relies on human guidance, such as data preprocessing, feature selection, algorithmic models, and the determination of hyperparameters. With the complexity of machine learning algorithms, the number of available algorithms and processes is increasing. AutoML is a spatial search optimization method that can find the optimal solution in a finite spatial range in the shortest possible time, reducing time and labor costs while improving operational accuracy [11]. AutoML generally consists of two main components: an evaluator and a tuner. The evaluator is responsible for measuring the performance of the learning tool under the adoption number conditions provided by the tuner and for feeding the results to the tuner, while the tuner is responsible for making the feedback information for the learning tool to update the configuration information. In addition to algorithm selection, hyperparameter optimization [20], and neural network architecture searches, AutoML can also cover automatic data preparation, automatic feature selection, automatic flowline construction, automatic model selection, and integrated learning [21]. One of the solutions for AutoML is AutoGluon–Tabular architecture [22], which is very suitable for structured data regression problem solving with more accurate model robustness. The analytical modeling of NIR spectral data is a typical structured data regression problem, and the screening of feature wavelengths is also a variable selection problem in machine learning, so it is very suitable for AutoGluon–Tabular architecture, which is used in this paper for the analytical modeling and model-based feature wavelength screening of NIR spectral data. This architecture differs from the common hyperparameter search-based technique architecture in that it relies on fusing multiple models that do not require a hyperparameter search, thus avoiding hyperparameter search and increasing the number of trained models in a prescribed time. The operational steps and the flow chart for data feature wavelength selection of NIR spectral data using AutoGluon–Tabular architecture are as shown in Figure 1.

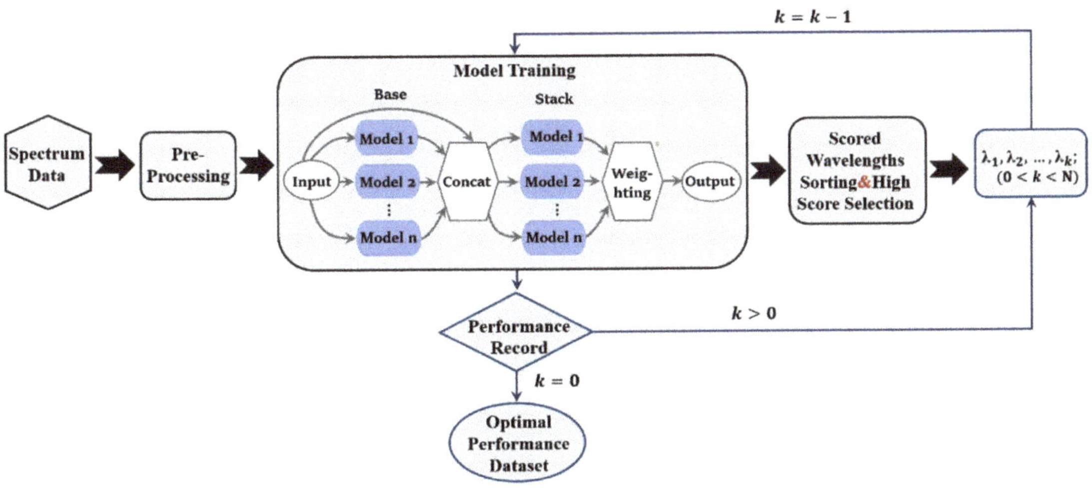

Figure 1. Feature wavelength selection using the AutoGluon-Tabular architecture.

(1) Data pre-processing

Firstly, the acquired transmission/reflection spectral data are converted into absorbance data, and the data are loaded into AutoGluon–Tabular. AutoGluon–Tabular will automatically check the label columns, determine the type of problem, and distinguish whether it is a classification problem or a regression problem; then the data are pre-processed, the feature data are classified, and the useless feature data are discarded.

(2) Model training

After the data pre-processing, the data will be trained by a series of machine learning models, using integration and stacking techniques to combine multiple models. AutoGluon–Tabular will select models for training in a unique order, firstly selecting the models with reliable performance and then gradually selecting the more computationally intensive but less reliable models. The current AutoGluon–Tabular architecture supports algorithms such as Random Forest [23], Super Random Tree, k Nearest Neighbor, LightGBM boosted tree, CatBoost boosted tree, AutoGluon–Tabular deep neural network, etc. In this paper, we will train all of the above algorithms. AutoGluon–Tabular ensembles multiple models and stacks them in multiple layers, which offers better use of allocated training time than seeking out the best. We set 'root_mean_squared_error' as "eval_metric". AutoGluon–Tabular tunes factors such as hyperparameters, early-stopping, ensemble-weights, etc. in order to improve this metric on validation data.

(3) Wavelength importance ranking

We calculated the permutation importance which measures the importance of a feature and ranked the wavelength variables according to this. The higher the permutation importance, the higher the contribution of the wavelength to the model and the more representative the wavelength is of the characteristics of the detection target.

(4) Feature wavelength combination screening

After ranking the importance of wavelengths, we are not sure which wavelength combinations can achieve the best performance of the model. In order to filter the best combination of feature wavelengths, we select the wavelengths with the top X permutation importance to retrain the model, record the evaluation index of the model performance trained by X wavelengths, reduce the number of X by 1, and repeat the retraining process until X equals 0, then stop the training. Based on the number of different X's and the evaluation metrics of the model performance, we introduce the precision criterion to determine the optimal number of wavelength variables. If a precision value is entered, the model suggested according to the precision criterion is the model with the lowest number of variables among all models, whose evaluation metric differs from the minimum evaluation metric by no more than the precision.

Finally, we determined the optimal wavelength combination to achieve the minimum number of measurements for a more effective model.

3. Instrument Design and Working Principle

After filtering the feature wavelengths of spectral data by AutoML, the construction of the excellent performance model and the accurate prediction results can be achieved via the collection of spectral data at specific wavelengths; however, the extraction of the feature wavelengths of the current conventional spectrometer spectral data requires us to obtain the full-band spectrum by scanning in advance, which results in sampling time waste and resource consumption. Thus, the above problems can be effectively solved if the wavelength filtering results can be targeted and finely sampled, while the AOTF near-infrared spectrometer has the characteristic of freely adjustable wavelength, which allows it to conduct discrete sampling and perform targeted and finely spectral sampling for specific filtering results.

AOTF is an electrically tunable filter based on the acousto-optic effect, which is mainly composed of a birefringent acousto-optic crystal (the most widely used material TeO_2), a piezoelectric transducer, and an acoustic wave absorber, as shown in Figure 2a. In practical applications, the RF drive signal is converted to ultrasonic waves inside the crystal by a piezoelectric transducer fixed on the crystal surface, and the diffraction wavelengths are selected by changing the frequency of the RF signal, with a wide tuning range and fast scanning speed. Spectrometers based on AOTF spectroscopy have the advantages of small size, light weight, all-solid state, and strong environmental adaptability and have been widely used in many fields such as food inspection [24], environmental monitoring [25,26], and deep space exploration [27,28]. According to the condition of momentum matching, the tuning relationship between the driving frequency $f(\lambda)$ of AOTF and the output diffraction wavelengths is shown as Equation (1) [29,30]:

$$f_a(\lambda) = V_a[n_i^2 + n_d^2 - 2n_i n_d \cos(\theta_i - \theta_d)]^{1/2}/\lambda, \quad (1)$$

where λ is the incident light wavelength, V_a is the ultrasonic propagation velocity in the crystal, $f_a(\lambda)$ is the corresponding ultrasonic frequency, θ_i is the incident angle for a device, θ_d is the output beam angle, n_i is the refractive index for the incident light, and n_d is the refractive index for the diffracted light. The relationship curve between the diffraction wavelength of the crystal and the driving frequency is shown in Figure 2b. The curve data were obtained by measurement. A HORIBA iHR 320 spectrometer was used to generate monochromatic light. The wavelength of the monochromatic light was first set, and then the scanning was performed by changing the driving frequency in fine steps. The frequency at which the most intense diffracted light was measured is the driving frequency corresponding to that wavelength. This procedure was repeated for the entire operating spectral range to obtain all data. It is possible to realize both sequences in the time-fine scanning of wide spectrum bands by wavelength point by point and the high-precision diffraction center wavelength picking of specific wavelength bands by coding control of RF driving frequency when using a good correspondence between RF driving frequency and diffraction wavelength, with simple wavelength picking and good repeatability.

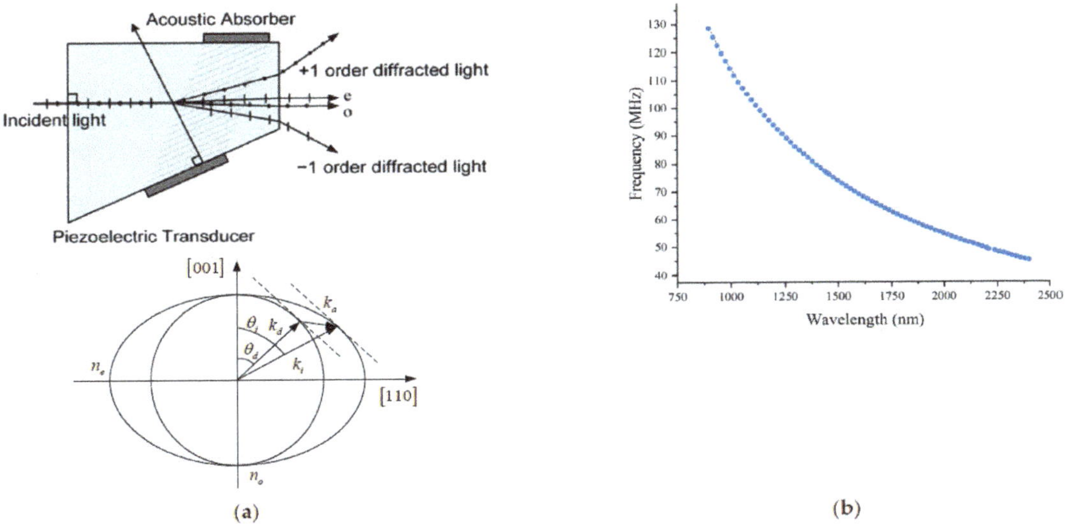

Figure 2. AOTF spectroscopy schematic. (**a**) AOTF spectroscopy and wave vector diagram; (**b**) curve of AOTF driving frequency with diffraction wavelength.

According to the characteristics of the AOTF spectrometer, the optical path structure of this system is shown in Figure 3a by combining an AutoML autonomous feature wavelength filtering algorithm and AOTF spectrometer-specific band sampling. The use of a dual optical path and a dual detector design, the use of an adjustable beam splitter to introduce the reference optical path, comprehensive data processing, and compensation of light intensity fluctuations caused by the instability of the light source and errors caused by environmental interference (in order to improve the accuracy of the instrument) allowed us to achieve a wide spectral scan range of 900–2400 nm. The actual spectra are sampled by a short-wave infrared AOTF, using both high-frequency and low-frequency drivers to achieve a wide spectral sampling range of 900–2400 nm, with a spectral resolution of 3.75–9.6 nm. The main performance parameters of the AOTF are shown in Table 1.

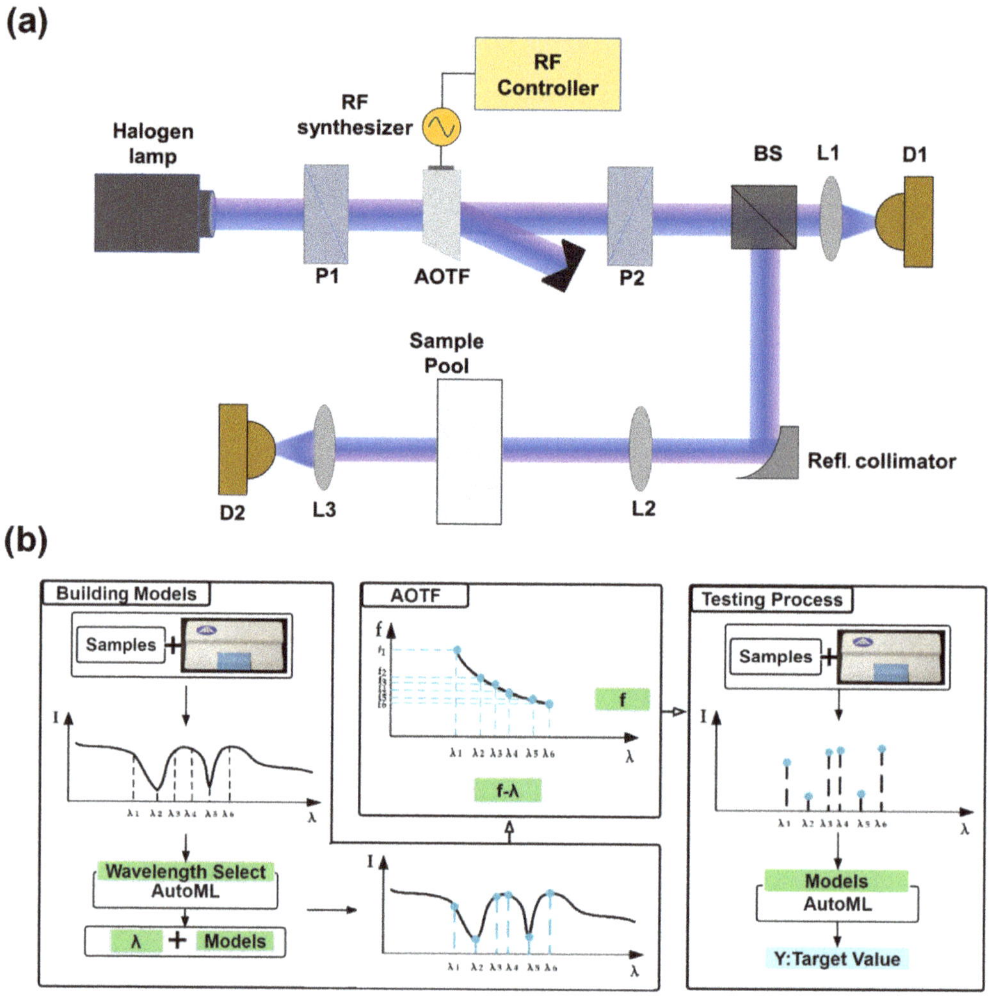

Figure 3. AOTF detection system and testing process. (**a**) schematic diagram of AOTF dual optical path and dual detector NIR spectrometer; P1, P2, polarizer; BS, adjustable beam splitter; D1, D2, InGaAs photodetector; (**b**) flow chart of AOTF detection based on AutoML.

Table 1. The main performance parameters of the AOTF.

Parameters	
Material	TeO_2
Spectral coverage/nm	900–2400
FWHM/nm	3.75–8.4 @ < 1380 nm 4.2–9.6 @ > 1380 nm
RF/MHz	45.25–128.55
Angular aperture/°	>8
Diffraction angle/°	>7.5
Power/W	~2

The flow chart of the system is shown in Figure 3b. Firstly, the full spectrum data of the substance is obtained through the AOTF spectrometer, and the filter modeling of the feature wavelength of the target is completed using the AutoML algorithm to extract the feature wavelength of the sample. After that, the RF drive function of the feature wavelengths is generated using the selected feature wavelength combinations combined with the drive frequency function of AOTF, and the drive function is stored in the RF controller. When the target is measured again, the system can take advantage of the flexible adjustment of the AOTF central wavelength at the subnanometer level to accurately locate the desired band and achieve the adjustable, high-precision, and fast acquisition of the feature spectrum. By transferring the obtained data to the model established by the combination of feature wavelengths, the detection and analysis of the characteristic components or contents of the sample can be realized.

4. Application Results

To verify the effectiveness of the system, two different types of liquid samples (milk and baijiu) were selected, and validation experiments were performed according to the detection system and testing process shown in Figure 3. The experimental results of the two measurement methods for fat content (%) in milk samples and alcohol by volume (ABV, %) in baijiu samples are shown.

4.1. Sample Preparation

For the milk samples, we chose nine different brands of liquid pure milk, and five samples of each milk brand were prepared for a total of 45 samples. The nine milk samples had different fat contents, as shown in Table 2, to establish the relationship between spectral data and fat content in the milk model using the nominal value of fat content.

Table 2. The fat contents of the milk samples.

Brand	Number of Samples	Fat (%)
Item1	5	1.6
Item2	5	0
Item3	5	3.5
Item4	5	3.9
Item5	5	3.5
Item6	5	3.0
Item7	5	3.1
Item8	5	3.0
Item9	5	3.1
all	45	–

For the liquor samples, we selected two brands and five types of liquor divided into 209 samples marked and sealed for storage. The determination of alcoholic content was entrusted to the Shenzhen Institute of Measurement and Quality Inspection using the national standard (GB 5009.225-2016) alcoholometer method, and the measurement result was the volume fraction of alcohol, i.e., alcoholic content (%vol). The range of alcoholic content of the samples was 42.0–56.2, with an accuracy error of 0.1, as shown in Table 3.

Table 3. ABV of baijiu samples.

Brand	ABV [1]	Number of Samples	ABV Measured
Item1	42	14	42.0
	50	49	50.2
	53	77	53.1
	56	47	56.2
Item2	56	22	55.8
all	–	209	–

[1] ABV = alcohol by volume.

The NIR transmittance spectra of all samples were collected using this system with a spectral sampling range of 900–2400 nm and a sampling interval of 5 nm, and the NIR spectral data for each sample was 301 wavelength points. For the milk samples, a 0.5 mm optical path length quartz cuvette was used to hold the samples, and the transmittance spectra of 45 samples were obtained. For the baijiu samples, the transmittance spectra of 209 samples were obtained using a quartz cuvette with a 2 mm optical path length.

The absorbance spectra of all samples were obtained by taking log(1/T) of the transmittance spectra of all samples. Savitzky-Golay (SG) [31,32] smoothing was used to reduce the random noise. The absorbance spectra were processed using Multiplicative Scatter Correction (MSC) [33] to remove the influence of non-concentration factors such as baseline shift and granularity on the spectra. The spectral data with low signal-to-noise ratio at the head and tail were removed, and the final data were in the wavelength range of 1250–2300 nm and 5 nm sampling interval, with a total of 211 data points. The absorbance spectral data obtained by processing are shown in Figure 4.

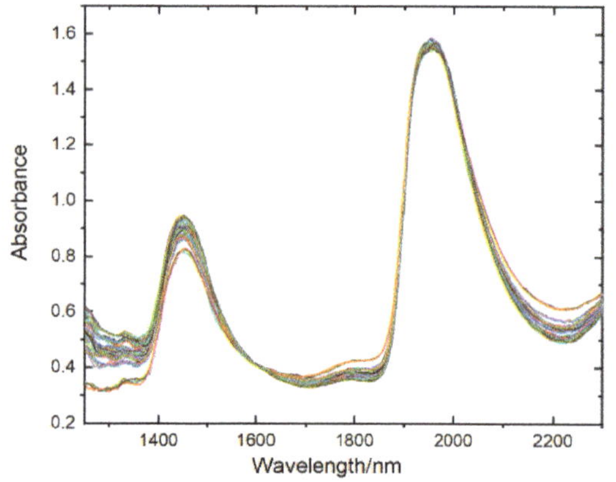

(a)

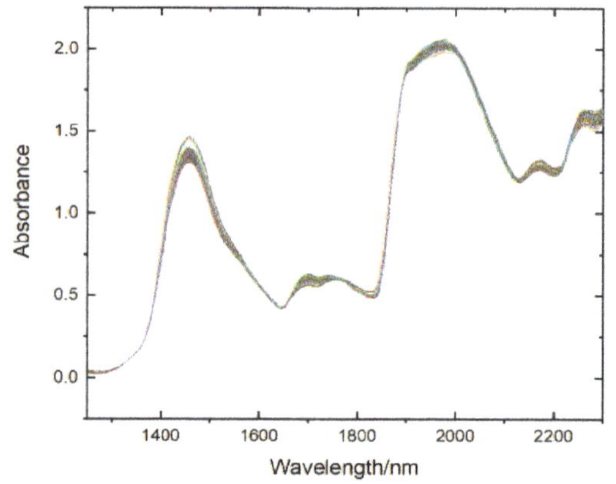

(b)

Figure 4. Absorbance spectra of different samples. (**a**) milk samples; (**b**) alcohol samples.

4.2. Modeling and Feature Wavelength Selection

The spectra of milk and alcohol samples obtained by AOTF spectrometer were converted into spectral absorbance values as input features, and fat content and alcohol content were used as model target values. The data sets were randomly divided, with 70% (31 samples) of milk used as training data and the remaining 30% (14 samples) used as test data; 70% (146 samples) of baijiu was used as training data and the remaining 30% (63 samples) was used as test data, as shown in Tables 4 and 5. Only the training data was provided to AutoML framework at training time, while the test data was only provided at prediction time.

Table 4. Statistics on fat of milk samples.

Data Sets	Number of Samples	Min	Max	Mean	STD
Total samples	45	0	3.9	2.99	1.12
Calibration set	31	0	3.9	2.96	1.15
Prediction set	14	0	3.9	2.76	1.32

Table 5. Statistics on ABV of baijiu samples.

Data Sets	Number of Samples	Min	Max	Mean	STD
Total samples	209	42	56.2	52.66	3.60
Calibration set	146	42	56.2	52.49	3.80
Prediction set	63	42	56.2	53.05	3.07

The fit application programming interface (API) of AutoGluon–Tabular was used to train these models. Within the call to fit, AutoGluon automatically preprocesses the raw data, identifies what type of prediction problem this is (binary, multi-class classification, or regression), partitions the data into various folds for model-training vs. validation, individually fits various models, and finally creates an optimized model ensemble that outperforms any of the individual trained models.

The performance of these models was adjudged based on root mean square errors (RMSE) of validation (RMSEV) and prediction (RMSEP). RMSEV is obtained from AutoGluon which is named as "score val," and RMSEP is calculated based on the prediction results of the test set. This is as shown in Figure 5a,c.

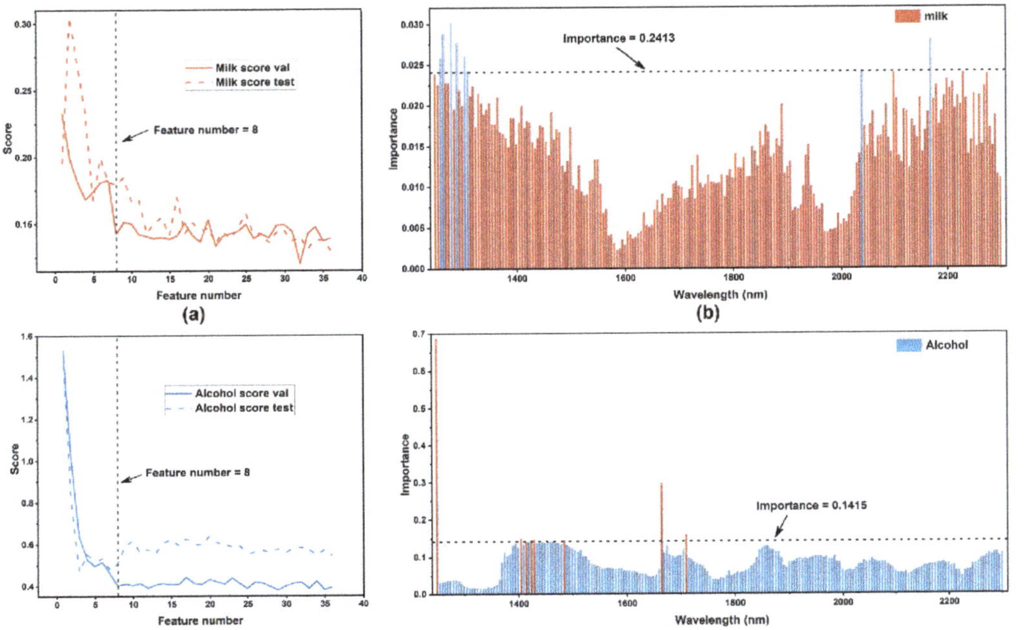

Figure 5. Results of the "score val" (RMSEV) and permutation importance of the two samples. (**a**) "score val" (RMSEV) of milk samples. We also calculated RMSEP as "score test". (**b**) degrees of permutation importance of the milk samples; (**c**) "score val" (RMSEV) of alcohol samples. We also calculated RMSEP as "score test". (**d**) permutation importance of alcohol samples.

The importance the ranking [34] of data variables was performed based on permutation importance for milk and baijiu samples, as shown in Figure 5b,d. Higher importance indicates that the feature variables have more influence on the model performance.

Based on the accuracy criteria, we calculated the difference between the RMSEV and the minimum RMSEV for all models, where the model with a difference not higher than 20% and with the least number of variables was selected. According to our precision criterion, eight characteristic wavelengths were selected for milk, and eight characteristic wavelengths were selected for baijiu. The actual results of the selected wavelengths are shown in Table 6.

Table 6. Results of feature wavelength selection.

Samples	Selected Wavelengths (nm)
Milk	1280, 1265, 2170, 1290, 1305, 1260, 1310, 2040
Baijiu	1250, 1665, 1710, 1405, 1430, 1425, 1485, 1415

4.3. Experimental Results of the System

The AOTF-NIR spectroscopy system combining the model and the corresponding sampling strategy was deployed to a real application scenario to perform experiments on the measurement of samples with unknown target value content. Using this instrumentation system with models, we were able to perform 211-point and 8-point measurements on the samples, respectively. The data obtained from the measurements could then be used to invoke the corresponding model for component content prediction. The analytical performance of the system was evaluated when the full-spectrum model and the characteristic wavelength model were applied separately. The performance of the prediction could be evaluated by calculating new RMSEPs.

The RMSEV shown in Table 7 is the "score val" obtained during the training phase. We also performed 211-point and 8-point spectral acquisitions for the above test set of samples, respectively. After data acquisition, the average spectrum calculated in the model building phase is used as the true spectrum for MSC processing. No smoothing operation is performed at this stage. The data are then input to the model to derive predicted values, which are displayed on the software interface. The RMSEP metric is obtained by calculating the error between the predicted and true values obtained from the inference of the input model after the spectral measurement; the inference time is automatically recorded by the program in the host computer; the sampling time is the estimated time taken by the instrument to acquire data at the desired wavelength point.

Table 7. The performance of the system for two datasets.

Data Sets	Number of Variables	RMSEV	RMSEP	Sampling Duration
milk	211	0.161	0.190	13.3 s
	8	0.143	0.180	0.6 s
baijiu	211	0.529	0.614	13.3 s
	8	0.403	0.507	0.6 s

As shown in Table 7 above, the use of the characteristic wavelength model and the corresponding sampling strategy achieves a slight improvement in prediction performance, while the sampling time consumption is ~22-fold smaller. The prediction performance improvement indicates that there are many irrelevant wavelengths throughout the spectrum that do not contribute to effective prediction of the target values and negatively affect the performance of the prediction model.

Combined with the above experimental results, the automatic machine learning framework can accurately and robustly filter the wavelength combinations with the best prediction performance for the target values; combined with the eight-wavelength models and the corresponding new sampling mechanism that only collects data at the eight wavelength points, higher prediction performance and considerable improvement in detection efficiency can be obtained compared with the full-spectrum model and the full-spectrum sampling mechanism. This advantage is particularly critical for the application of AOTF-NIR analysis systems for online inspection. The combination of AOTF's flexibility and automatic machine learning provides a good reference case for the application of intelligent optical inspection in automated production by applying an automatic machine learning framework to easily build a well-performing model for different application objectives and adaptively change the sampling strategy.

5. Conclusions

In this study, an NIR detection system based on AOTF with AutoML feature wavelength screening was proposed and validated. Combining various chemometric algorithms, spectral detection based on the characteristic wavelengths instead of the full spectrum achieves the improvement of detection efficiency. Meanwhile, the model is simplified, interpretability is improved, and storage and computational resource usage are reduced. Taking milk and alcohol as examples, the validation experiments achieve a reduction in the number of sampling wavelengths while achieving accurate and nondestructive rapid determination of fat and alcohol content.

Author Contributions: Conceptualization, Z.J. and Z.H.; methodology, Z.J.; software, Z.J.; validation, Y.G., J.L. and Z.H.; formal analysis, Z.J.; investigation, Z.J. and Y.G.; resources, Y.T. and R.X.; data curation, Z.J. and B.W.; writing—original draft preparation, Z.J. and Y.G.; writing—review and editing, Z.H. and J.L.; visualization, Y.G., Y.T. and B.W.; supervision, J.W. and Z.H.; project administration, Z.H. and J.W.; funding acquisition, Z.H. All authors have read and agreed to the published version of the manuscript.

Funding: This research was funded by the National Natural Science Foundation of China (Grant No. 61605231), the Program of Shanghai Academic Research Leader (Grant No. 19XD1424100), and the National Science Foundation for Distinguished Young Scholars (Grant No. 62125505).

Institutional Review Board Statement: Not applicable.

Informed Consent Statement: Not applicable.

Data Availability Statement: Not applicable.

Conflicts of Interest: The authors declare no conflict of interest.

References

1. Osborne, B.G. Near-infrared spectroscopy in food analysis. In *Encyclopedia of Analytical Chemistry: Applications, Theory and Instrumentation*; BRI Australia Ltd.: North Ryde, NSW, Australia, 2006.
2. Pu, H.; Kamruzzaman, M.; Sun, D.-W. Selection of feature wavelengths for developing multispectral imaging systems for quality, safety and authenticity of muscle foods-a review. *Trends Food Sci. Technol.* **2015**, *45*, 86–104. [CrossRef]
3. Scafi, S.H.F.; Pasquini, C. Identification of counterfeit drugs using near-infrared spectroscopy. *Analyst* **2001**, *126*, 2218–2224. [CrossRef] [PubMed]
4. Wang, D.-J.; Zhou, X.-Y.; Jin, T.-M.; Hu, X.; Zhong, J.; Wu, Q. Application of near-infrared spectroscopy to agriculture and food analysis. *Guang Pu Xue Yu Guang Pu Fen Xi = Guang Pu* **2004**, *24*, 447–450.
5. Yang, J.; Li, J.; Hu, J.; Yang, W.; Zhang, X.; Xu, J.; Zhang, Y.; Luo, X.; Ting, K.C.; Lin, T.; et al. An interpretable deep learning approach for calibration transfer among multiple near-infrared instruments. *Comput. Electron. Agric.* **2022**, *192*, 106584. [CrossRef]
6. Lorente, D.; Aleixos, N.; Gómez-Sanchis, J.; Cubero, S.; Blasco, J. Selection of optimal wavelength features for decay detection in citrus fruit using the ROC curve and neural networks. *Food Bioprocess Technol.* **2013**, *6*, 530–541. [CrossRef]
7. Yun, Y.-H.; Li, H.-D.; Deng, B.-C.; Cao, D.-S. An overview of variable selection methods in multivariate analysis of near-infrared spectra. *TrAC Trends Anal. Chem.* **2019**, *113*, 102–115. [CrossRef]
8. Li, H.; Liang, Y.; Xu, Q.; Cao, D. Key wavelengths screening using competitive adaptive reweighted sampling method for multivariate calibration. *Anal. Chim. Acta* **2009**, *648*, 77–84. [CrossRef]
9. Li, H.-D.; Xu, Q.-S.; Liang, Y.-Z. Random frog: An efficient reversible jump Markov Chain Monte Carlo-like approach for variable selection with applications to gene selection and disease classification. *Anal. Chim. Acta* **2012**, *740*, 20–26. [CrossRef]
10. Leardi, R.; Gonzalez, A.L. Genetic algorithms applied to feature selection in PLS regression: How and when to use them. *Chemom. Intell. Lab. Syst.* **1998**, *41*, 195–207. [CrossRef]
11. Gijsbers, P.; LeDell, E.; Thomas, J.; Poirier, S.; Bischl, B.; Vanschoren, J. An Open Source AutoML Benchmark. *arXiv* **2019**, arXiv:1907.00909.
12. Amsaraj, R.; Mutturi, S. Real-coded GA coupled to PLS for rapid detection and quantification of tartrazine in tea using FT-IR spectroscopy. *Lwt-Food Sci. Technol.* **2021**, *139*, 8. [CrossRef]
13. Holroyd, S.E. The use of near infrared spectroscopy on milk and milk products. *J. Near Infrared Spectrosc.* **2013**, *21*, 311–322. [CrossRef]
14. Moscetti, R.; Raponi, F.; Ferri, S.; Colantoni, A.; Monarca, D.; Massantini, R. Real-time monitoring of organic apple (var. Gala) during hot-air drying using near-infrared spectroscopy. *J. Food Eng.* **2018**, *222*, 139–150. [CrossRef]

15. Bellincontro, A.; Taticchi, A.; Servili, M.; Esposto, S.; Farinelli, D.; Mencarelli, F. Feasible application of a portable NIR-AOTF tool for on-field prediction of phenolic compounds during the ripening of olives for oil production. *J. Agric. Food Chem.* **2012**, *60*, 2665–2673. [CrossRef]
16. Cirilli, M.; Bellincontro, A.; Urbani, S.; Servili, M.; Esposto, S.; Mencarelli, F.; Muleo, R. On-field monitoring of fruit ripening evolution and quality parameters in olive mutants using a portable NIR-AOTF device. *Food Chem.* **2016**, *199*, 96–104. [CrossRef]
17. Fernández-Espinosa, A.J. Combining PLS regression with portable NIR spectroscopy to on-line monitor quality parameters in intact olives for determining optimal harvesting time. *Talanta* **2016**, *148*, 216–228. [CrossRef]
18. Allouche, Y.; López, E.F.; Maza, G.B.; Márquez, A.J. Near infrared spectroscopy and artificial neural network to characterise olive fruit and oil online for process optimisation. *J. Near Infrared Spectrosc.* **2015**, *23*, 111–121. [CrossRef]
19. Naganathan, G.K.; Cluff, K.; Samal, A.; Calkins, C.R.; Jones, D.D.; Lorenzen, C.L.; Subbiah, J. A prototype on-line AOTF hyperspectral image acquisition system for tenderness assessment of beef carcasses. *J. Food Eng.* **2015**, *154*, 1–9. [CrossRef]
20. Bergstra, J.; Bardenet, R.; Bengio, Y.; Kégl, B. Algorithms for hyper-parameter optimization. *Adv. Neural Inf. Processing Syst.* **2011**, *24*, 2546–2554.
21. Feurer, M.; Klein, A.; Eggensperger, K.; Springenberg, J.; Blum, M.; Hutter, F. Efficient and robust automated machine learning. *Adv. Neural Inf. Processing Syst.* **2015**, *28*, 2944–2952.
22. Erickson, N.; Mueller, J.; Shirkov, A.; Zhang, H.; Larroy, P.; Li, M.; Smola, A. AutoGluon-Tabular: Robust and Accurate AutoML for Structured Data. *arXiv* **2020**, arXiv:2003.06505.
23. Calhoun, P.; Su, X.; Spoon, K.M.; Levine, R.A.; Fan, J. Random Forest. In *Wiley StatsRef: Statistics Reference Online*; John Wiley & Sons: Hoboken, NJ, USA, 2014; pp. 1–20.
24. Bellincontro, A.; Cozzolino, D.; Mencarelli, F. Application of NIR-AOTF spectroscopy to monitor Aleatico grape dehydration for Passito wine production. *Am. J. Enol. Vitic.* **2011**, *62*, 256–260. [CrossRef]
25. Dekemper, E.; Vanhamel, J.; Kastelik, J.-C.; Pereira, N.; Bolsée, D.; Cessateur, G.; Lamy, H.; Fussen, D. New AOTF-based instrumental concepts for atmospheric science. In Proceedings of the 14th School on Acousto-Optics and Applications, Torun, Poland, 24–27 June 2019; p. 112100S.
26. Dekemper, E.; Vanhamel, J.; Van Opstal, B.; Fussen, D. The AOTF-based NO_2 camera. *Atmos. Meas. Tech.* **2016**, *9*, 6025–6034. [CrossRef]
27. Korablev, O.I.; Belyaev, D.A.; Dobrolenskiy, Y.S.; Trokhimovskiy, A.Y.; Kalinnikov, Y.K. Acousto-optic tunable filter spectrometers in space missions. *Appl. Opt.* **2018**, *57*, C103–C119. [CrossRef] [PubMed]
28. Li, J.; Gui, Y.; Xu, R.; Zhang, Z.; Liu, W.; Lv, G.; Wang, M.; Li, C.; He, Z. Applications of AOTF Spectrometers in In Situ Lunar Measurements. *Materials* **2021**, *14*, 3454. [CrossRef]
29. Gass, P.A.; Sambles, J.R. Accurate design of a noncollinear acousto-optic tunable filter. *Opt. Lett.* **1991**, *16*, 429–431. [CrossRef]
30. Georgiev, G.; Glenar, D.A.; Hillman, J.J. Spectral characterization of acousto-optic filters used in imaging spectroscopy. *Appl. Opt.* **2002**, *41*, 209–217. [CrossRef]
31. Rinnan, Å.; Berg, F.V.D.; Engelsen, S.B. Review of the most common pre-processing techniques for near-infrared spectra. *TrAC Trends Anal. Chem.* **2009**, *28*, 1201–1222. [CrossRef]
32. Savitzky, A.; Golay, M.J. Smoothing and differentiation of data by simplified least squares procedures. *Anal. Chem.* **1964**, *36*, 1627–1639. [CrossRef]
33. Geladi, P.; MacDougall, D.; Martens, H. Linearization and scatter-correction for near-infrared reflectance spectra of meat. *Appl. Spectrosc.* **1985**, *39*, 491–500. [CrossRef]
34. Importance Ranking. Available online: https://explained.ai/rf-importance/ (accessed on 1 March 2022).

materials

MDPI

Article

Analysis on the Influence of Incident Light Angle on the Spatial Aberrations of Acousto-Optical Tunable Filter Imaging

Kai Yu [1,*] and Huijie Zhao [1,2,3,*]

[1] School of Instrumentation Science & Opto-Electronics Engineering, Beihang University, No. 37 Xueyuan Road, Haidian District, Beijing 100191, China
[2] Key Laboratory of "Precision Opto-Mechatronics Technology", Ministry of Education, No. 37 Xueyuan Road, Haidian District, Beijing 100191, China
[3] Institute of Artificial intelligence, Beihang University, No. 37 Xueyuan Road, Haidian District, Beijing 100191, China
* Correspondence: yukai@buaa.edu.cn (K.Y.); hjzhao@buaa.edu.cn (H.Z.); Tel.: +86-152-0144-1833 (K.Y.); +86-010-8233-8728 (H.Z.)

Abstract: Acousto-optical tunable filter (AOTF) does not conform to the pinhole model due to the acousto-optic interaction. A calculation method of AOTF aberrations under the condition of incident light with a large arbitrary angle is proposed to solve the problem of coordinate mapping between object space and image space of the AOTF system without refractive index approximation. This approach can provide accurate pointing information of the interested targets for the tracking and searching system based on AOTF. In addition, the effect of cut angle values of the paratellurite crystal on aberrations was analyzed to optimize the design of AOTF cutting according to different application requirements. Finally, distribution characteristics and quantitative calculation results of AOTF aberrations were verified by experiments with different targets, respectively. The experimental results are in good agreement with the simulations.

Keywords: AOTF spectrometer; aberrations; coordinate mapping

Citation: Yu, K.; Zhao, H. Analysis on the Influence of Incident Light Angle on the Spatial Aberrations of Acousto-Optical Tunable Filter Imaging. *Materials* **2022**, *15*, 4464. https://doi.org/10.3390/ma15134464

Academic Editors: Alexander S. Machikhin and Vitold Pozhar

Received: 10 May 2022
Accepted: 22 June 2022
Published: 24 June 2022

1. Introduction

Photoelectric tracking and searching system is the most important way of passive detection at present. With the development of infrared technology, all-day and all-weather detection can be realized. At the same time, more advanced jamming technologies have emerged accordingly. These jamming techniques can make the target and the decoy highly similar in geometry, kinematics, and radiance within a wide spectral range. The consequence is the failure of traditional detection and identification methods. Therefore, it is necessary to introduce the characteristics of the spectral dimension. For example, the apparent temperature, the gradient of temperature, or the emissivity of the target can be obtained by multispectral detection [1–3]. The features of these kinds are very stable and different from the bait, so they can be used to distinguish the target from the bait. Thus, spectral detection plays an important role in anti-jamming imaging systems now and in the future.

Spectral imaging detection includes optical system design, mechanical design, photoelectric conversion technology, target detection, recognition algorithm, and many other fields. However, to adapt to the rapid changes in the scene, there is an urgent need for an imaging method with a flexible spectrum. The acousto-optic tunable filter (AOTF) is a kind of narrowband filter based on birefringent crystal material. The wavelength, direction, and intensity of diffracted light can be modified by modulating the sound field with a high-frequency sinusoidal signal. Moreover, it has many advantages, such as being electronically controlled with high speed, wide spectral range, and all-solid-state. As a result, AOTF staring imaging spectrometer is commonly used in spectral analysis [4], laser shaping [5,6], spectral microscopic [7,8], and many other fields [9–13]. Similarly, the advantages of AOTF make it very suitable for tracking and searching system.

As we know, AOTF is mostly applied to the spectral imaging detection of static targets due to the different application requirements before. In addition, the field angle of view of the AOTF system is generally designed to be small (approximately equal to $\pm 2^{\circ}$) because of the constraint of the separation angle. Additionally, it is considered that the imaging aberration caused by AOTF is very small and can be ignored. Therefore, within a small angle range, many scholars only considered the impact of incident light angle on device performance. Zhang Zhonghua analyzed the influence of the incident light angle on the spectral bandwidth, diffraction efficiency, and the separation angle inside and outside the crystal; however, their analysis is limited to the interaction plane of AOTF [14–17]. Yushkov gave the analytical expressions of the central wavelength and the tuning frequency under oblique incidence [18], but for some special applications, the precise geometric information of the target is required. At this time, the aberration caused by AOTF can not be ignored even under the condition of a small field angle of view. Pozhar and Machikhin focused on the quantitative description of AOTF spatial aberrations. In their previous works, an approximation of small birefringence was made, and they only analyzed the aberrations in the direction of the optical axis and the direction perpendicular to the optical axis. Then they extended the analysis of AOTF spatial aberrations from two-dimensional space to three-dimensional space [19–23] and proposed a method of correcting spectral and spatial aberrations by AOTF pre-calibration for applications in which geometric characteristics shall be attained at the same time, such as microimaging [24]. Later, they used the developed DLL to combine the ray-tracing model with ZEMAX and applied it to the optical design of the stereoscopic spectral imaging system [25–27]. In addition, they used a single AOTF to obtain the three-dimensional information while obtaining the spectral information of the target [28]. Nonetheless, no quantitative experimental results of AOTF aberrations were given. In a word, the influence of incident light angle on AOTF can be summarized into several stages. From the analysis within the two-dimensional plane (acousto-optic interaction plane) to three-dimensional space, from the approximate analytical solution neglecting the change in refractive index to the accurate iterative solution. Until now, the acousto-optic interaction model based on momentum matching conditions has been relatively perfect.

However, if AOTF is used for tracking and searching systems, different issues need to be considered. The number of pixels occupied by the detected target on the image plane is relatively few, and the angle of light entering AOTF may no longer be constrained by the separation angle because there is no overlap between the zero-order light imaging and the diffracted light imaging. In other words, the zero-order light and the first-order diffracted light are allowed to be imaged at the same time under the condition that the imaging is not saturated. Moreover, zero-order light imaging can be eliminated through image processing because it always exists stably, and the intensity of diffracted light fluctuates with the change in driving frequency. Therefore, for AOTF used for tracking and searching system, there are two main problems to be solved: (1) How can the problem of central wavelength drift, that is, to ensure the accuracy of the spectral data of the target, be solved? (2) How can the accurate pointing information (azimuth and pitch angle) of the target be obtained? As photoelectric detection equipment, it is one of the most basic functions to give the target pointing information accurately. Otherwise, additional optoelectronic equipment is required for assistance.

As the field of view increases, the drift of the central wavelength of AOTF becomes more serious. Some scholars corrected the drift through the computational technique to achieve the purpose of expanding the field of view [29,30]. It means that problem (1) mentioned above can be solved for the AOTF system with a large field of view. On the other hand, the expansion of the field of view also means that the aberration of AOTF is more severe. The AOTF cannot use the pinhole model in ray tracing. We first conducted an in-depth study of AOTF aberration to solve the problem (2). If the calculation and analysis of aberration of AOTF can ensure sufficient accuracy, AOTF can be extended to the tracking and searching system of moving targets. Therefore, it is necessary to analyze the influence of different incident angles on imaging aberrations in a large range of angles. As far as we know, the study of this kind is not complete enough and lacks experimental verification. In this article, we analyzed the influence of the angle of incident light on AOTF spatial aberrations in three-dimensional space by the ray-tracing method and show quantitative results of the experiments. Moreover, the influence of cut angle values of the paratellurite crystal on aberrations was simulated.

The main contribution of this paper was to deeply analyze the aberration of AOTF with a large field of view so that it can give the target pointing information accurately when AOTF is used for the detection of moving targets and put forward a new application direction of AOTF. For the AOTF imaging spectrometer with a small field of view, the aberration is very small. AOTF can be considered to conform to the pinhole model, and the method in this paper is not required for ray tracing. In addition, the algorithm needs to be optimized in the real-time system due to the use of the iterative algorithm.

2. Method

At present, the AOTF imaging spectrometer can be classified into two types of structures according to the light entering the AOTF. One is with parallel entering light, and the other is with the convergent entering light. Both have their advantages; in theory, the lateral chromatic aberration of the confocal structure is less, while the axial chromatic aberration of the collimated optical path is less. Considering that lateral chromatic aberration can be corrected by the wedge angle of the AOTF rear surface (the rear wedge angle of AOTF we used is 4.83°), so the collimated light structure is used for the simulations and the experimental design.

The analysis of aberration is essential to analyze the relationship between the incident light angle of AOTF and the exit direction of diffracted light. As a spectrometer, the ray-tracing method can no longer be limited to the crystal but needs to be converted to the world coordinate system for analysis. The advantage of this method is that it has better adaptability. It can still analyze and guide the design of the optical system when changing the cut angle values of AOTF or the angle of Bragg's diffraction.

2.1. Ray Tracing of Arbitrary Light in Three-Dimensional Space

In the design and performance analysis of AOTF devices, monochromatic light is usually employed as the light source. It should be pointed out that in many analyses of AOTF device modeling, the incident light is usually vertical, and a certain diffracted light is selected by changing the driving frequency that matches the wavelength of the incident light. Momentum mismatch occurs when the incident light is not perpendicular to the AOTF front surface, or the wavelength of the incident light does not match the driving frequency. However, in actual applications of spectral imaging detection, the incident light is mostly polychromatic, and the diffracted light also essentially satisfies the momentum matching condition in another wave vector diagram in which the wavelength of incident light drifts.

Take e-polarized incident light and o-polarized diffracted light as an example. When the incident light is white, the driving frequency is fixed, and the angle of the incident light is changed, then the original incident wavelength (λ_1) no longer satisfies the momentum matching condition, as the blue triangle shown in Figure 1a. Additionally, there is another wavelength (λ_2) of incident light and the sound vector to meet the new momentum matching condition, such as the red triangle shown in Figure 1b, the wavelength and direction of the diffracted light can be calculated accordingly. In Figure 1a,b, the length and direction of L_{AB} and L_{DE} do not change, which means the sound vector remains unchanged. In other words, when the incident light is polychromatic, the diffracted light of a certain wavelength can always be found to meet the momentum matching condition. The momentum mismatch is only an artificially defined quantity to calculate the diffraction efficiency or spectral bandwidth. If the intensity of the sound field is constant, the amount of momentum mismatch is only related to the direction and wavelength of the diffracted light. Momentum mismatch is not considered in the analysis of AOTF imaging aberrations. Therefore we need to find the wavelength and exit direction of diffracted light that meet the conditions.

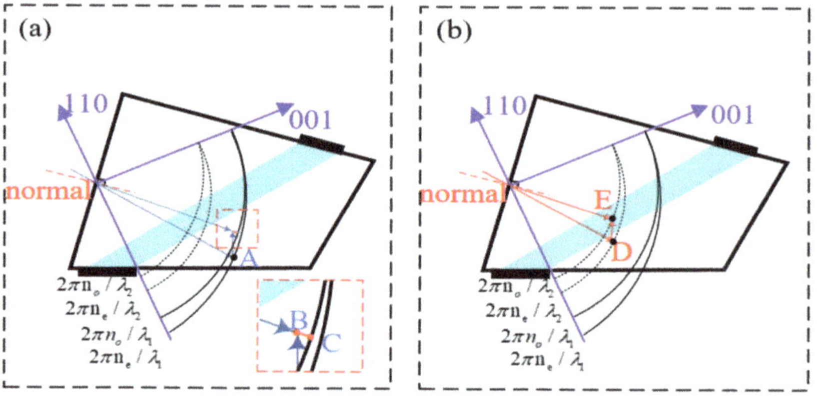

Figure 1. Wave diagram of anisotropic AO diffraction. Momentum mismatching (**a**) and momentum matching (**b**) under oblique incidence.

In the world coordinate system, assuming the direction cosine of the incident light is (L_0, M_0, N_0), in the crystal coordinate system, the direction cosine of the incident light is converted to $(L_1, M_1 \cos\theta_i^* - N_1 \sin\theta_i^*, M_1 \sin\theta_i^* + N_1 \cos\theta_i^*)$ and θ_i^* represents the angle between the front surface of AOTF and the 110 axes. The first surface [31] adopts the refraction theorem and the equation for calculating the refractive index of extraordinary light, as shown in Equation (1).

$$\begin{cases} n_{air} \sin\theta_{in} = n_e(\theta)\sin(\theta - \theta_i{}^*), \\ n_e^2(\theta) = \frac{n_o^2 n_e^2}{n_o^2 \sin^2\theta + n_e^2 \cos^2\theta}, \end{cases} \quad (1)$$

where n_{air} is the refractive index of the air medium, θ_{in} is the incident angle in the air medium, $n_e(\theta)$ is the refractive index of e light corresponding to the refracted light in the crystal, θ and $\theta_i{}^*$ is defined as shown in the partially enlarged view of Figure 2, θ is the angle between the refracted light in the crystal and the crystal axis in the polar coordinate system, $\theta - \theta_i{}^*$ is the angle between the refracted light in the crystal and the normal of the interface, and n_o and n_e represent the refractive index of o light and e light, respectively. Then, the phase velocity direction of the incident light in the crystal can be obtained. Point A (x_a, y_a, z_a) is the intersection point of the incident light, and the ellipsoid of the light wave vector shown in Figure 1. $|\mathbf{K_a}|$ can be obtained with driving frequency corresponding to the wavelength of the incident light and the direction of the sound vector $\mathbf{K_a}$ under the condition of momentum matching when the incident light is perpendicular to the front surface of AOTF. Since the driving frequency has not been tuned and $\mathbf{K_a}$ remains the same, even if the incident light angle has been changed, the coordinates of point B $(x_a - k_{ax}, y_a - k_{ay}, z_a - k_{az})$ can be calculated from the coordinates of point A. The coordinate of point C (x_c, y_c, z_c) is the intersection of the extension line of OB and o-wave vector sphere, and the length of the BC line, L_{BC}, can be obtained.

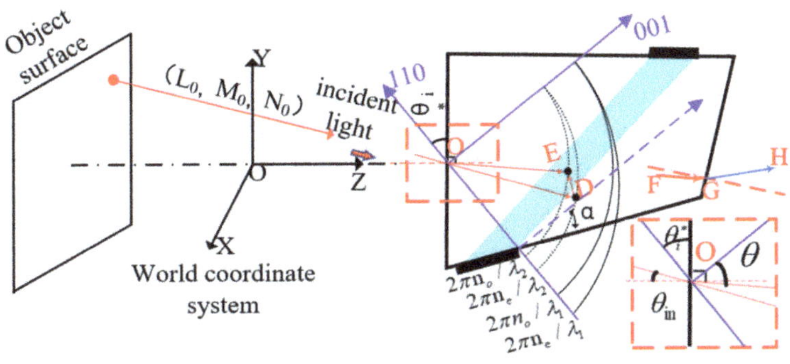

Figure 2. Ray tracing of AOTF in three-dimensional space.

It can be found that the value of L_{BC} is related to the incident wavelength. It can be considered that the incident light vector, diffracted light vector, and sound vector satisfy the momentum matching condition when L_{BC} is close to 0 with the change in incident light wavelength. Therefore, the corresponding central wavelength at an arbitrary incident angle is expressed by Equation (2). The "argmin" in Equation (2) means that the center wavelength corresponds to the smallest L_{BC} when the wavelength changes within a certain range.

$$\lambda_{\text{center}} = \underset{\lambda_{\min} \le \lambda \le \lambda_{\max}}{\operatorname{argmin}} \left(L_{BC}(\lambda) \right). \tag{2}$$

The intersection point of the incident light and the E-light ellipsoid can be calculated, i.e., D (x_d, y_d, z_d), after we obtain λ_{center}. It can be understood that in the process of wavelength change under the condition of a certain angle of the incident light, point A moves to point D. Thus far, the intersection point E (x_e, y_e, z_e) of the diffracted light and the o-light sphere, and the exit direction of diffracted light in the crystal coordinate system can be obtained simultaneously. After the diffracted light passes through the AOTF exit surface and then changes from the crystal coordinate system to the world coordinate system, the directional cosine of the diffracted light($\overrightarrow{\mathbf{GH}}$ in Figure 2) can be obtained. The iterative method must be used to obtain more accurate ray tracing because the refractive index of tellurium dioxide changes with the wavelength, and the analytical solution is no longer applicable.

Through this method, the exit direction of diffracted light corresponding to incident polychromatic light at any angle can be calculated, and further its imaging position, then the corresponding relationship between AOTF spatial aberrations and the angle of incident light can be calculated. This method is also applicable to the cases where the incident light is o-polarized, the diffracted light is e-polarized, and cases about the different cut angle values of the paratellurite crystal.

2.2. Aberration Simulation of AOTF at Different Incident Angles

Through the method above, it can be known that another wavelength of the incident satisfies the momentum matching condition with the original sound vector when the white light is incident obliquely. Select the incident light wavelength λ = 632 nm, driving frequency f = 73.25 MHz, and calculate the matching center wavelength under the condition of oblique incidence within the range of $\pm 4.0^\circ$ polar angle and azimuth angle. The ultrasonic cut angle of AOTF is 6.5°. The rear wedge angle is 4.83°. These parameter settings are consistent with the actual design parameters of the AOTF cutting.

As shown in Figure 3, the central wavelength drifts with the increase in the incident light angle. In the azimuth plane, the drift of the central wavelength is symmetrically distributed. In the polar angle plane, the drift of the central wavelength is asymmetric, and the drift of the central wavelength near the AOTF crystal axis is greater than that away from the crystal axis when the incident light is inclined by the same angle from the vertical incidence to both sides.

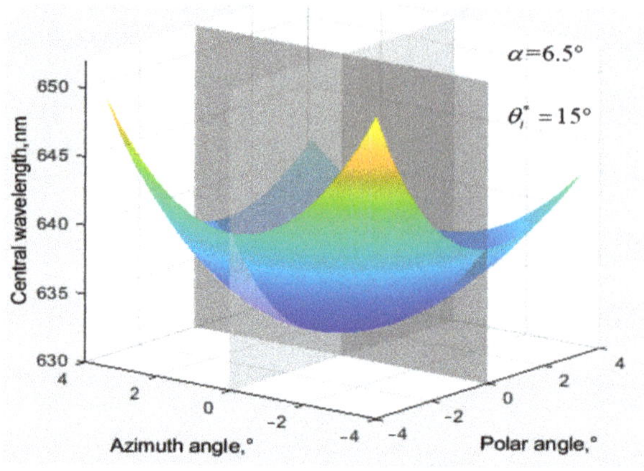

Figure 3. Central wavelength shift at ±4° angle of incident light.

The direction of diffracted light can be further calculated and compared with the image plane position of diffracted light under the condition of pinhole imaging after the central wavelength is obtained, and the aberrations of AOTF can be obtained accordingly. The focal length of the rear objective lens is 50 mm, and the pixel size is 4.5 μm.

The $L_{(AC)}$ in Figure 4c corresponds to the ordinate in Figure 4b, and the $L_{(BC)}$ in Figure 4c corresponds to the ordinate in Figure 4a. AOTF aberrations are different in polar direction and azimuth direction significantly. In the azimuth direction, the aberration is 180° rotational symmetry, and the aberration is less than two pixels within a ±4° incidence angle. In the polar direction, the aberration follows the symmetrical distribution of the azimuth plane, and the aberration caused by oblique incidence is considerable. The larger the tilt angle is, the greater the aberration is, up to 30 pixels within ±4° incidence angle. It can be discovered from the comparison between Figures 3 and 4 that the greater the center wavelength drifts, the more serious the AOTF aberration is.

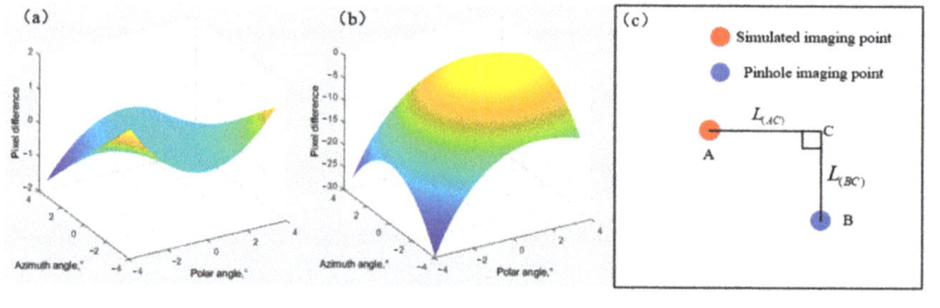

Figure 4. The difference between the imaging position of diffracted light at different incident angles of AOTF and pinhole imaging: (**a**) the difference in azimuth direction and (**b**) in polar direction, and (**c**) the image plane.

The accuracy of the simulation mainly considers the difference between the design value and the theoretical value of the ultrasonic cut angle.

As shown in Figure 5, the error of edge field aberration does not exceed two pixels when the error of ultrasonic cut angle is ±0.2°. The aberration error in the acousto-optic action plane is mainly considered here because the error in the azimuth plane affected by the ultrasonic cut angle is very small, so it is not shown. In the simulation, the principle of parallel tangent is no longer satisfied due to the error of the ultrasonic cut angle.

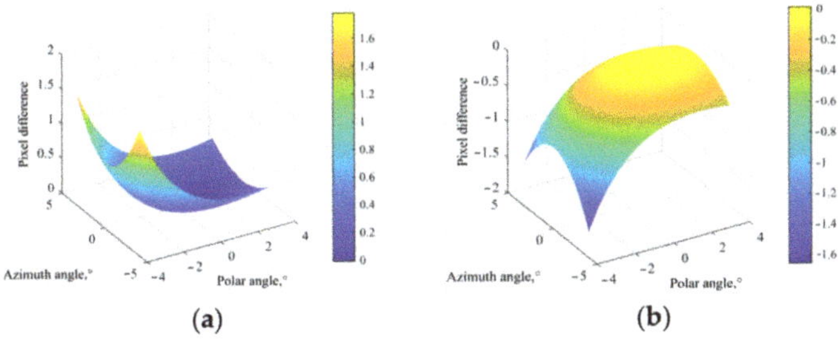

Figure 5. The accuracy of the simulation (**a**) $\alpha = 6.3°$ (**b**) $\alpha = 6.7°$.

In addition, cut angle values of the paratellurite crystal, i.e., α, affect the aberrations of AOTF. The aberrations in the polar direction of AOTF with different α are simulated, as shown in Figure 6. The results show that the aberrations of AOTF decrease with the increase in α within the range of $(0°, 18.9°)$.

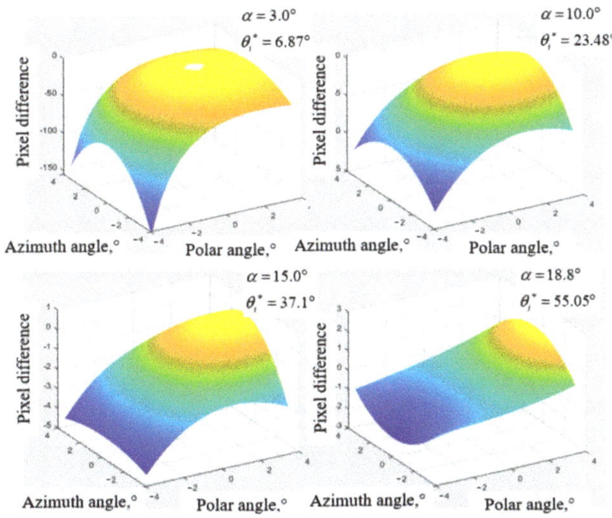

Figure 6. Aberrations of different ultrasonic cut angles of AOTF.

3. Results

3.1. Experiment of the Aberration Distribution Characteristics with Integrating Sphere as the Target

We chose a small area target, the integrating sphere, as the light source to analyze the distribution characteristics of AOTF distortion more intuitively. The integrating sphere had a diameter of 12 inches. Such target selection can make the diffracted light imaging not overlap with the zero-order light imaging so that the influence of incident angle on aberrations of AOTF can be analyzed in a wider range.

The experimental imaging setup of different incident angles of the integrating sphere is shown in Figure 7. φ means the half-angle aperture in the azimuth direction (plane perpendicular to the acousto-optic interaction plane). The distance between the integrating sphere and AOTF was 2.8 m, the outlet diameter of the integrating sphere was 4 inches, the aperture of AOTF was 20 mm, the ultrasonic cut angle was 6.5°, the rear wedge angle was

4.83°, the spectral range was 400–1000 nm, the change in polarization was $e \rightarrow o$, the focal length of the rear objective lens was 50 mm, and the pixel size was 4.5 μm. AOTF was fixed on the precision turntable, the driving frequency of the AOTF was adjusted by the PC, and the camera could obtain images of diffracted light with different wavelengths. Images of diffracted light at different incident angles can be obtained through the precision turntable.

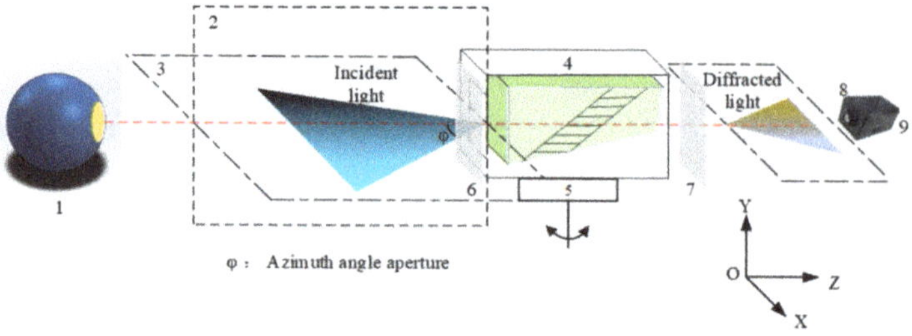

Figure 7. Integrating experimental sphere setup of AOTF based on different incident light angles: 1—integrating sphere; 2—polar plane; 3—azimuth plane; 4—AOTF; 5—rotatable stage; 6 and 7—polarizers; 8—imaging lens; 9—camera.

The distance between the target and AOTF is far in the experimental setting so that the pre-collimating mirror group can be omitted. It is only necessary to measure the aberration of the imaging mirror group and add the aberration parameters to the simulation model.

In Figure 8, the diameter of the large dots is 15 mm, the smaller is 8 mm, and the distance between the calibration points is 40 mm. The coordinates of the calibration point were obtained by the method of geometric center extraction, and the unsatisfactory calibration points were eliminated. Finally, the radial distortion coefficient of the imaging lens group could be obtained by Zhang Zhengyou's calibration method, and the result is shown below as Equation (3).

$$\begin{pmatrix} k_1 \\ k_2 \\ k_3 \end{pmatrix} = \begin{pmatrix} 0.37667 \\ -3.40922 \\ -0.00195 \end{pmatrix}. \tag{3}$$

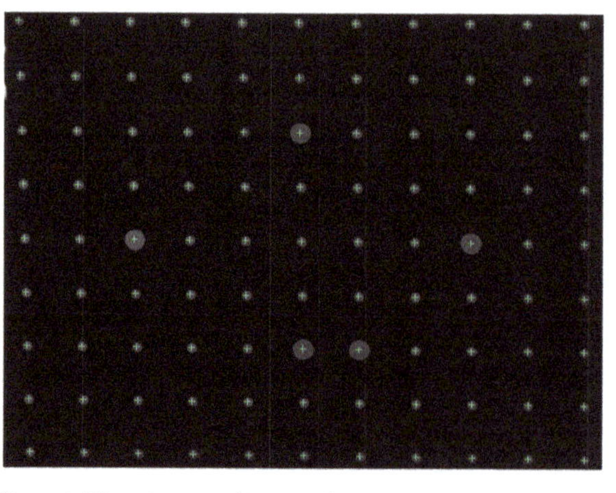

Figure 8. Distortion test of imaging lens group.

As shown in Figure 9, the angle of incident light from left to right gradually approaches the AOTF crystal axis, and the image of diffracted light of the integrating sphere has changed. In general, the aberrations present an asymmetric distribution in the polar direction. By taking the image of *polar angle* = 0° (vertical incidence) as the benchmark, the imaging changes are small when the incident polar angle gradually increases in the direction away from the crystal axis; thus, the spatial aberrations of the AOTF in this direction are small. The imaging of the integrating sphere changed significantly when the incident polar angle increased in the direction close to the crystal axis; thus, the spatial aberrations of the AOTF in this direction are tremendous.

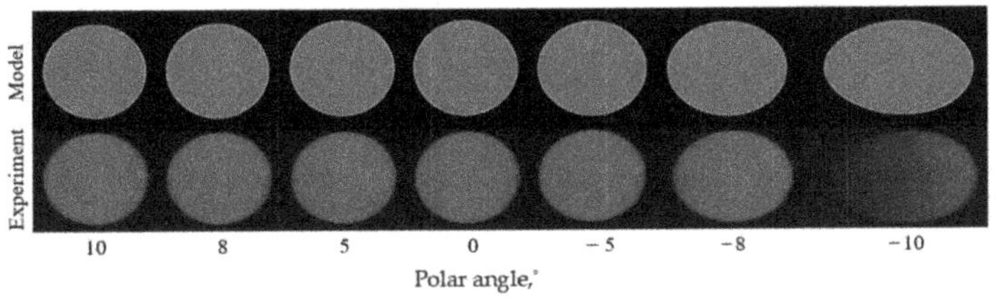

Figure 9. Simulated (upper row) and experimental (lower row) images with the AOTF tuned to λ = 632.8 nm (f = 72.84 MHz) for different angles of incident light in the polar plane.

The influence of incident azimuth angles on the aberrations of AOTF can be obtained with the same experimental method by rotating the AOTF by 90° around the optical axis. From Figure 10, the aberrations caused by the change in the azimuth incidence angle still exist mainly in the polar direction and are distributed symmetrically up and down based on the image of *azimuth angle* = 0°.

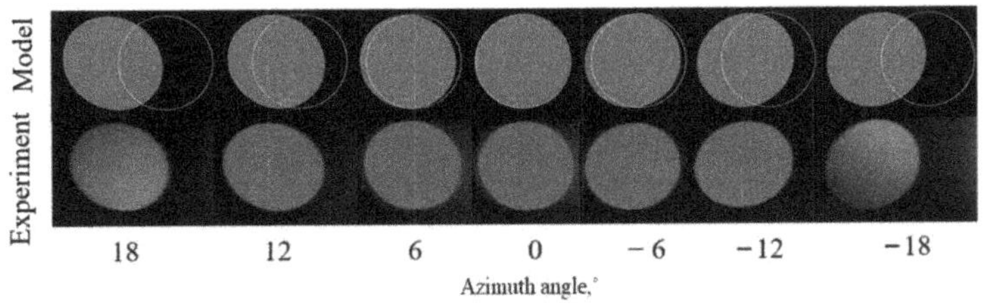

Figure 10. Simulated (upper row) and experimental (lower row) images with the AOTF tuned to λ = 632.8 nm (f = 72.84 MHz) for different angles of incident light in the azimuth plane. The solid red line in the upper row indicates the ideal positions of the integrating sphere with the assumption of pinhole imaging.

Therefore, we can conclude that the aberrations of AOTF mainly exist in the polar direction, the amount of aberrations in the azimuth direction is small, the aberrations are affected by the incident angle in the polar direction greatly, and the closer the incident angle to the AOTF crystal axis is, the greater the aberrations are.

3.2. *Quantitative Experiment with the Square Grid as the Target*

We chose to quantitatively analyze the influence of incident angle on AOTF aberrations in the polar direction. The experimental target is vertically placed at a distance of 5 m from AOTF, and other experimental conditions are consistent with those in Section 3.1. The imaging of the target center point, (x_{55}, y_{55}) in Figure 11a, is located at the center of the image plane. The size of the target is 350 mm × 350 mm, the length of the small square grid is 9 mm, and the distance between adjacent squares is 40 mm.

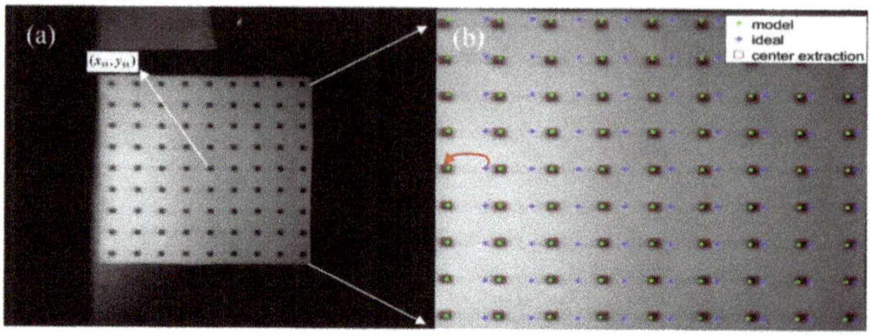

Figure 11. The experiment of the square target: (**a**) actual image of diffracted light at a polar angle of −5.3°; (**b**) the partially enlarged view of the square target.

In Figure 11b, the blue star points represent the ideal points of pinhole imaging, the solid green dots represent the simulated imaging points, and the red boxes indicate the actual imaging areas of the square grid. The coordinate of the actual image point is obtained by extracting the geometric center of the sampling points. We found that the simulated imaging points were located near the actual imaging areas. The reason for the errors may include the surface of the target is not strictly perpendicular to AOTF, there exists an error in distance measurement and target center extraction, and the sound field of AOTF is not ideal. The $error_{RMS}$ can be calculated by Equation (4).

$$error_{RMS} = \sqrt{\frac{1}{81}\sum_{i=1}^{9}\sum_{j=1}^{9}\left[(x_{actual,ij} - x_{model,ij})^2 + (y_{actual,ij} - y_{model,ij})^2\right]}. \quad (4)$$

where $x_{actual,ij}$ is the row coordinate of the diffracted light imaging position. $x_{model,ij}$ is the row coordinate of the calculated imaging position. y is the column coordinate. The $error_{RMS}$ between the actual imaging points and the ideal imaging points is 39.70 $\approx$ 40 pixels, and the deviation mainly exists in the polar direction. The $error_{RMS}$ between the actual imaging points and the modeled imaging points is 2.14 $\approx$ 2 pixels, in which the coefficient of fixing errors is fitted by the least square method. The pixel difference of each point is shown in Figure 12. The sequence of the labels of the sampling points is to take the upper left point as the first sampling point. The sampling is in the row direction.

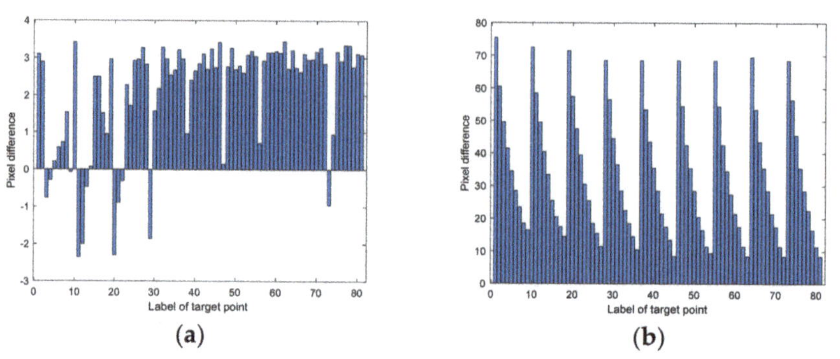

Figure 12. The pixel difference of each point: (**a**) pixel difference between the actual target position and simulated target position; (**b**) pixel difference between the ideal target position and actual target position.

As can be seen from Figure 12a, the pixel difference of all 81 sampling points between the actual target position and the simulated target position is less than 3.5 pixels. In

Figure 12b, the closer the sampling point is to the right of the image, the smaller the aberration is, which is consistent with the previous analysis.

We made the polar angle closer to the AOTF crystal axis, and the result is shown in Figure 13.

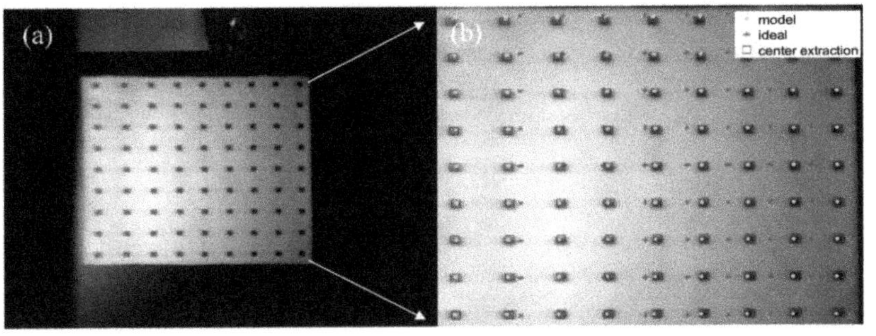

Figure 13. The experiment of the square target: (**a**) actual image of diffracted light at a polar angle of −7°; (**b**) the partially enlarged view of the square target.

By comparing Figures 11b and 13b, it was found that the aberrations of AOTF became more severe with the increase in polar angle, and the black squares on the left were deformed, which is consistent with the results of the previous analysis and simulations.

4. Conclusions and Future Directions

The angle of incident light influences the spatial aberrations of AOTF imaging, and the aberrations mainly exist in the direction of the polar angle. In addition, the influence of cut angle values of the paratellurite crystal on aberrations was analyzed. In the range $(0°, 18.9°)$, the larger the ultrasonic cut angle is, the smaller the spatial aberrations of AOTF imaging are.

In this article, we verified the simulation results through quantitative experiments. The results show that the $error_{RMS}$ between the simulation results and the experimental results is about two pixels when the incident angle in the polar direction is −5.3° under the experimental parameter settings above. Moreover, it provides some guidance for the design of the AOTF system with a large field of view. The computational accuracy of AOTF aberration guarantees that AOTF has the potential to be used in tracking and searching system, which is still blank in the previous research.

Compared with other spectroscopic methods, the biggest advantage of AOTF lies in the staring spectral imaging. Therefore, AOTF can be used not only for the spectral detection of static targets but also for the detection of moving targets, which can expand the application range of AOTF in the future.

Author Contributions: Conceptualization, H.Z. and K.Y.; methodology, K.Y.; software, K.Y.; validation, K.Y. and H.Z.; formal analysis, K.Y.; investigation, K.Y.; resources, H.Z.; data curation, K.Y.; writing—original draft preparation, K.Y.; writing—review and editing, H.Z.; visualization, H.Z.; supervision, H.Z.; project administration, H.Z.; funding acquisition, H.Z. All authors have read and agreed to the published version of the manuscript.

Funding: This work was supported by the National Key Research and Development Program of China (2016YFB0500505) and the National Natural Science Foundation of China (NSFC) (Grant No. 61227806).

Institutional Review Board Statement: Not applicable.

Informed Consent Statement: Not applicable.

Data Availability Statement: Not applicable.

Acknowledgments: We are very grateful to Zefu Xu for his dedication to the related discussions.

Conflicts of Interest: The authors declare no conflict of interest.

References

1. Machikhin, A.S.; Zinin, P.V.; Shurygin, A.V. Imaging system based on a tandem acousto-optical tunable filter for In Situ measurements of the high temperature distribution. *Opt. Lett.* **2016**, *41*, 901–904. [CrossRef] [PubMed]
2. Scharf, V. Four-band fiber optic radiometry for true temperature measurements during an exothermal process. *Opt. Eng.* **2002**, *41*, 1502–1506. [CrossRef]
3. Campbell, A.J. Measurement of temperature distributions across laser heated samples by multispectral imaging radiometry. *Rev. Sci. Instrum.* **2008**, *79*, 21731. [CrossRef]
4. Zhao, H.; Wang, Z.; Jia, G. Chromatic aberrations correction for imaging spectrometer based on acousto-optic tunable filter with two transducers. *Opt. Express* **2017**, *25*, 23809–23825. [CrossRef] [PubMed]
5. Yushkov, K.B.; Molchanov, V.Y.; Ovchinnikov, A.V.; Chefonov, O.V. Acousto-optic replication of ultrashort laser pulses. *Phys. Rev. A* **2017**, *96*, 043866. [CrossRef]
6. Molchanov, V.Y.; Yushkov, K.B.; Kostryukov, P.V. Measurement of amplified binary-modulated chirped laser pulses generated by different acousto-optic pulse shaping algorithms. *Opt. Laser Technol.* **2021**, *142*, 107220. [CrossRef]
7. Zhang, C.; Wang, H.; Huang, J. The visible to the near infrared narrow band acousto-optic tunable filter and the hyperspectral microscopic imaging on biomedicine study. *J. Opt.* **2014**, *16*, 125303. [CrossRef]
8. Yushkov, K.B.; Molchanov, V.Y.; Belousov, P.V. Contrast enhancement in microscopy of human thyroid tumors by means of acousto-optic adaptive spatial filtering. *J. Biomed. Opt.* **2016**, *21*, 016003. [CrossRef] [PubMed]
9. Yushkov, K.B.; Molchanov, V.Y. Acousto-optic filters with arbitrary spectral transmission. *Opt. Commun.* **2015**, *355*, 177–180. [CrossRef]
10. Gupta, N.; Voloshinov, V.B. Hyperspectral imaging performance of a TeO_2 acousto-optic tunable filter in the ultraviolet region. *Opt. Lett.* **2005**, *30*, 985–987. [CrossRef] [PubMed]
11. Gupta, N.; Suhre, D.R. Notch filtering using a multiple passband AOTF in the SWIR region. *Appl. Opt.* **2016**, *55*, 7855–7860. [CrossRef] [PubMed]
12. Chanover, N.J.; Glenar, D.A.; Voelz, D.G.; Xiao, X.; Tawalbeh, R.; Boston, P.J.; Brinckerhoff, W.B.; Mahaffy, P.R.; Getty, S. An AOTF-LDTOF spectrometer suite for In Situ organic detection and characterization. In Proceedings of the IEEE Aerospace Conference, Big Sky, MT, USA, 5–12 March 2011.
13. Glenar, D.A.; Hillman, J.J.; Saif, B.; Bergstralh, J. Acousto-optic imaging spectropolarimetry for remote sensing. *Appl. Opt.* **1994**, *33*, 7412–7424. [CrossRef] [PubMed]
14. Wang, P.; Zhang, Z. Effect of direction of incident light on the basic performance of a TeO_2 acousto-optic tunable filter. *Appl. Opt.* **2016**, *55*, 5860–5867. [CrossRef] [PubMed]
15. Zhang, C.; Zhang, Z.; Yang, Y.; Wang, H. Design and analysis of a noncollinear acousto-optic tunable filter. *Opt. Lett.* **2007**, *32*, 2417–2419. [CrossRef] [PubMed]
16. Zhang, C.; Wang, H.; Qiu, Y. Analysis of the Spectral Resolution of a TeO_2 based Noncollinear Acousto-Optic Tunable Filter. *Engineering* **2011**, *3*, 233–235. [CrossRef]
17. Zhang, C.; Wang, H. Design of an acousto-optic tunable filter based on momentum mismatching together with the rotatory property. *Appl. Opt.* **2018**, *57*, 8595–8599. [CrossRef] [PubMed]
18. Yushkov, K.B.; Chizhikov, A.I.; Makarov, O.Y.; Molchanov, V.Y. Optimization of noncollinear AOTF design for laser beam shaping. *Appl. Opt.* **2020**, *59*, 8575–8581. [CrossRef] [PubMed]
19. Machikhin, A.S.; Pozhar, V.E. Image aberrations in an acousto-optical tunable filter. *Opt. Quant. Electron.* **2010**, *55*, 1490–1496. [CrossRef]
20. Machikhin, A.; Batshev, V.; Pozhar, V. Aberration analysis of AOTF-based spectral imaging systems. *J. Opt. Soc. Am. A* **2017**, *34*, 1109–1113. [CrossRef] [PubMed]
21. Machikhin, A.S.; Pozhar, V.E. Spatial and spectral image distortions caused by diffraction of an ordinary polarised light beam by an ultrasonic wave. *Quantum Electron.* **2015**, *45*, 161–165. [CrossRef]
22. Pozhar, V.E.; Machikhin, A.S. Image aberrations caused by light diffraction via ultrasonic waves in uniaxial crystals. *Appl. Opt.* **2012**, *51*, 4513–4519. [CrossRef] [PubMed]
23. Gorevoy, A.V.; Machikhin, A.S.; Martynov, G.N. Spatiospectral transformation of noncollimated light beams diffracted by ultrasound in birefringent crystals. *Photonics Res.* **2021**, *9*, 687–693. [CrossRef]
24. Machikhin, A.S.; Shurygin, A.V.; Pozhar, V.E. Spatial and spectral calibration of an acousto-optical spectrometer. *Instrum. Exp. Tech.* **2016**, *59*, 692–697. [CrossRef]
25. Batshev, V.; Naumov, A.; Pozhar, V.E. Improved optical design of AOTF-based stereoscopic system for 3D imaging spectroscopy. In Proceedings of the Twelfth International Conference on Digital Image Processing, Osaka, Japan, 19–22 May 2020.
26. Batshev, V.I.; Gorevoy, A.V.; Pozhar, V.E.; Machikhin, A.S. Aberration analysis of AOTF-based stereoscopic spectral imager using optical design software. *J. Phys. Conf. Ser.* **2021**, *2127*, 012035. [CrossRef]
27. Khokhlov, D.; Gorevoy, A.V.; Machikhin, A.S. Applying computer simulation using optical design software to multi-spectral geometrical calibration of stereoscopic measurement systems. In Proceedings of the Automated Visual Inspection and Machine Vision III, Munich, Germany, 21 June 2019.

28. Batshev, V.; Machikhin, A.S.; Pozhar, V.E.; Naumov, A. Single-volume dual-channel acousto-optical tunable filter. *Opt. Express* **2019**, *28*, 1150–1157.
29. Gorevoy, A.V.; Machikhin, A.S.; Martynov, G.N. Computational technique for field-of-view expansion in AOTF-based imagers. *Opt. Lett.* **2022**, *47*, 585–588. [CrossRef] [PubMed]
30. Xu, Z.; Zhao, H.; Jia, G. Optical schemes of super-angular AOTF-based imagers and system response analysis. *Opt. Commun.* **2021**, *498*, 127204. [CrossRef]
31. Zhao, H.; Li, C.; Zhang, Y. Three-surface model for the ray tracing of an imaging acousto-optic tunable filter. *Appl. Opt.* **2014**, *53*, 7684–7690. [CrossRef]

materials

MDPI

Article

Calibration of Acousto-Optic Interaction Geometry Based on the Analysis of AOTF Angular Performance

Hao Zhang [1], Huijie Zhao [1,2], Qi Guo [1,*] and Yan Xuan [1]

1 School of Instrumentation Science and Opto-Electronics Engineering, Beihang University, Beijing 100191, China; hao2017@buaa.edu.cn (H.Z.)
2 Institute of Artificial Intelligence, Beihang University, Beijing 100191, China
* Correspondence: qguo@buaa.edu.cn

Abstract: Acousto-optic interaction geometry determines the spectral and spatial response of an acousto-optic tunable filter (AOTF). The precise calibration of the acousto-optic interaction geometry of the device is a necessary process before designing and optimizing optical systems. In this paper, we develop a novel calibration method based on the polar angular performance of an AOTF. A commercial AOTF device with unknown geometry parameters was experimentally calibrated. The experimental results show high precision, in some cases falling within 0.01°. In addition, we analyzed the parameter sensitivity and Monte Carlo tolerance of the calibration method. The results of the parameter sensitivity analysis show that the principal refractive index has a large influence on the calibration results, while other factors have little influence. The results of the Monte Carlo tolerance analysis show that the probability of the results falling 0.1° using this method is greater than 99.7%. This work provides an accurate and easy-to-perform method for AOTF crystal calibration and can contribute to the characteristic analysis of AOTFs and the optical design of spectral imaging systems.

Keywords: AOTF; acousto-optic interaction geometry; polar angular analysis; tolerance analysis

Citation: Zhang, H.; Zhao, H.; Guo, Q.; Xuan, Y. Calibration of Acousto-Optic Interaction Geometry Based on the Analysis of AOTF Angular Performance. *Materials* **2023**, *16*, 3708. https://doi.org/10.3390/ma16103708

Academic Editors: Virgil-Florin Duma, Alexander S. Machikhin and Vitold Pozhar

Received: 26 February 2023
Revised: 26 April 2023
Accepted: 9 May 2023
Published: 13 May 2023

1. Introduction

An AOTF device is a spectral-splitting device based on the acousto-optic effect [1]. When compared with traditional optical splitters, such as prism, grating and interferometer, it has advantages in terms of a large-angle aperture, high spectral resolution, arbitrary wavelength configuration, fast tuning speed and it does not require moving elements [2,3]. Therefore, it is widely used in many applications, such as spectral imaging [4,5], polarization analysis [6,7], stereoscopic imaging [8] and notch filtering [9], etc. In addition, in the case of monochromatic input light, AOTFs also have the abilities of spatial filtering and edge enhancement [10,11]. There are two basic configurations, collinear and noncollinear, for AOTFs. Furthermore, in the collinear configuration, the interacting optical and acoustic waves propagate in identical directions, while the directions of the optical and acoustic waves are different in the noncollinear configuration. The first AOTF with a collinear design was reported by Harris and Wallace in 1969 [12]. Subsequently, Chang described noncollinear AOTFs in 1974 [13], which are commonly used today. Noncollinear AOTFs have some advantages, as a larger angular aperture and more materials with a large acousto-optic figure of merit can be chosen. Among them, TeO_2 crystals are very suitable for noncollinear AOTFs that cover the spectral range of 350–4500 nm [14,15].

Acousto-optic interaction geometry is a key inherent property of an AOTF device, which greatly affects the spectral and spatial response of the device. Previously, Voloshinov analyzed the acousto-optic effect under three kinds of acoustic cut angles, and the results showed that acousto-optic interaction geometry affects the angular apertures and spectral resolution of an AOTF device [16]. In addition, acousto-optic interaction geometry will also affect the tuning relationship, sound field distribution, aberration and chromatic aberration characteristics of the device, thus greatly affecting all aspects of the AOTFs in characteristic analysis and optical system design [4,17]. Therefore, AOTFs must be designed in detail to meet the special output requirements. Chang first proposed the parallel tangent condition for the design of noncollinear AOTFs with large angular apertures in the 1970s [13]. Yano discussed some properties of AOTFs using the simplified treatment [18]. Moreover, Gass corrected the birefringence approximation for the accurate design of the acousto-optic interaction geometry in the analysis [19]. These theories are all typically under the parallel tangent condition. For the non-parallel tangent condition, Yushkov expressed an exact phase-matching calculation equation at an arbitrary incident angle and recently proposed an alternative method for analyzing the Bragg angle curve in wide-angle AOTFs [10,20]. Zhang discussed the function of the phase mismatching condition and proposed a new tuning method with a non-radio-frequency signal [21]. After the processes of analysis and design, an AOTF will be manufactured using a series of fabrication technologies. Many steps are involved in the fabrication of AOTFs, including X-ray orientation, cutting, polishing, transducer orientation and fabrication, mounting and grinding [22]. The fabrication technologies of an AOTF are so complex that they easily lead to machining tolerances between the designed and actual device. If the designed values were used for device characteristic analysis and optical system design, it would lead to inaccurate results. Therefore, it is vital to calibrate the acousto-optic interaction geometry of an actual AOTF before use. In the past, our team proposed an acousto-optic interaction geometry calibration method by the tested tuning frequency curve under the parallel tangent condition [23]. However, this multi-wavelength method is not conducive to calibration accuracy as it relies on more constant parameters. In addition, to find the parallel tangent condition, as shown in Figure 1a, the incident angles must be adjusted accurately, which requires high precision in the experiment.

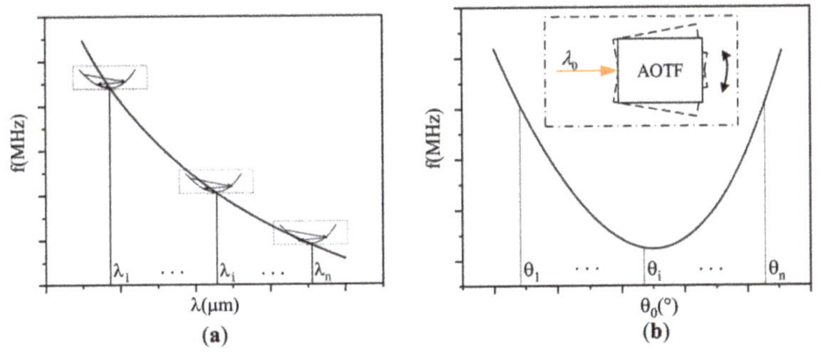

Figure 1. Comparison of operating principles. (**a**) Calibration method by the tested tuning curves under the parallel tangent condition. (**b**) Calibration method using polar angular analysis, which has no specific requirement of incident angles.

To overcome these shortcomings, here we develop a calibration method based on the polar angular performance of an AOTF. Firstly, we establish an AOTF angular frequency relationship model that can be solved analytically. Moreover, based on this model, a novel method is developed to calibrate the acousto-optic interaction geometry of an actual AOTF device. It does not introduce principal refractive index errors between multiple wavelengths and works with a single monochromatic light source. Furthermore, this method does not depend on determining characteristic incident angles, as shown in Figure 1b. Finally, using the principle of the minimum root mean square error (RMSE) between the measured and theoretical data, the acousto-optic interaction geometry of the actual AOTF device can be calculated through the use of the parameter traversal method. This method is an improvement of the calibration process in terms of simplicity and robustness and has been tested with high precision in experiments. Simultaneously, the method analyzes the influence of crystal constants on calibration results in the visible range. This work is significant and provides a database for a range of research related to AOTF devices.

2. Methods

The acousto-optic interaction geometry of an AOTF refers to the front facet angle (θ_i^*), the acoustic cut angle (θ_α) and the back facet angle (θ_β), respectively. As shown in Figure 2, this is the top view of the AOTF device and corresponds to the polar plane. In the AOTF, an acoustic wave is generated by a transducer and absorbed by an absorber. By switching the radio frequency signals applied to the transducer, the AOTF can scan the spectral regions of interest [24]. Given that the polarization state of the incident light (L_0) is inconsistent with one of the eigenwave modes in TeO_2, four types of emitted light are produced, namely diffraction ordinary polarized light (L_d^o), transmission extraordinary polarized light (L_t^e), transmission ordinary polarized light (L_t^o) and diffraction extraordinary polarized light (L_d^e).

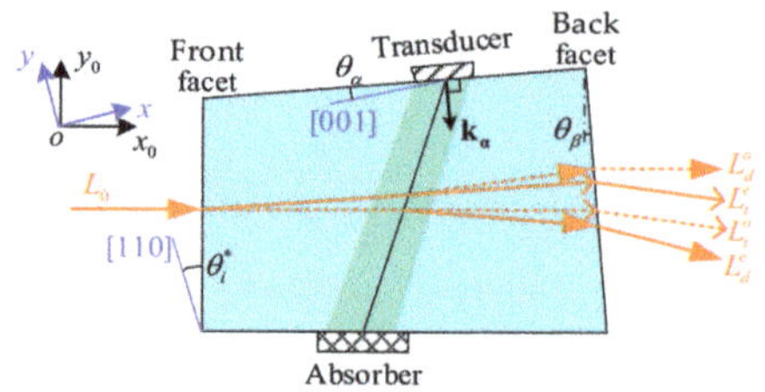

Figure 2. Acousto-optic interaction geometry of an AOTF. The θ_i^* is the front facet angle between the front facet and the crystal axis [110]. The θ_α is the acoustic cut angle between the transducer surface and the crystal axis [001]. The θ_β is the back facet angle between the back and front facets.

Two coordinate systems have been established for analysis in this paper: the optical axes coordinate system (x_0oy_0) and the crystal axes coordinate system (xoy), as shown in Figure 2. The optical axes coordinate system is a rectangular coordinate system, where the y_0 axis is the intersection line of the incident surface and polar plane, while the x_0 axis is perpendicular to the incident surface. The crystal axes coordinate system is also a rectangular coordinate system, wherein the y axis is the crystal axis [110], while the x axis is the crystal axis [001].

The model between acousto-optic interaction geometry and polar angular performance for AOTFs involves two processes: (a) the calculation of the refraction at the plane of incidence, shown in Section 2.1, and (b) wave vector analysis of acousto-optic interaction, shown in Section 2.2. In addition, the relationship between the incident polar angles and matching frequencies is independent of the back facet, which will be analyzed in Section 2.1. Therefore, we need two measurements to obtain complete calibration results by swapping the front and back facets, as shown in Figure 3. For the exchange of the input and output facets, the AOTF device needs to be rotated about 180° around the axis perpendicular to the xoy plane.

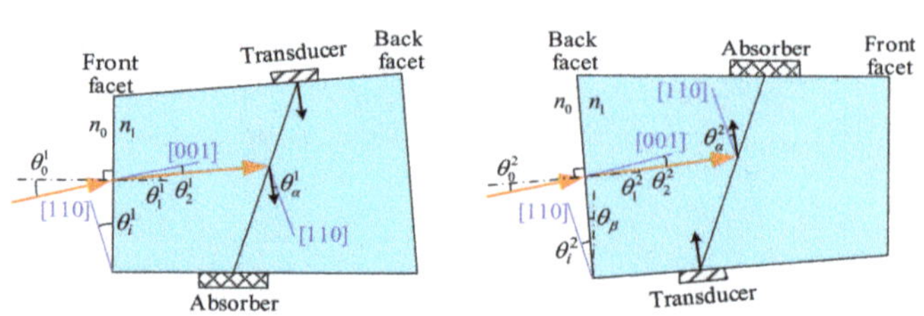

Figure 3. Two modes for analysis. (**a**) Positive mode, lights enter AOTF from the front facet. (**b**) Reverse mode, lights enter AOTF from the back facet.

2.1. Refraction at the Plane of Incidence

Firstly, the refraction of light in the plane of incidence obeys Snell's law as follows:

$$n_0 \sin\theta_0 = n_1 \sin\theta_1 \quad (1)$$

where θ_0 is the incident polar angle between the incident light and the normal of the incident plane. θ_1 is the refraction angle in the crystal between the refracted light and the normal to the plane of incidence. n_0 and n_1 are the refractive indices in the air and crystal, respectively. Moreover, it is known that n_0 is equal to 1 in the air. TeO_2 crystal has the anisotropy and n_1 can be solved by:

$$\begin{cases} n_1^o = n^o \\ n_1^e = \frac{n^o n^e}{\sqrt{(n^e)^2 \cos^2\theta_2^e + (n^o)^2 \sin^2\theta_2^e}} \end{cases} \quad (2)$$

with superscripts (o and e) used to distinguish ordinary (o-polarized) from extraordinary (e-polarized) light. n^o and n^e, related to the wavelength λ, are the principal refractive indices of TeO_2. θ_2^o and θ_2^e are the angles between refracted light and the crystal axis [001] for o-polarized and e-polarized lights, respectively. The difference between θ_1 and θ_2 is as follows:

$$\theta_2 = \theta_1 - \theta_i^c \quad (3)$$

where θ_i^c is the angle between the incident plane and the crystal axis [110], for which $c = 1$ corresponds to the positive mode, as shown in Figure 3a, while $c = 2$ corresponds to the reverse mode, as shown in Figure 3b. When switching from the positive mode to the reverse mode, the AOTF device must be rotated for swapping the input and output facets. Furthermore, the relationships between θ_i^c and the acousto-optic interaction geometry (θ_i^*, θ_α and θ_β) of an AOTF are as follows:

$$\begin{cases} \theta_i^1 = \theta_i^* \\ \theta_i^2 = \theta_i^* - \theta_\beta \end{cases} \quad (4)$$

which means that both the positive and reverse modes are necessary to obtain the complete acousto-optic interaction geometry of an AOTF device. From Equations (1)–(3), we can obtain:

$$\begin{cases} F^e(x) = \left(a_1a_3 - a_2a_3a_5^2\right)x^2 - 2a_2a_3a_4a_5x + a_1a_2 - a_2a_3a_4^2 = 0 \\ F^o(x) = \left(a_1 - a_3a_5^2\right)x^2 - 2a_3a_4a_5x + a_1 - a_3a_4^2 = 0 \end{cases} \quad (5)$$

where F^e and F^o correspond to the conditions under which the incident lights are the e-polarized and o-polarized, respectively. Furthermore, $a_1 = \sin^2\theta_0$, $a_2 = (n^e)^2$, $a_3 = (n^o)^2$, $a_4 = \sin\theta_i^c$, $a_5 = \cos\theta_i^c$ and $x = \tan\theta_2$. Equation (5) contains both the quadratic equations and can easily be solved. Then, using Equations (1)–(5), the relationship between θ_0 and θ_2 can be expressed as follows:

$$\theta_2 = F_1\left(\theta_0, \lambda, \theta_i^*, \theta_\beta\right) \quad (6)$$

where F_1 is an implicit function.

2.2. Wave Vector Analysis of Acousto-Optic Interaction

Acousto-optic interaction in the AOTF is usually analyzed through a wave vector diagram [25]. When the momentum-matching condition is satisfied, the incident wave vector $\mathbf{k_i}$, the acoustic wave vector $\mathbf{k_\alpha}$ and the diffraction wave vector $\mathbf{k_d}$ constitute a closed triangle (Figure 4), shown as $\mathbf{k_i} \pm \mathbf{k_\alpha} = \mathbf{k_d}$. Some are dependent on the incident wavelength and refraction indices, as follows:

$$\begin{cases} |\mathbf{k_i}| = \frac{2\pi n_i}{\lambda} \\ |\mathbf{k_d}| = \frac{2\pi n_d}{\lambda} \end{cases} \quad (7)$$

where n_i and n_d are the refractive indices of the incident and diffracted light in TeO_2. They satisfy:

$$\begin{cases} \frac{x^2}{(n^o)^2} + \frac{y^2}{(n^o)^2} = \left(\frac{2\pi}{\lambda}\right)^2 \\ \frac{x^2}{(n^o)^2} + \frac{y^2}{(n^e)^2} = \left(\frac{2\pi}{\lambda}\right)^2 \end{cases} \tag{8}$$

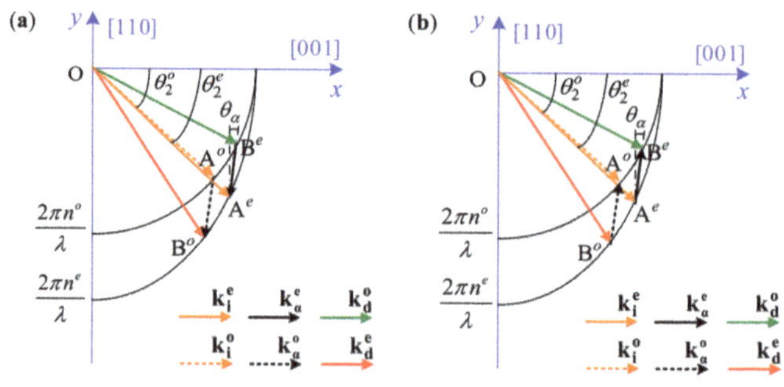

Figure 4. Wave vector diagrams of the noncollinear AOTF. (**a**) Positive mode. (**b**) Reserve mode.

In addition, acoustic wave vector $\mathbf{k}_\alpha$ satisfies:

$$|\mathbf{k_a}| = \left|\overrightarrow{AB}\right| = \frac{2\pi f_\alpha}{V_\alpha} \tag{9}$$

where f_α is the acoustic frequency and V_α is the acoustic wave velocity. In the crystal, V_α is given by [26]:

$$V_\alpha^2 = V_{110}^2 \cos^2\theta_\alpha + V_{001}^2 \sin^2\theta_\alpha \tag{10}$$

where V_{001} and V_{110} are the acoustic wave velocities along the respective crystal axes. According to Equation (9), in order to solve acoustic frequency f_α, we need to calculate the $\left|\overrightarrow{AB}\right|$. As shown in Figure 4, point A (x_A, y_A)satisfies $\tan\theta_2 = \frac{y_A}{x_A}$ and Equation (8), which can be solved by following:

$$\begin{cases} A^o = (1, \tan\theta_2^o)\frac{2\pi n^o}{\lambda\cdot\sqrt{1+\tan^2\theta_2^o}} \\ A^e = (1, \tan\theta_2^e)\frac{2\pi n^o n^e}{\lambda\cdot\sqrt{(n^e)^2+(n^o)^2\tan^2\theta_2^e}} \end{cases} \tag{11}$$

Point B (x_B, y_B) satisfies the linear equation of $\overrightarrow{AB}$ and Equation (8). For both positive and reverse modes, the linear equations of $\overrightarrow{AB}$ are the same one as:

$$\frac{x_B - x_A}{y_B - y_A} = \tan\theta_\alpha \tag{12}$$

Therefore, point B can be solved by following:

$$\begin{cases} F^{o\to e}(y_B) = \left(\frac{\tan^2\theta_\alpha}{(n^o)^2} + \frac{1}{(n^e)^2}\right){y_B}^2 + \frac{2}{(n^o)^2}(x_A\tan\theta_\alpha - y_A\tan^2\theta_\alpha)y_B \\ \qquad + \left(\frac{{x_A}^2+{y_A}^2\tan^2\theta_\alpha - 2x_Ay_A\tan\theta_\alpha}{(n^o)^2} - \frac{2\pi}{\lambda}\right) = 0 \\ F^{e\to o}(y_B) = \left(\frac{\tan^2\theta_\alpha}{(n^o)^2} + \frac{1}{(n^o)^2}\right){y_B}^2 + \frac{2}{(n^o)^2}(x_A\tan\theta_\alpha - y_A\tan^2\theta_\alpha)y_B \\ \qquad + \left(\frac{{x_A}^2+{y_A}^2\tan^2\theta_\alpha - 2x_Ay_A\tan\theta_\alpha}{(n^o)^2} - \frac{2\pi}{\lambda}\right) = 0 \end{cases} \tag{13}$$

which are all quadratic equations and easy to be solved exactly. $F^{o\to e}$ corresponds to the condition wherein the incident light is o-polarized and the diffraction light is e-polarized, while $F^{e\to o}$ corresponds to the condition wherein the incident light is e-polarized and the diffraction light is o-polarized. From Equations (9)–(13), the relationship between f_α and θ_2 can be expressed as follows:

$$f_\alpha = F_2(\theta_2, \lambda, \theta_\alpha) \tag{14}$$

where F_2 is an implicit function.

In summary, with Equations (6) and (14), the model between the acousto-optic interaction geometry and the incident polar angular frequencies of AOTFs can be established as follows:

$$f_\alpha = F\left(\theta_0, \lambda, \theta_i^*, \theta_\alpha, \theta_\beta\right) \tag{15}$$

where F is an implicit function. This means that, for an actual AOTF device with inherent acousto-optic interaction geometry (θ_i^*, θ_α and θ_β), acoustic frequencies and incident polar angles are correlated when the wavelength (λ) of the incident lights is fixed. Therefore, the acousto-optic interaction geometry of the AOTF device can be calibrated by analyzing the incident polar angles and corresponding acoustic frequencies.

The numerical values of the constants used in the calculations in this paper are provided in Table 1 [27].

Table 1. The numerical value of the constants used in the calculations.

Parameters	Values
λ	632.8 nm
n^o	2.2597 at 632.8 nm
n^e	2.4119 at 632.8 nm
V_{001}	616 m/s
V_{110}	2104 m/s

3. Experiments and Discussions

3.1. Experimental Setup

The schematic diagram of our experimental setup is shown in Figure 5. The monochromatic source we used was a 632.8 nm He-Ne laser (DH-HN250), from which the linearly polarized light was generated. A group of frosted glasses was used to reduce light intensity, and the effect of the frosted glasses can be replaced by multiple polarizers. The commercial AOTF used in the experiment was manufactured by China Electronics Technology Group Corporation (CETC) and is referred to as SGL100-400/850-20LG-K. The polarizer, located ahead of the AOTF, was used to adjust the polarization state of incident lights so that o-polarized and e-polarized components were close and convenient for measurements. The turntable (GCM-1107M) was used to accurately control the incident polar angles of the incident light into the AOTF, and here the accuracy of the rotation angle was 2′. A Basler acA640-120gm camera with an 8 mm focal length lens was used as the detector in the experiments. Compared with the optical power meter, a camera can simultaneously detect the intensities of transmitted and diffracted lights (Figure 6) to effectively avoid the measurement error caused by the instability of the power intensity and the polarization state of the laser.

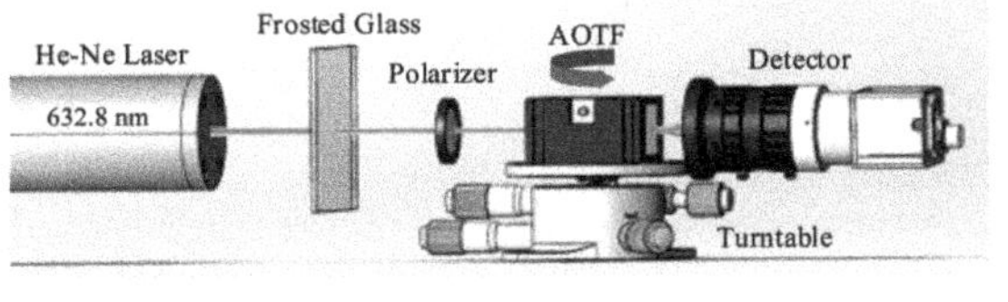

Figure 5. Experimental setup for the polar angular frequency tests.

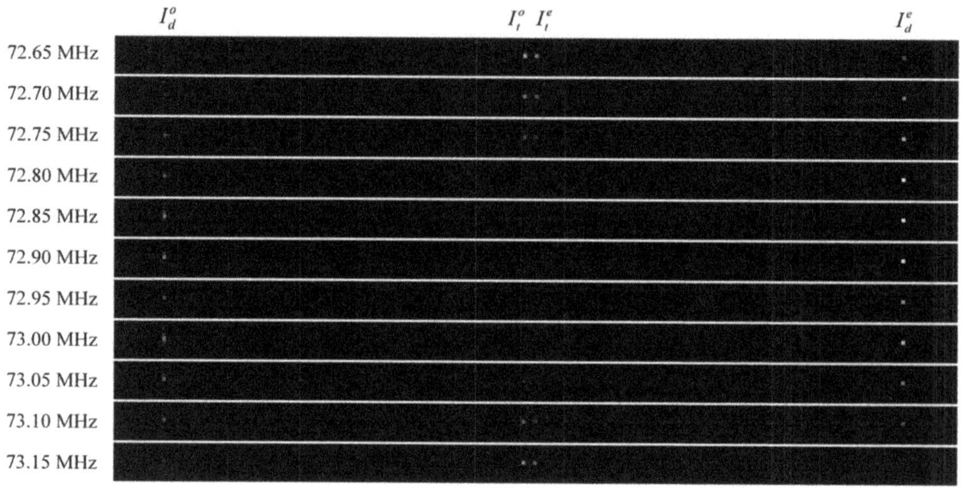

Figure 6. A montage of 11 images taken at different frequencies in sequence. The incident polar angle is 2.19° in the positive mode.

3.2. Results and Discussions

For incident polar angle analysis, we needed to measure the matching frequency at each incident polar angle, which corresponds to the peak diffraction intensity. In practice, the potential range of the matching frequency can be estimated from Equation (15) using the design geometry parameters from the AOTF manufacturer or through the direct observation of the maximum diffraction intensity. It should be noted that in some special applications, the changing ultrasonic signal and the angle of the incident light will greatly change the shape of the AOTF transfer function, which needs further discussion [28,29]. The processes of the tests are organized as follows:

- Step 1: Adjust the polar angle of the AOTF by using the turntable and make sure that the incident plane of the AOTF is perpendicular to the incident light. This step can be judged by whether the reflected laser point coincides with the exit point. We recorded the scale of the turntable at this point as the "0" scale, and the other incident polar angles were able to be adjusted with this scale.
- Step 2: After adjusting the AOTF incident polar angle, the laser, AOTF and detector must be switched on. Then a montage of images, including transmitted and diffracted light, can be taken by scanning the acoustic frequencies, as shown in Figure 6. For each image, both transmitted and diffracted light can be captured, or only o-polarized and e-polarized light can be measured separately by adjusting the polarizer. Given that, in some cases, the AOTFs do not have the wedge angle compensation, the directions of transmitted o-polarized and e-polarized light are coincident. In these experiments, the frequency step was 0.05 MHz.
- Step 3: To find the matching frequency corresponding to the peak diffraction intensity, use the relative diffraction efficiency to evaluate as:

$$\eta = \frac{I_d}{I_d + I_t} / \eta_{\max} \tag{16}$$

where I_t and I_d are the light intensity values for the transmitted and diffracted light from the same incident light. The intensity values are quantified by the digital number (DN) values with 8-bit digitization. For each order of emitted light, we used the sum of DN values in the effective area, where nine adjacent pixels were selected for calculation, as shown in Figure 7a.

- Step 4: The matching frequencies were able to be solved by quartic polynomial fitting, as shown in Figure 7b, and at least five frequency points are required for each incident polar angle.
- Step 5: In order to ensure that the temperature of each measurement is close to the room temperature, the AOTF needs to be switched off for a few minutes because the temperature of the AOTF rises during operation, which would affect its polar angular performance.
- Step 6: Adjust another incident polar angle of the AOTF, switch on the AOTF and repeat Steps 2–6 again.

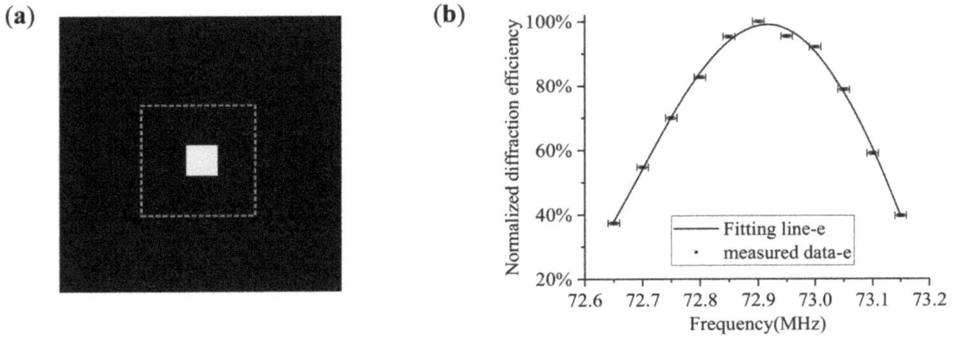

Figure 7. Experimental data for incident polar angular frequency analysis. (**a**) The effective DN value is the sum of the DN values within 9 adjacent pixels, and the 9 adjacent pixels are here surrounded by red dotted lines here; (**b**) The data fitting result by quartic polynomial fitting, and the measured data can be found in Figure 6.

The coefficients of determination (R^2) of all the results were better than 0.98, and the fitting residuals were less than 0.003. Some other data fitting results at different incident polar angles are shown in Figure 8. From these results, we found that the matching frequencies of o-polarized and e-polarized lights are generally not consistent under the same incident polar angles. In other words, at the same acoustic frequency, the diffracted wavelengths of the o-polarized and e-polarized lights are not consistent. However, under a specific incident polar angle, we obtained the same diffracted wavelengths of the o-polarized and e-polarized lights at the same acoustic frequency. In some research, they named this condition the equivalent point, wherein the matching frequencies are the same for o-polarized and e-polarized light at the same incident polar angle [30,31]. Here, we obtained this condition by adjusting the incident polar angle. Moreover, it would be exactly calculated with an exact acousto-optic interaction geometry of the AOTF. Therefore, we will discuss it after the acousto-optic interaction geometry calibration.

In this paper, a total of 21 incident polar angles were sampled in the positive mode, while 17 incident polar angles were sampled in the reverse mode. In the experiments, the minimum angle sampling step was 0.5°. All of the matching frequencies in positive and reserve modes can be found in Figure 9c,d. After sampling all of the measured data, we used the parameter traversal method to calculate the acousto-optic interaction geometry of the AOTF with the principle of minimum RMSE. For each input of the geometry parameters, the RMSE between the theoretical data and measured data is as follows:

$$RMSE = \sqrt{\frac{1}{N}\sum_{i=1}^{N}\left|F\left(\theta_0(i), \lambda_0, \theta_i^*, \theta_\alpha, \theta_\beta\right) - F_m(\theta_0(i))\right|^2} \tag{17}$$

where N is the number of sampling points for each mode, and there are two types of sampling points for both o-polarized and e-polarized lights at some incident polar angles.

N was 31 for the positive mode in this paper, and 26 for the reserve mode. The $\theta_0(i)$ is the incident polar angle, and the incident wavelength (λ_0) was 632.8 nm. F_m is the measured data of the matching frequency at each incident polar angle. As shown in Figure 9, we obtained the RMSE distributions in both positive and reserve modes. Then, the acousto-optic interaction geometry of the AOTF device, corresponding to the minimum RMSE, was able to be obtained. As shown in Figure 9a, we obtained $\theta_i^1 = 15.074°$ and $\theta_\alpha^1 = 6.484°$ in the positive mode with the minimum RMSE (0.032 MHz). Meanwhile, we also obtained $\theta_i^2 = 10.435°$ and $\theta_\alpha^2 = 6.486°$ in the reserve mode with the minimum RMSE (0.042 MHz) in Figure 9b. The difference of θ_α between the two measurements was 0.002°, which means this calibration method has a high precision and can be better than 0.01°. We took the average of two results as the calibration value that $\theta_\alpha = 6.485°$. As shown in Figure 9c,d, the measured data were very close to the theoretical data. In summary, we obtained the calibration results of $\theta_i^* = 15.074°$, $\theta_\alpha = 6.485°$ and $\theta_\beta = 4.639°$ with Equation (4). In addition, according to the reference [19], we calculated the acousto-optic interaction geometry, meeting the parallel tangent condition, whereby $\theta_i^* = 15.074°$ and $\theta_\alpha = 6.548°$ under normal incidence of e-polarized light at 632.8 nm. Therefore, we found that the calibration result of the actual AOTF device was close but did not meet the parallel tangent condition at 632.8 nm.

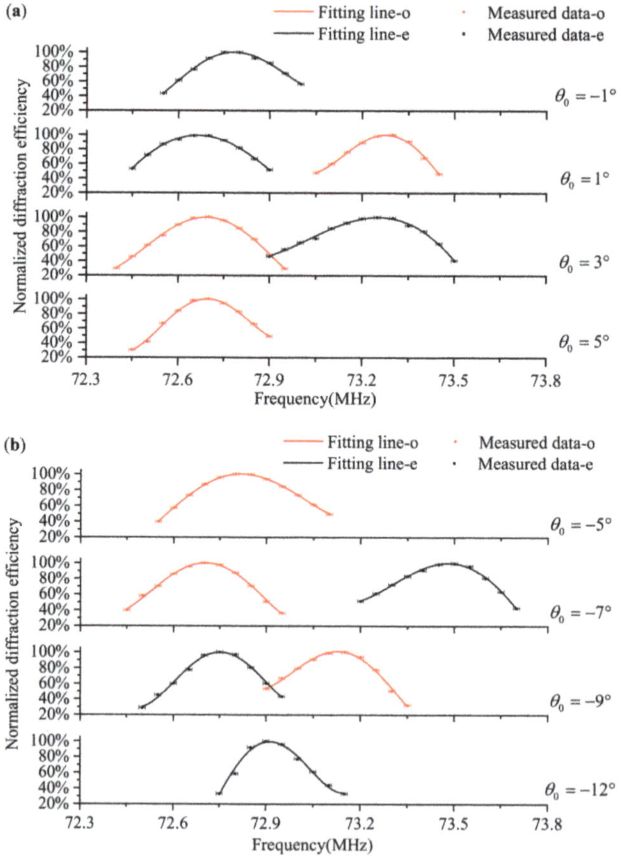

Figure 8. Some other data fitting results. (**a**) Positive mode, normalized diffraction efficiencies by frequency scanning at some incident polar angles (−1°, 1°, 3°, 5°); (**b**) reverse mode, normalized diffraction efficiencies by frequency scanning at some incident polar angles (−5°, −7°, −9°, −12°).

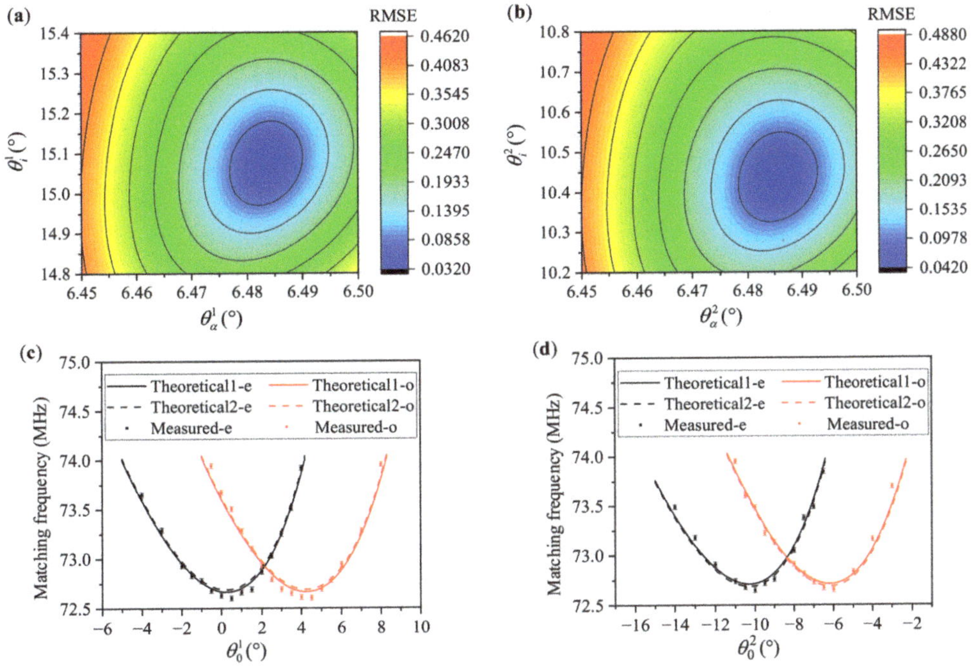

Figure 9. Calibration results of the actual AOTF device. (**a**) RMSE distribution in the positive mode; (**b**) RMSE distribution in the reserve mode; (**c**) measured data and theoretical simulation curves in the positive mode. The solid line marked "Theoretical-1e" is the theoretical simulation curve under which $\theta_i^1 = 15.074°$ and $\theta_\alpha^1 = 6.484°$, while the dashed line marked "Theoretical-2e" is the theoretical simulation curve under which $\theta_i^1 = 15.074°$ and $\theta_\alpha^1 = 6.485°$; (**d**) measured data and theoretical simulation curves in the reserve mode. The solid line marked "Theoretical-1e" is the theoretical simulation curve under which $\theta_i^2 = 10.435°$ and $\theta_\alpha^2 = 6.486°$, while the dashed line marked "Theoretical-2e" is the theoretical simulation curve under which $\theta_i^2 = 10.435°$ and $\theta_\alpha^2 = 6.485°$.

From Figure 9a,b, we found that the acoustic cut angle of the AOTF was more sensitive than the front facet angle. Therefore, we further analyzed the angular frequency relationship under the different acoustic cut angles and front facet angles, and the results are shown in Figure 10. The deviation caused by the change of acoustic cut angles (±0.01°) was higher than that caused by the change of front facet angles (±0.1°). These results confirm that changing the acoustic cut angle has a greater influence. Moreover, changing the acoustic cut angle makes the angular frequency curves shift up and down, and the larger acoustic cut angle corresponds to the state of shifting up. In comparison, changing the front facet angle makes the angular frequency curves shift left and right, and the larger front facet angle corresponds to the state of shifting right.

In order to further verify the accuracy of the calibration result, the equivalent points in two modes are calculated and tested here. According to the calibration results and Equation (15), the equivalent points can be solved as the incident polar angle is 2.19° in the positive mode and −8.34° in the reserve mode. The measured results are shown in Figure 11a,b, respectively. The results show that the matching frequencies are approximately the same at the same incident polar angle when ignoring the bandwidth for o-polarized and e-polarized lights. Furthermore, the differences in the peak diffraction efficiency are both less than 0.01 MHz. This work is significant for non-polarization AOTF applications.

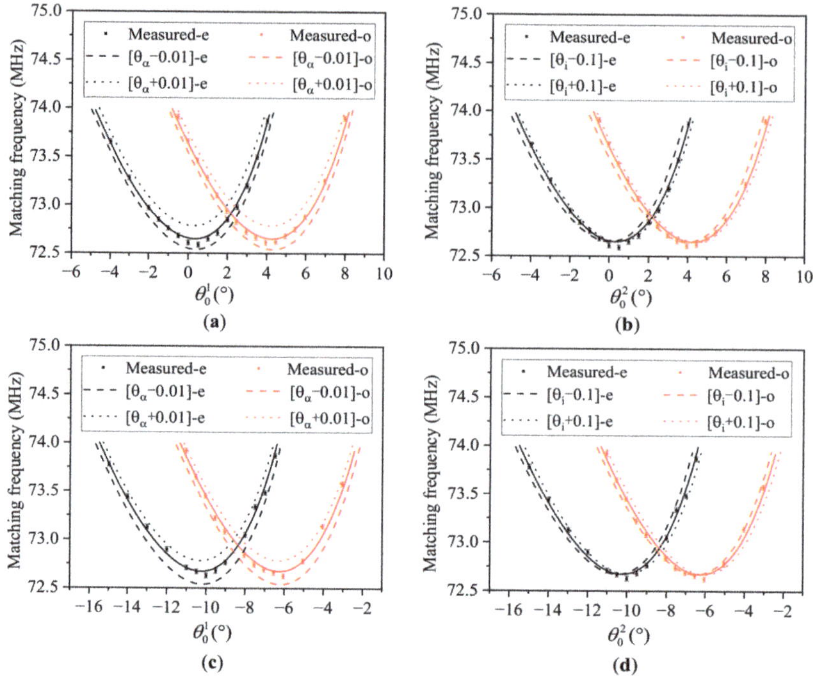

Figure 10. Polar angular frequency analysis under the different acoustic cut angles and front facet angles. (**a**,**b**) are both in the positive mode, while (**c**,**d**) are both in the reserve mode. The solid lines in (**a**,**c**) are the theoretical simulation curves under the calibration result whereby $\theta_i^1 = 15.074°$ and $\theta_\alpha^1 = 6.485°$; while the solid lines in (**b**,**d**) are the theoretical simulation curves under the calibration result whereby $\theta_i^2 = 10.435°$ and $\theta_\alpha^2 = 6.485°$. The dashed lines in (**a**,**c**) marked "$[\theta_\alpha - 0.01°]$" are the theoretical simulation curves under which $\theta_\alpha = 6.475°$, while the dotted lines in (**a**,**c**) marked "$[\theta_\alpha + 0.01°]$" are the theoretical simulation curves under which $\theta_\alpha = 6.495°$. The dashed lines in (**b**) and (**d**) marked "$[\theta_i - 0.1°]$" are the theoretical simulation curves under which $\theta_i^1 = 14.974°$ and $\theta_i^2 = 10.335°$, respectively, while the dotted lines in (**b**,**d**) marked "$[\theta_i + 0.1°]$" are the theoretical simulation curves under which $\theta_i^1 = 15.174°$ and $\theta_i^2 = 10.535°$, respectively.

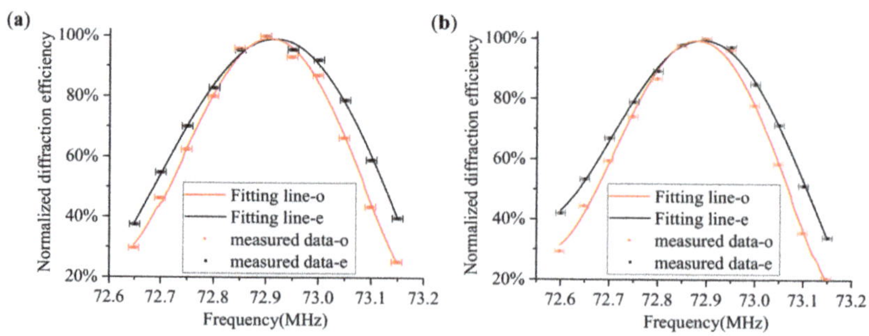

Figure 11. Measurement results at the equivalent points. (**a**) Incident polar angle at 2.19° in the positive mode. (**b**) Incident polar angle at −8.34° in the reserve mode.

3.3. Tolerance Analysis

The constant parameters and measurement parameters involved in the calibration method may have some tolerances, which have not been taken into account above. The constant parameters mainly include the principal refractive index tolerance and the acoustic wave velocity tolerance. The principal refractive index tolerance was taken from reference [27], and the acoustic wave velocity tolerance was set to ±0.5 m/s here. The measurement parameters mainly include the rotational accuracy of the precision turntable (2′) and the sampling step of the tuning frequency (0.05 MHz). The specific range of the tolerance setting can be found in Table 2, and the tolerance distribution of all parameters assumes the uniform distribution probability.

Table 2. Tolerance setting and standard deviation of calibration results by the variable-controlled method.

Parameters		Tolerances	Standard Deviation of Calibration Results		
			θ_i^*	θ_α	θ_β
Principle refractive index	n^o n^e	±0.0006 ±0.0007	0.066°	0.032°	0.001°
Acoustic wave velocity	V_{001} V_{110}	±0.5 m/s	0.002°	0.001°	0.001°
Incident polar angle		±1′	0.003°	<0.001°	0.006°
Matching frequency		±0.025 MHz	0.004°	<0.001°	0.005°

We performed a tolerance analysis on the calibration method. The tolerance analysis includes two aspects: parameter sensitivity analysis and Monte Carlo analysis [32]. The variable-controlled method was used to analyze the parameter sensitivity, wherein only one of the parameters varied for 100 times at a time. The standard deviations of the calibration results were used as the parameter sensitivity analysis index. The results show that the principal refractive index tolerance had the greatest influence on the calibration results, especially for the front facet angle and the acoustic cut angle. In comparison, other parameter tolerances had very little effect. Monte Carlo analysis was then used to analyze the statistical tolerance of the entire calibration process. In this paper, a total of 1000 simulated calibrations were performed as the statistical sample in Figure 12. The statistical results of the error distribution of the calibration results are shown in Table 3, and the cumulative probability of a result within than 0.1° was greater than 99.7%. Moreover, the cumulative probability of the front facet angle falling within 0.01° was greater than 18.4%. The cumulative probability of an acoustic cut angle falling within 0.01° was greater than 35.3%. The cumulative probability of a back facet angle falling within 0.01° was greater than 83.0%. Furthermore, the cumulative probability of maximum a cut-angle deviation falling within 0.01° was greater than 15.0%.

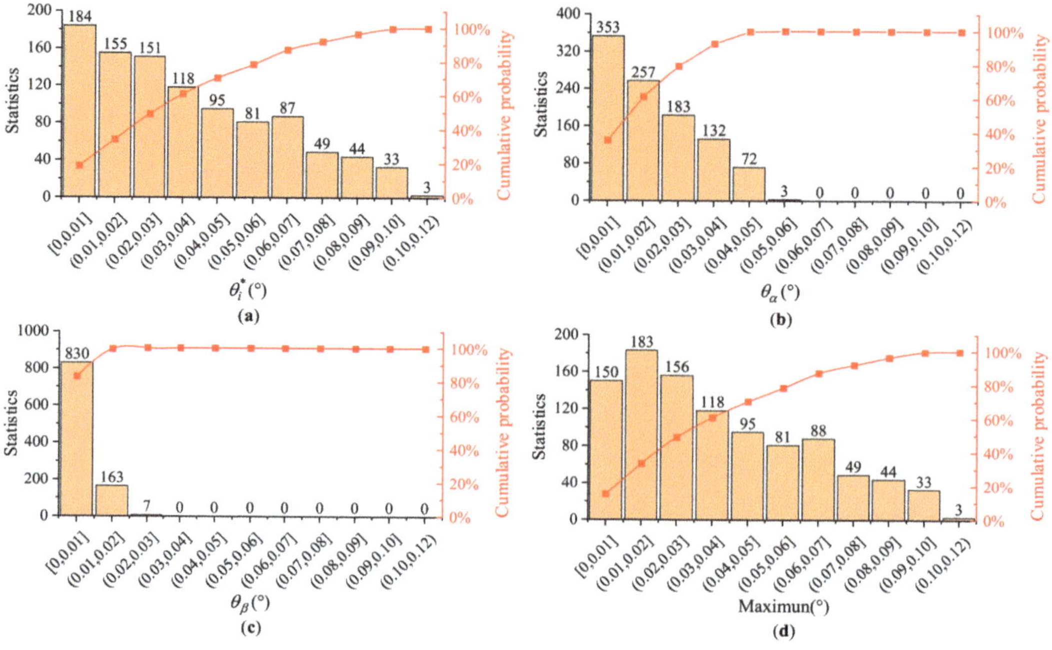

Figure 12. Monte Carlo simulation results. (**a**) Statistical distribution of the front facet angle deviations. (**b**) Statistical distribution of the acoustic cut angle deviations; (**c**) Statistical distribution of the back facet angle deviations. (**d**) Statistical distribution of the maximum cut angle deviations.

Table 3. Cumulative probability of the Monte Carlo analysis result.

Parameters	≤0.01°	≤0.1°
θ_i^*	18.4%	99.7%
θ_α	35.3%	100%
θ_β	83.0%	100%
Maximum	15.0%	99.7%

4. Conclusions

In summary, we proposed a method for calibrating the acousto-optic interaction geometry of AOTFs based on polar angular analysis. Moreover, based on this method, we obtained the complete acousto-optic interaction geometry of an actual AOTF device, including the front facet angle (θ_i^*), the acoustic cut angle (θ_α) and the back facet angle (θ_β). Specifically, we carried out the following research:

(a) We established a model of the AOTF angular frequency relationship that can be solved analytically.

(b) We proposed a novel and easy-to-perform method for calibrating the acousto-optic interaction geometry of an actual AOTF device. Furthermore, the experimental results showed a high precision with the acoustic cut angle, within results falling within 0.01°.

(c) We analyzed the polar angular performance with the acousto-optic interaction geometry of the AOTF and the results showed that the acoustic cut angle of the AOTF is more sensitive than the front facet angle. Specifically speaking, changing the acoustic cut angle makes the angular frequency curves shift up and down, and the larger acoustic cut angle corresponds to the state of shifting up. In comparison, changing the front facet angle makes the angular frequency curves shift left and right, and the larger front facet angle corresponds to the state of shifting right.

(d) We calculated and tested the equivalent points for the o-polarized and e-polarized lights in both positive and reserve modes, which is vital to the non-polarization applications of AOTFs.

(e) We analyzed the parameter sensitivity and Monte Carlo tolerance of the calibration method. The results of the parameter sensitivity analysis showed that the principal refractive index of the crystal has a large influence on the calibration results, while other factors have little influence. The results of the Monte Carlo tolerance analysis showed that the cumulative probability of the results falling within 0.1° with this method is greater than 99.7%. Moreover, the probability of falling within 0.01°, for the front facet angle is greater than 18.4%. In comparison, the acoustic cut angle and the back facet angle are greater than 35.3% and 83.0%, respectively.

These works are of great significance for the studies of AOTFs, such as ray tracing, characteristic analysis and optical system design.

Author Contributions: Conceptualization, H.Z. (Hao Zhang) and H.Z. (Huijie Zhao); methodology, H.Z. (Hao Zhang); investigation, H.Z. (Hao Zhang); validation, H.Z. (Hao Zhang) and Y.X.; formal analysis, H.Z. (Hao Zhang); resources, H.Z. (Huijie Zhao) and Q.G.; writing—original draft preparation, H.Z. (Hao Zhang); writing—review and editing, H.Z. (Huijie Zhao) and Q.G.; supervision, H.Z. (Huijie Zhao) and Q.G.; project administration, H.Z. (Huijie Zhao); funding acquisition, H.Z. (Huijie Zhao) and Q.G. All authors have read and agreed to the published version of the manuscript.

Funding: This research was funded by the National Natural Science Foundation of China (61875004, 62075202).

Institutional Review Board Statement: Not applicable.

Informed Consent Statement: Not applicable.

Data Availability Statement: Not applicable.

Acknowledgments: We would like to thank the Nanofabrication facility in Beihang Nano for supplies for the darkroom and some devices.

Conflicts of Interest: The authors declare no conflict of interest.

References

1. Dixon, R. Acoustic diffraction of light in anisotropic media. *IEEE J. Quantum Electron.* **1967**, *3*, 85–93. [CrossRef]
2. Chang, I.C. Tunable Acousto-Optic Filters: An Overview. *Opt. Eng.* **1977**, *16*, 455–460. [CrossRef]
3. Tran, C.D. Principles and analytical applications of acousto-optic tunable filters, an overview. *Talanta* **1997**, *45*, 237–248. [CrossRef] [PubMed]
4. Calpe-Maravilla, J. 400– to 1000–nm imaging spectrometer based on acousto-optic tunable filters. *J. Electron. Imaging* **2006**, *15*, 023001. [CrossRef]
5. Li, J.; Gui, Y.; Xu, R.; Zhang, Z.; Liu, W.; Lv, G.; Wang, M.; Li, C.; He, Z. Applications of AOTF Spectrometers in In Situ Lunar Measurements. *Materials* **2021**, *14*, 3454. [CrossRef]
6. Gupta, N.; Suhre, D.R. Acousto-optic tunable filter imaging spectrometer with full Stokes polarimetric capability. *Appl. Opt.* **2007**, *46*, 2632–2637. [CrossRef]
7. Anchutkin, V.S.; Bel'skii, A.B.; Voloshinov, V.B.; Yushkov, K.B. Acoustooptical method of spectral-polarization image analysis. *J. Opt. Technol.* **2009**, *76*, 473–477. [CrossRef]
8. Machikhin, A.; Batshev, V.; Pozhar, V.; Naumov, A. Single-volume dual-channel acousto-optical tunable filter. *Opt. Express* **2020**, *28*, 1150–1157. [CrossRef]
9. Gupta, N.; Suhre, D.R. Notch filtering using a multiple passband AOTF in the SWIR region. *Appl. Opt.* **2016**, *55*, 7855–7860. [CrossRef] [PubMed]
10. Yushkov, K.B.; Molchanov, V.Y.; Belousov, P.V.; Abrosimov, A.Y. Contrast enhancement in microscopy of human thyroid tumors by means of acousto-optic adaptive spatial filtering. *J. Biomed. Opt.* **2016**, *21*, 16003. [CrossRef] [PubMed]
11. Chu, Y.; Chen, L.; Wang, H.; ZHang, C.; Liu, W.; Wang, P.; Fan, S.; Yan, B.; Li, Z. Research on edge enhancement of optical image based on acousto-optic filtering. In Proceedings of the Twelfth International Conference on Information Optics and Photonics, Xi'an, China, 23–26 July 2021; pp. 1084–1088.
12. Harris, S.E.; Wallace, R.W. Acousto-Optic Tunable Filter*. *J. Opt. Soc. Am.* **1969**, *59*, 744–747. [CrossRef]

13. Chang, I.C. Noncollinear acousto-optic filter with large angular aperture. *Appl. Phys. Lett.* **1974**, *25*, 370–372. [CrossRef]
14. Gupta, N. Materials for imaging acousto-optic tunable filters. In Proceedings of the Image Sensing Technologies: Materials, Devices, Systems, and Applications, Baltimore, MD, USA, 7–8 May 2014; pp. 41–50.
15. Gupta, N.; Voloshinov, V. Hyperspectral imaging performance of a TeO_2 acousto-optic tunable filter in the ultraviolet region. *Opt. Lett.* **2005**, *30*, 985–987. [CrossRef]
16. Voloshinov, V.B. Anisotropic light diffraction on ultrasound in a tellurium dioxide single crystal. *Ultrasonics* **1993**, *31*, 333–338. [CrossRef]
17. Voloshinov, V.B.; Yushkov, K.B.; Linde, B.B.J. Improvement in performance of a TeO_2 acousto-optic imaging spectrometer. *J. Opt. A-Pure Appl. Opt.* **2007**, *9*, 341–347. [CrossRef]
18. Yano, T.; Watanabe, A. Acoustooptic TeO_2 tunable filter using far-off-axis anisotropic Bragg diffraction. *Appl. Opt.* **1976**, *15*, 2250–2258. [CrossRef]
19. Gass, P.A.; Sambles, J.R. Accurate design of a noncollinear acousto-optic tunable filter. *Opt. Lett.* **1991**, *16*, 429–431. [CrossRef] [PubMed]
20. Yushkov, K.B.; Kupreychik, M.I.; Obydennov, D.V.; Molchanov, V.Y. Acousto-optic k-space filtering for multifrequency laser beam shaping. *J. Opt.* **2023**, *25*, 014002. [CrossRef]
21. Zhang, C.; Wang, H.; Zhang, Z.; Yuan, J.; Shi, L.; Sheng, Z.; Zhang, X. Non-radio-frequency signal tuned acousto-optic tunable filter. *Opt. Express* **2018**, *26*, 1049–1054. [CrossRef]
22. Knuteson, D.; Singh, N.; Gupta, N.; Gottlieb, M.; Suhre, D.; Berghmans, A.; Thomson, D.; Kahler, D.; Wagner, B.; Hawkins, J. Design and fabrication of mercurous bromide acousto-optic filters. In Proceedings of the Infrared and Photoelectronic Imagers and Detector Devices, San Diego, CA, USA, 31 July–1 August 2005; pp. 98–105.
23. Li, C.; Zhao, H.; Zhang, Y.; Zhou, P. Crystal geometry measurement of an acousto-optic tunable filter using the tested tuning curves. In Proceedings of the 2012 8th IEEE International Symposium on Instrumentation and Control Technology (ISICT), London, UK, 11–13 July 2012; pp. 267–272.
24. Georgiev, G.; Georgieva, E.; Konstantinov, L. Angular and power characteristics of noncollinear acousto-optic tunable filters. *Opt. Lasers Eng.* **1999**, *31*, 1–12. [CrossRef]
25. Chang, I.C. Acousto-Optic Tunable Filters. *Opt. Eng.* **1981**, *20*, 824–829. [CrossRef]
26. Goutzoulis, A.P.; Pape, D.R. *Design and Fabrication of Acousto-Optic Devices*; Marcel Dekker Inc.: New York, NY, USA, 1994.
27. Uchida, N.; Ohmachi, Y. Elastic and Photoelastic Properties of TeO_2 Single Crystal. *J. Appl. Phys.* **1969**, *40*, 4692–4695. [CrossRef]
28. Gorevoy, A.V.; Machikhin, A.S.; Martynov, G.N.; Pozhar, V.E. Spatiospectral transformation of noncollimated light beams diffracted by ultrasound in birefringent crystals. *Photonics Res.* **2021**, *9*, 687–693. [CrossRef]
29. Pozhar, V.; Machihin, A. Image aberrations caused by light diffraction via ultrasonic waves in uniaxial crystals. *Appl. Opt.* **2012**, *51*, 4513–4519. [CrossRef]
30. Voloshinov, V.B.; Molchanov, V.Y.; Mosquera, J.C. Spectral and polarization analysis of optical images by means of acousto-optics. *Opt. Laser Technol.* **1996**, *28*, 119–127. [CrossRef]
31. Xue, B.; Xu, K.; Yamamoto, H. Discussion to the equivalent point realized by the two polarized beams in AOTF system. *Opt. Express* **1999**, *4*, 139–146. [CrossRef] [PubMed]
32. Liu, J.S. *Monte Carlo Strategies in Scientific Computing*; Springer: Berlin/Heidelberg, Germany, 2001; Volume 75.

materials

MDPI

Article

The Calibration Methods of Geometric Parameters of Crystal for Mid-Infrared Acousto-Optic Tunable Filter-Based Imaging Systems Design

Kai Yu [1], Qi Guo [1,2,*], Huijie Zhao [1,2] and Chi Cheng [1]

[1] School of Instrumentation Science & Opto-Electronics Engineering, Beihang University, No. 37 Xueyuan Road, Haidian District, Beijing 100191, China

[2] Key Laboratory of "Precision Opto-Mechatronics Technology", Ministry of Education, No. 37 Xueyuan Road, Haidian District, Beijing 100191, China

* Correspondence: qguo@buaa.edu.cn; Tel.: +86-178-0106-6063

Abstract: AOTF calibration is a complex topic that has various aspects. As far as geometric calibration is concerned, it includes not only processing errors and fixing errors in the optical system, but also the error of geometric parameters of crystal (GPC). GPC is the preset input in the optical design and optimization of Zemax, which determines the key parameters, including the spatial resolution, the field of view, and aberration. In particular, the compensation of aberration during the optical design requires accurate values of GPC. However, it is currently considered ideal. Therefore, two calibration methods based on the principle of parallel tangent are proposed: (1) the minimum-central wavelength method; (2) the minimum-frequency method. The deviation of the parallel tangent incident angle calibrated by the two methods is 0.03°. As a result, the tuning curve calculated in theory with the calibrated geometric parameters of AOTF is consistent with the tuning curve measured in practice.

Keywords: AOTF spectral imager; geometric calibration of AO crystal; spectral response; imaging aberrations; tuning curve

Citation: Yu, K.; Guo, Q.; Zhao, H.; Cheng, C. The Calibration Methods of Geometric Parameters of Crystal for Mid-Infrared Acousto-Optic Tunable Filter-Based Imaging Systems Design. *Materials* **2023**, *16*, 2341. https://doi.org/10.3390/ma16062341

Academic Editors: Alexander S. Machikhin and Vitold Pozhar

Received: 14 February 2023
Revised: 28 February 2023
Accepted: 7 March 2023
Published: 14 March 2023

1. Introduction

The acousto-optic tunable filter (AOTF) has been widely applied in spectral imaging systems for remote sensing [1,2], spectral microscopy [3,4], three-dimensional detection [5], notch filtering [6,7], and edge enhancement [8] due to its advantages of a narrow bandpass, electric tunability, all-solid-state, fast response, large angular aperture, and wide spectral range. In particular, the flexible electric tunability makes AOTF spectral imagers the most promising candidates for identifying and tracking targets. The central wavelength can be arbitrarily changed by modulating the frequency of the acousto-optic grating generated in a birefringent crystal material with an ultrasonic transducer. Until now, novel structures of AOTF imaging spectrometers have been developed for the purpose of performance enhancement—for instance, to improve the resolution of the spectrum or of space [9,10]. Moreover, some new crystalline materials [11] have emerged in the field of AOTF devices to improve the spectral range and efficiency. The low price and high spectral resolution of AOTF indicate its potential application in the field of plasmonic sensing [12].

The calibration of AOTF is inseparable from the study of device characteristics and system design. In the area of modeling for AOTF devices, Zhang was the first to propose a noncollinear AOTF operating mode [13] and a design scheme according to the principle of parallel tangent [14]. This principle is the criterion for the design of a wide-aperture AOTF, which needs to be considered before designing. Yushkov [15,16] put forward the analytical expressions of the central wavelength and tuning frequency at oblique incidence. The spectral response of AOTF depends not only on the GPC but also on the direction of incident light. Before the AOTF spectrometer is processed, the imaging characteristics need to be simulated by modeling. In traditional methods of the optical system design of an AOTF spectrometer, the AOTF device is considered as a transparent slab, which is far from sufficient in terms of aberration evaluation and optical design optimization. Batshev and Machikhin [17] summarized and tested four typical optical schemes in common use: confocal, collimating, tandem, and double-path. As a special component in an optical system, AOTF cannot be represented by the existing surface of Zemax. The method of using the user-defined surface (UDS) function in Zemax and computer dynamic link library (DLL) technique to design an optical system was introduced [18]. The GPC is the input of UDS. Subsequently, Machikhin solved the limitation of the field of view of AOTF with a computational technique [19]. It can be found the device characteristics of AOTF are crucial for system optimization, and GPC directly affects the device characteristics. However, little attention was paid to the accuracy of GPC.

AOTF needs to be calibrated in practical applications. Calibrations mainly include spectral, geometric, radiometric, thermal, etc. For spectral calibration, it is necessary to establish the mapping relationship between spectrum and frequency [20], which is usually characterized by the tuning curve [21,22]. For geometric calibration, scholars are mostly committed to the research of system-level calibration methods. For instance, Pozhar, Machikhin [23–25], and Liu Hong [26] focused on the quantitative description of the AOTF spatial aberrations. As mentioned earlier, the aberrations of the system are caused by the coupling of many factors. The influence of GPC is not considered separately. Shi [27] proposed a multiplane camera calibration model (MPM). Together with the fringe-phase marking method and the neural network algorithm, MPM established the mapping between the image points and the corresponding space lines. Katrašnik [28,29] proposed some fully automatic geometrical calibration methods for the hyper-spectral imaging systems based on AOTF. However, they could only aim at imaging samples on a 2D plane due to the defocusing of 3D objects within a broad spectral range. For radiometric calibration, the accuracy of the radiometric response of AOTF hyperspectral imaging systems is crucial for obtaining reliable measurements. Katrašnik analyzed the influence of the noise of the detector on the radiometric calibration accuracy [30]. The charge-coupled device (CCD) imaging sensor was thoroughly analyzed in [31], while the complementary metal–oxide–semiconductor (CMOS) imaging sensor was analyzed in [32]. The difference in spectral value caused by the inhomogeneity of temperature in the crystal, such as anomalous sidelobes [33], and the drift of the central wavelength, was often considered in the thermal calibration [34]. As a part of AOTF calibration, the geometric calibration of the system is deservedly important, but it cannot guide the optimization of an optical system. The geometric aberrations are the result of the joint influence of the mirror group and AOTF in the optical system. The vertical incidence angle (θ_i) and ultrasonic cut angle (α) determine the GPC, which ultimately affects the performance of the system. The influence of the wedge angle of the AOTF rear surface in the mid-infrared band is very small and generally not considered. Normally, most devices follow the design rules proposed by Zhang, where θ_i and α are constrained under the principle of parallel tangent [13]. In this case, θ_i can be calculated once α is fixed. However, for devices not designed under this principle, there is no effective method to obtain α and θ_i accurately. In addition, since the GPC may shift from the designed values due to the sophistication during the fabrication process of AOTF devices, a feasible and accurate calibration of GPC is necessary.

Table 1 summarizes the works related to AOTF in recent years. The table includes calibration, optical design and optimization, operating modes, and ray tracing.

The different GPC directly affect the distribution of relative diffraction efficiency, spectral response, and frequency response of the device. On the contrary, we can use the characteristics of the test results to obtain the desired GPC. Therefore, we propose two methods: (1) the minimum-central wavelength method; (2) the minimum-frequency method. The calibration of GPC should be completed before the design and optimization of the optical system based on AOTF. Ultimately, the accuracy of the simulation of spectral resolution, imaging distortion, spectral bandwidth, and other performance indexes can be guaranteed.

The proposed methods can also be used in a polarizer-free AOTF-based spectral imaging system [35], as well as the installation and adjustment of AOTF imaging systems.

Table 1. The works related to AOTF in recent years.

AOTF-Related Work	Recent Research
Calibration	Bürmen, Li, Vila-Francés, spectral calibration [21,22]; Pozhar, Machikhin [23–25] and Liu [26], geometric calibration; Katrašnik, Healey, Tian, radiometric calibration [30–32]; Trutna, Yano, thermal calibration [33,34]; Shi [27], 3D calibration; Katrašnik [28,29], automatic geometrical calibration methods.
Optical design and optimization	Batshev and Machikhin [17], typical optical schemes; Batshev and Gorevoy [18], Zemax integration.
Operating mode	Zhang [13], noncollinear AOTF operating mode, the principle of parallel tangent, tuning curve.
Ray tracing	Yushkov [15], tuning frequency at oblique incidence; Pozhar and Machikhin [23], 2D to 3D ray tracing.

2. Background

At present, TeO_2 is widely used as an acousto-optic material in the spectral span from visible light to mid-infrared. For non-collinear anisotropic acousto-optic interaction in TeO_2, there are two working modes. The arbitrarily polarized incident beam is split into two orthogonally polarized propagation modes: the extraordinary "e" mode and the ordinary "o" mode. We can limit the polarization direction of incident light by adding a polarizer in front of the AOTF. For instance, when the incident light is extraordinary, diffracted light will be shifted to normal light. This working mode can be recorded as e→o mode. Correspondingly, the other working mode is o→e mode.

The main equations describing AO interaction may be derived from the laws of the conservation of energy and momentum for photons and a phonon [36]:

$$\begin{cases} \omega_d = \omega_i \pm f \\ \mathbf{K}_d = \mathbf{K}_i \pm \mathbf{K}_a \end{cases}, \tag{1}$$

where (f, K_a),(ω_i, K_i), and (ω_d, K_d) are the frequencies and wavevectors of sound, incident light, and diffracted light. The sign "+" corresponds to the absorption of the phonon, and the sign "−" corresponds to its stimulated birth. $|\mathbf{K}_i| = 2\pi n_i(\lambda)/\lambda$, $|\mathbf{K}_d| = 2\pi n_d(\lambda)/\lambda$, $|\mathbf{K}_a| = 2\pi f_a/\mathbf{V}_a$, $n_i(\lambda)$, and $n_d(\lambda)$ represent the refractive index of incident and diffracted light, respectively; f_a and $\mathbf{V}_a$ represent the ultrasonic frequency and acoustic phase velocity, respectively. The actual device will adjust the power of the ultrasonic transducer at different incident wavelengths to achieve high diffraction efficiency in each wave band.

Spectral imaging applications require that the Bragg phase matching condition $\mathrm{K}_d = \mathrm{K}_i + \mathrm{K}_a$ for the wave vectors of ultrasound K_a, incident K_i, and diffracted K_d light must be approximately satisfied in a wide range of incident light angles. This wide-aperture configuration of acousto-optic (AO) interaction [13] is achievable when the tangents to the wave surfaces for incident and diffracted light are parallel (Figure 1).

In Figure 1, λ is the wavelength of incident light. $n_o(\lambda)$ and $n_e(\lambda)$ represent the main refractive index of ordinary light and extraordinary light in the crystal. θ_i is the vertical incidence angle; α is the ultrasonic cut angle. The length of vectors K_{ei} and K_{od} depends on the refractive indices $n_o(\lambda)$ and $n_e(\lambda)$; the wavenumber k_0 corresponds to the light wavelength in a vacuum: $k_0 = 2\pi/\lambda$. AB is parallel to CD under the constraint of the principle of parallel tangent. In this case, the incident light angle is the parallel tangent incident angle ($\theta_i(\alpha)$). $\delta\theta$ is the offset between θ_i and $\theta_i(\alpha)$. According to the equations for calculating the refractive index of ordinary light and extraordinary light in anisotropic crystal, the refractive index of the refracted light in the crystal can be obtained once the wavelength of the incident light is determined, as shown in Equation (2) [9]. In addition, the geometric relationship can be deduced—that is, Equation (3) needs to be satisfied at the same time. In the mid-infrared band, Berny and Georgiev proposed the following dispersion equations and coefficients [37,38].

$$\begin{cases} n_i = \left(\cos^2\theta_i(\alpha)/n_o^2(\lambda) + \sin^2\theta_i(\alpha)/n_e^2(\lambda)\right)^{-1/2} \\ n_d = n_o(\lambda) \end{cases}, \tag{2}$$

$$\begin{cases} \tan\theta_d = (n_o(\lambda)/n_e(\lambda))^2 \tan\theta_i(\alpha) \\ \tan(\alpha) = (n_d\cos\theta_d - n_i\cos\theta_i(\alpha))/(n_i\sin\theta_i(\alpha) - n_d\sin\theta_d) \end{cases}', \tag{3}$$

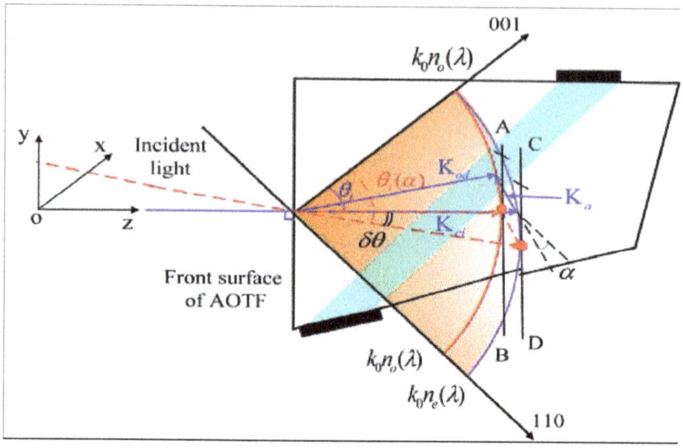

Figure 1. Wave vector layout under the parallel tangent principle (e→o mode). $\mathbf{K}_{ei}$, incident light is extraordinary; $\mathbf{K}_{od}$, diffracted light is ordinary; $\mathbf{K}_a$, acoustic vector; **oy**, the direction of acousto-optic diffraction.

Once the wavelength of incident light is determined, α and $\theta_i(\alpha)$ will present a one-to-one correspondence. At the same time, the value of the ultrasonic cut angle α is also limited. $\theta_i(\alpha)$ is 55.53° when α is equal to 18.9°. If α continues to increase, the incident light angle that meets the constraint of the principle will not be found in the acousto-optic plane, losing practical significance. Therefore, the value range of α should be considered in the cutting design of AOTF.

In the past, for AOTF with wide-aperture geometry, the tuning curve was mostly used to correct the GPC. When the wavelength of incident light is determined, the length of the acoustic vector is only related to the GPC, and the matching frequency f_a of the driver can be obtained eventually, which can be expressed as follows [21]:

$$f_a = \frac{|\mathbf{K}_a|\mathbf{V}_a}{2\pi} = \mathbf{F}(\lambda, \alpha, \theta_i), \tag{4}$$

where $\mathbf{V}_a$ is the acoustic phase velocity, and f_a is the acoustic frequency. However, Equation (4) is workable only if the default condition ($\theta_i = \theta_i(\alpha)$) is met. Then, the modified ultrasonic cut angle can be obtained by using Equation (5) [21].

$$\alpha_m = \underset{\alpha}{\operatorname{argmin}} \sum_{i=1}^{M} |f_{ai} - f_{mi}|^2, \tag{5}$$

where f_{ai} is the matching frequency calculated theoretically. f_{mi} is the measured matching frequency. α_m is the modified ultrasonic cut angle. Generally, a small correction to α can make the theoretically calculated tuning curve very close to the measured tuning curve. If the difference between the corrected value of α and the nominal value of α given by the manufacturer is greater than 0.1°, the device does not strictly meet the parallel tangent principle, and the GPC needs to be calibrated separately. Of course, we can continue optimizing the value of α to match the tuning curve, but the optimization results at this time are not the real geometric parameters. The large deviation of α will directly affect the simulation of the spectral resolution, spatial resolution, aberrations, and other indicators, also bringing large errors to the optical simulation and design.

3. Simulation and Method

The mid-infrared AOTF is used as the simulation object, and the ray tracing method is a three-surface model that we proposed in our previous work [39]. This model can calculate the spectral response and frequency response at different incident light angles by numerical solution. The simulated optical scheme of the system is shown in Figure 2.

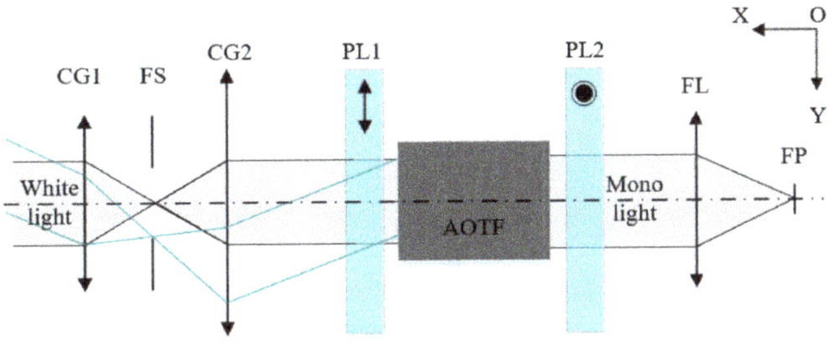

Figure 2. The simulation optical scheme of the system. CG1 and CG2: collimator group; FS: field stop; PL1 and PL2: polarizer; FL: focal lens; FP: focal plane; OY: the direction of acousto-optic diffraction and the column direction.

A parallel light structure is applied to the AOTF imaging spectrometer, the incident light is polychromatic, the driving frequency of AOTF is set to be 13.27 MHz, and the corresponding central wavelength of diffracted light is 4000 nm, calculated via the tuning curve. The column direction of the image plane corresponds to the direction of acousto-optic diffraction (within the color plane in Figure 3a). The central wavelength drift of the image plane can be obtained through simulation; the result is shown in Figure 3a. Specifically, when the frequency of the driver is fixed, the wavelength of diffracted light will shift with the change in the angle of incident light. Similarly, if the wavelength of incident light is fixed at 4000 nm, the central frequency drift of the image plane can be obtained, as shown in Figure 3b.

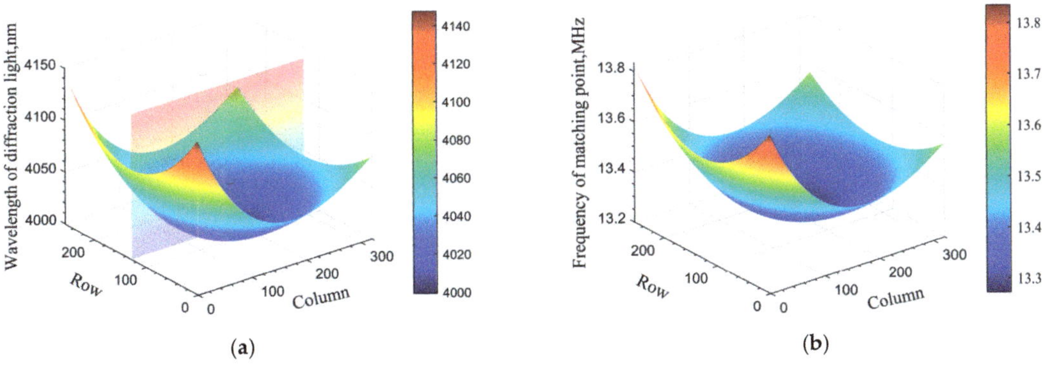

Figure 3. Mid-infrared AOTF image plane drift. (**a**) Central wavelength drift and (**b**) central frequency drift.

Here, we take the central wavelength shift as an example. Different positions of the image plane correspond to different incident light angles. From Figure 3a, it can be seen that the central wavelength drift on the image plane is symmetrically distributed in the row direction. In the column direction, the drift on one side is more serious than that on the other side. The wavelength of diffracted light reaches the lowest point ideally when the angle of incident light is perpendicular to the front surface of the crystal, which means that the central wavelength of the momentum matching point reaches a minimum when the angle of incidence satisfies the principle of parallel tangent ($\theta_i = \theta_i(\alpha)$). If the GPC does not meet this principle, the spectral response will change. Assuming that the vertical incidence angle (θ_i) deviates from the parallel tangent incidence angle ($\theta_i(\alpha)$) by $\pm 1°$, the simulation results are as shown in Figure 4.

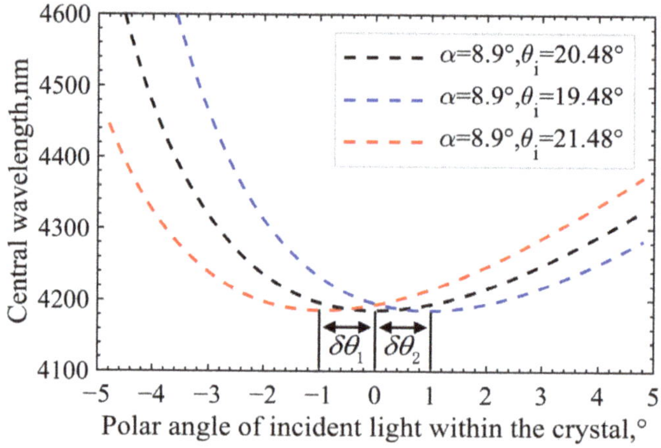

Figure 4. Influence of the GPC on spectral response.

In Figure 4, the polar angle of incident light (θ_p) within the crystal represents the angle between the incident light within the crystal and the optical axis. $\theta_i(\alpha)$ corresponding to the three GPCs remains unchanged and can be calculated from α directly. Therefore, in turn, an accurate GPC can be obtained through the angle difference ($\delta\theta_1$ or $\delta\theta_2$) between $\theta_i(\alpha)$ and θ_i. For the AOTF device with undefined design parameters, accurate parameters can be obtained by seeking the lowest point of matching central wavelength or central frequency. The differences between the calibration optical scheme of the device and the optical scheme of the system lie in the following aspects: (1) because of the asymmetry of spectral response in the direction of acousto-optic diffraction, it is necessary to accurately change the angle of incident light in this direction; (2) the radiance of the infrared laser is too strong, so an optical power meter should be used to measure the diffraction light intensity instead of a detector; (3) during the calibration process, the position of the diffracted light will follow the change in the incidence angle and the wavelength (or the driving frequency), which cannot be obtained accurately. Moreover, the receiving aperture of an optical power meter is limited, so the method of measuring the radiance of the non-diffracting light is used to obtain the intensity of the diffracted light indirectly. It is noteworthy that the angle processing accuracy of AOTF is $\pm 1°$; therefore, an order of magnitude higher is needed for the accuracy of our angle control so as to meet the calibration requirements. The angle of the incident light is not easy to control accurately, so a precise turntable is used to change the angle of the AOTF device instead. The schematic diagram of the AOTF geometric parameter calibration is shown in Figure 5.

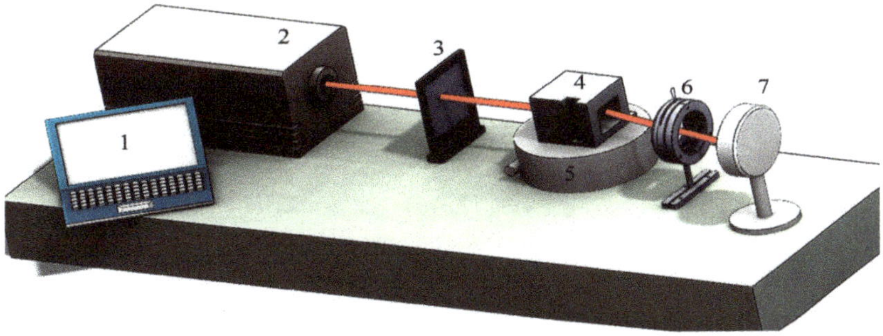

Figure 5. The schematic diagram of AOTF geometric parameter calibration. 1—computer; 2—mid-infrared laser; 3—polarizer; 4—AOTF; 5—precision turntable; 6—diaphragm; 7—optical power meter.

The AOTF is placed on a precision turntable, and the accuracy of the angle adjustment of the turntable is 0.01°. The light produced by the laser is elliptically polarized. A polarizer needs to be added before AOTF to ensure that the incident light is extraordinary. The power of zero-order light in the on and off states of the AOTF is obtained from the optical power meter, and the difference between the two is the power of the diffracted light. The deflection angle of the AOTF is positive when the AOTF rotates clockwise. At the same time, it is equal to the absolute value of the polar angle of the incident light, the sign opposite (the deflection direction is opposite). Here, the incident polar angle refers to the angle between the incident light outside the crystal and the optical axis.

The procedure of calibration of GPC is as follows:

- Ensure that the incident laser is perpendicular to the front surface of the AOTF;
- Adjust the rotation angle of the AOTF using the precision turntable;
- The spectral response at different angles can be obtained by adjusting the output wavelength of the laser (the minimum-central wavelength method);
- The frequency response at different angles can be obtained by adjusting the frequency of the transducer (the minimum-frequency method).

To improve the fitting accuracy of the final data, the angle step value is set as 1° for the large deflection angle of the AOTF, while the angle step value is selected as 0.1° when it is close to the parallel tangent incidence angle. Because the temperature of the AOTF has a certain influence on the spectral response, to shorten the acquisition time and to recover the spectra as completely as possible, the wavelength scan interval is set to be 10 nm when the diffraction efficiency is lower than 20%, 5 nm when the diffraction efficiency is 20~50%, and 1 nm when the diffraction efficiency is higher than 50%. During the test, the temperature change range of the crystal is controlled within ±0.1° by temperature control equipment.

4. Experiment and Results

4.1. The Minimum-Central Wavelength Method

According to the experimental setup in Figure 5, the frequency of the AOTF remains fixed at 14.0 MHz, and the wavelength of the laser needs to be tuned. The wavelength tuning range of the laser is 3674.0~4566.7 nm.

The test data of each spectral response are fitted with the least squares method to find the central wavelength corresponding to the point with the highest diffraction efficiency, as shown in Equation (6).

$$\begin{cases} DE_{ci} = \mathbf{x}(1) \times \sin c(\mathbf{x}(2) \times (\lambda_{mi} - \mathbf{x}(3)))^2 + \mathbf{x}(4) \\ L(DE_{mi}, DE_{ci}) = \sum\limits_{i=1}^{m} (DE_{mi} - DE_{ci})^2 \end{cases}, \quad (6)$$

where **x** is a matrix of 1×4, which represents the initial value of the least squares method. λ_{mi} represents the tested wavelength. DE_{ci} represents the calculated diffraction efficiency and DE_{mi} represents the tested diffraction efficiency. We minimize $L(DE_{mi}, DE_{ci})$ by iteration, and **x** is the fitting coefficient.

Spectral responses at some incidence angles are shown in Figure 6. Figure 6a,b represent spectral responses at different incident polar angles. Different incident polar angles indicate that the angle of incident light changes in the acousto-optic interaction plane. The angle of incident light can be accurately obtained from the precision turntable. The central wavelength of the matched curve is the extreme value of the fitted curve, in order to reduce the error caused by random sampling.

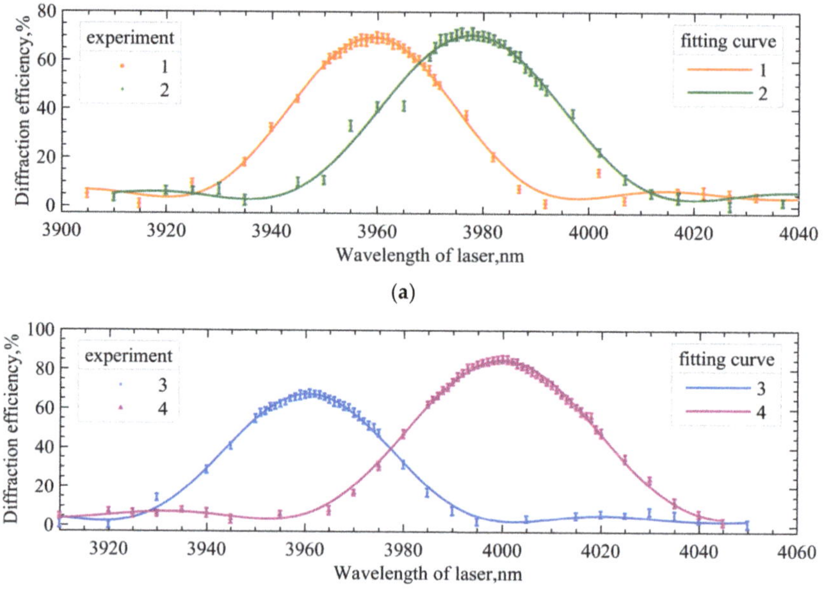

Figure 6. Spectral responses at some different incident angles. (**a**) 1: $\theta_p = 0°$; 2: $\theta_p = -3°$; (**b**) 3: $\theta_p = -1°$; 4: $\theta_p = -4°$.

Figure 7 shows the results of the minimum-central wavelength method. Figure 7a is a three-dimensional plot of the spectral responses at different angles. Figure 7b is the fitting curve of the central wavelengths; θ_p corresponding to the lowest point of the central wavelengths is 0.49°, and the corresponding central wavelength is 3960.6 nm. At this time, the angle of incident light is not perpendicular to the front surface of the AOTF. Thus, it can be seen that the principle of parallel tangent is not satisfied for this mid-infrared AOTF device, even presenting a large deviation.

Here, θ_p corresponding to the minimum central wavelength is marked as $\theta\prime$. The offset between θ_i and $\theta_i(\alpha)$, i.e., $\delta\theta$, can be obtained by the refraction law, as shown in Equation (7).

$$\sin(\theta\prime) = n_e(\lambda_s)\sin(\delta\theta), \tag{7}$$

where λ_s is the wavelength of the laser. In this experiment, the incident polar angle is 0.49°, $\delta\theta = 0.22°$ accordingly. According to the actual position of the transducer and the geometric relationship in the coordinate system, θ_i satisfies Equation (8).

$$\theta_i = \theta_i(\alpha) + \delta\theta, \tag{8}$$

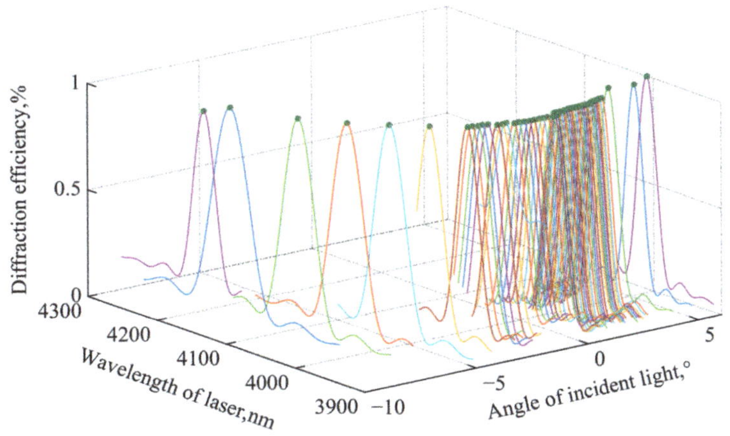

(**a**)

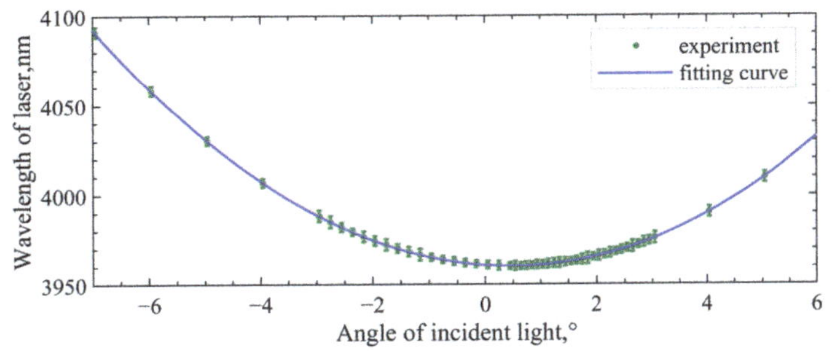

(**b**)

Figure 7. Measurement and fitting results of the minimum-central wavelength method. (**a**) Three-dimensional diagram of spectral responses at different angles; (**b**) fitted curve of central wavelengths.

4.2. The Minimum-Frequency Method

The difference between the two methods is that the fixed parameter in the minimum-frequency method is the output wavelength of the laser, while, in the minimum-central wavelength method, the frequency of the RF is relatively fixed. As shown in Figure 8a, f_1, f_2, and f_3 indicate the selection of three ultrasound frequencies. There will be two intersections with the spectral drift curve when the frequency is selected as f_1. The two points meet the momentum-matching condition and the diffraction efficiency is high. However, because the spectral drift in acousto-optic planes has no symmetry characteristic, the incident angle cannot be calculated through the deflection angle of AOTF corresponding to the two intersections. When the frequency is selected as f_2, the tuning curve cannot be accurately obtained due to the error in the actual design parameters, and the approximate value will also produce an error. When the frequency is selected as f_3, the extreme point of the curve corresponds to the highest diffraction efficiency under the premise of momentum mismatch, requiring no accurate tuning curve relationship. In practice, the actual frequency is 0.5 MHz lower than the frequency calculated by the ideal tuning curve. Finally, the incident polar angle corresponding to the highest diffraction efficiency point is the desired one. The experimental setup is the same as the method above. Figure 8b shows the measured data and the fitting results.

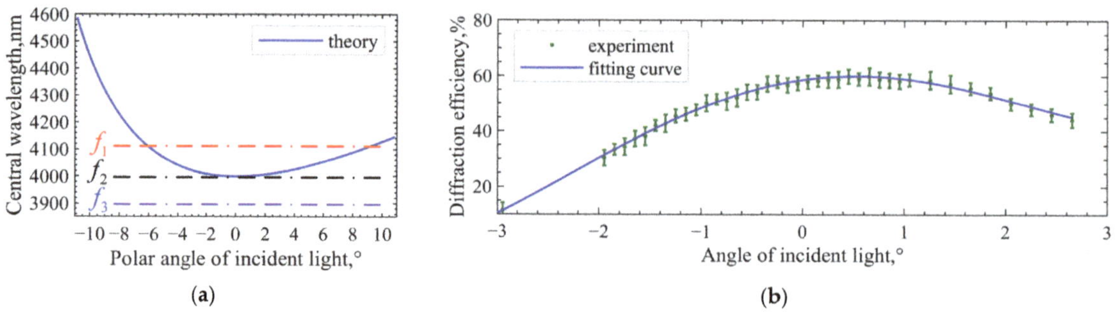

Figure 8. The minimum-frequency method. (**a**) Method of frequency setting; (**b**) measured data and fitting results.

Third-order polynomial fitting is performed based on the measured data. The diffraction efficiency corresponding to the extreme point of the fitting curve is 59.91%, and the incident polar angle corresponding to the extreme point is 0.52°. Compared with the minimum-central wavelength test result, there is an error of 0.03°. The results obtained by the two test methods are very close, indicating the feasibility and accuracy of the two. Finally, the AOTF design parameters obtained from the test are taken into the model to calculate the tuning curve. Before correction, $\alpha = 8.9°$, $\theta_i = 20.48°$. After correction, $\alpha = 8.9°$, $\theta_i = 20.70°$. The geometric layout in Figure 1 has changed, so the tuning curve (model calculation) has changed accordingly. The measured results of the tuning curve and the results of model calculation before and after parameter correction are shown in Figure 9a. The results show that the theoretical tuning curve can be more consistent with the measured data only after GPC is corrected. The relative difference in central frequency can be expressed by Equation (9). The results are shown in Figure 9b.

$$relative\ error = \frac{|f_c - f_t|}{f_t} \times 100\%, \tag{9}$$

where f_c is the theoretical value of the central frequency. f_t is the measured value of the central frequency. Within the test band, the frequency error can be controlled within 1% after the geometric parameter correction of the AOTF crystal.

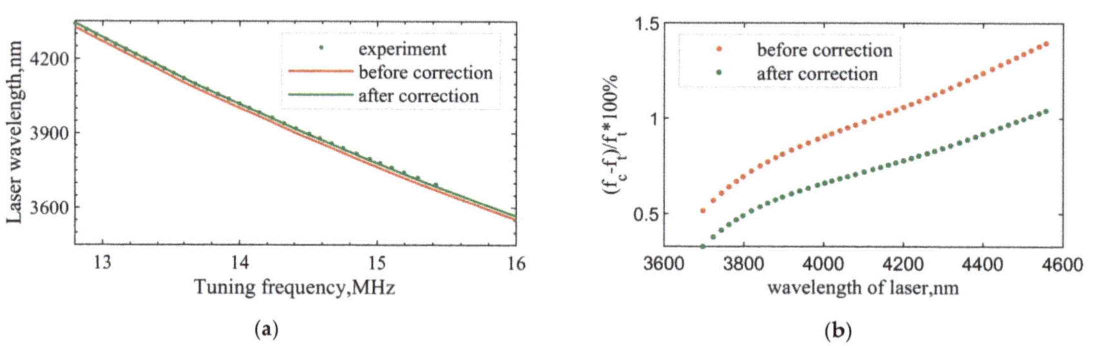

Figure 9. Correction result of the tuning curve. (**a**) Correction of tuning curve; (**b**) relative difference in central frequency.

Then, we performed five complete measurements with the two different methods. The uncertainty was calculated by Equation (10).

$$\begin{cases} U_A \approx \delta_x = \sqrt{\sum_{i=1}^{n} (x_i - \overline{x})^2 / (n-1)} \\ U_B \approx \Delta ins \\ U = \sqrt{U_A{}^2 + U_B{}^2} \end{cases} \tag{10}$$

where U_A is the measurement uncertainty; δ_x is the standard deviation of the measurement result of the vertical incidence angle (θ_i); Δins is the uncertainty of the measuring instruments. It is assumed that $\Delta ins = 0.01°$ to maintain the consistency of the two methods; U is the combined uncertainty. The calculation results are shown in Table 2. The advantages and disadvantages of the two methods are also given.

Table 2. The advantages and disadvantages of the two calibration methods.

Method	Calibration Time/Min	Calibration Results/°	Frequency Error
A: The minimum-central wavelength method	~270	$\alpha = 8.9, \theta_i = 20.70 \pm 0.04$	≤1%
B: The minimum-frequency method	~30	$\alpha = 8.9, \theta_i = 20.72 \pm 0.10$	≤1.05%

From Table 2, the time required for method A is much longer than for method B. In our case, method A is more precise because the transducer will produce a power fluctuation during frequency adjustment. Method B is greatly affected by the experimental conditions. Besides the above, the measured values of error in the tuning curves of both methods fall within the confidence intervals.

Finally, we carried out the mid-infrared spectral imaging experiment, and the experimental schematic diagram is shown in Figure 10.

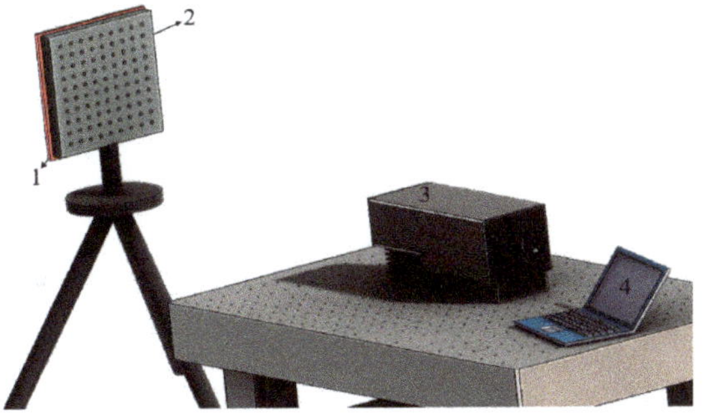

Figure 10. Schematic diagram of mid-infrared AOTF spectral imaging experiment for geometric calibration. 1—heating plate; 2—geometric calibration target; 3—AOTF spectrometer; 4—computer.

As seen in Figure 10, the distance between the target and AOTF spectrometer was far in the experimental setting, so the pre-collimating mirror group (telescope) could be omitted. It was only necessary to measure the aberration of the imaging mirror group and add the aberration parameters to the simulation model. The temperature of the heating plate could reach 100 °C. The diameter of the dots was 6 mm, and the distance between the calibration points was 20 mm. The coordinates of the calibration points were obtained by the method of geometric center extraction, and the unsatisfactory calibration points were eliminated. The size of the detector's target plane was 256 × 320 and the pixel size was 30 μm. The focal length of the imaging mirror group was 50 mm. The ultrasonic cut angle provided by the manufacturer was 8.9°. The luminous aperture of AOTF was 20 × 20 mm. The length of the piezotransducer was 21 mm. Different GPCs were brought into the model to calculate the theoretical imaging points, and the results are shown in Figure 11 without considering the alignment errors.

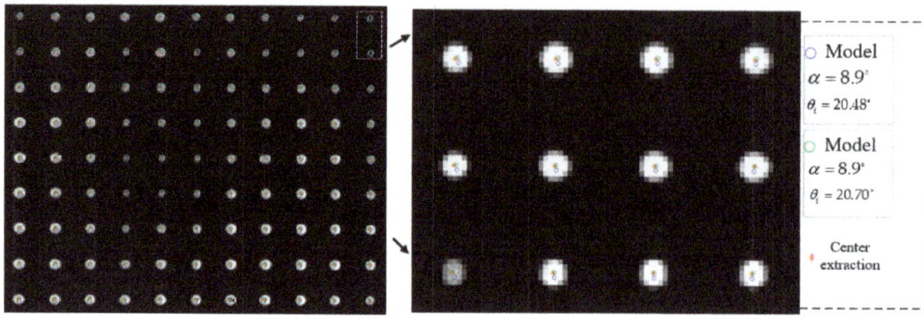

(a) (b)

Figure 11. Spectral imaging experiment based on mid-infrared AOTF. (**a**) Actual image of diffracted light; (**b**) the partially enlarged view of the target.

In Figure 11, the circles represent the simulated imaging points, and the red asterisks represent the actual imaging points. If we remove the two calibration points within the red dotted box, we can find that the simulated imaging points are all located near the actual imaging areas. After the geometric calibration of the AOTF crystal, the ideal imaging points are closer to the actual imaging points, but the effect is not obvious. The main reasons lie in the following aspects: (1) the aberration of mid-infrared AOTF is much smaller than that of the visible band; (2) the geometric calibration result of AOTF crystal has little difference from the reference value; (3) limited by the angular aperture, the field of view of the system is $\pm 3°$. In this angle range, the aberration is small. The $error_{RMS}$ can be calculated by Equation (11).

$$error_{RMS} = \sqrt{\frac{1}{N}\sum_{j=1}^{N}\left[(x_{actual,j} - x_{model,j})^2 + (y_{actual,j} - y_{model,j})^2\right]}. \quad (11)$$

where $x_{actual,j}$ is the row coordinate of the diffracted light imaging position. $x_{model,j}$ is the row coordinate of the calculated imaging position. y is the column coordinate. N is the total number of sampling points. Before the correction, $error_{RMS} = 2.01$. After correction, $error_{RMS} = 1.07$. The pixel difference for each point is shown in Figure 12. The fitting curve adopts cubic polynomial fitting. The pixel difference is controlled at approximately one pixel after correction.

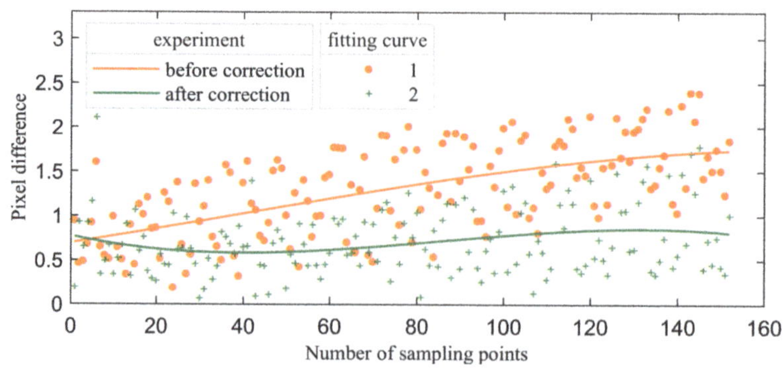

Figure 12. The pixel difference for each point.

5. Conclusions and Future Directions

Before designing and manufacturing a spectrometer based on AOTF, it is necessary to theoretically analyze the key parameters, such as the spectral bandwidth, imaging distortion, and so on. If the GPC cannot be obtained accurately, it will be impossible to conduct simulation analysis on the premise of accuracy, making the results meaningless. For the AOTF devices not strictly conforming to the principle of parallel tangent, two test methods to determine the design parameters are proposed, and the final test results of the two are highly consistent, with an error of 0.03°. Finally, the accuracy of the obtained design parameters is verified by the tuning curve and geometric calibration experiment. Results reveal that the accuracy of GPC is the premise on which the AOTF model can be used for ray tracing or analyzing the response of the device or the system. It is also the original intention of this work.

As for the advantages and disadvantages of the two methods, we believe that the data measured by the minimum-central wavelength method will have better stability. Because of the heating due to the operation of AOTF, the fluctuation in diffraction efficiency is influenced more seriously than the shift in the central wavelength.

The design of all AOTFs is based on the optical anisotropy of crystals; until now, however, only uniaxial crystals have been used. The method in this paper is also only applicable to the calibration of the geometric parameters of uniaxial crystals. The spectral response of AOTF based on biaxial crystal is more complex, and the calibration method remains to be determined.

Author Contributions: Conceptualization, Q.G. and K.Y.; methodology, Q.G.; software, C.C.; validation, Q.G. and H.Z.; formal analysis, K.Y.; investigation, K.Y.; resources, H.Z.; data curation, C.C.; writing—original draft preparation, C.C.; writing—review and editing, Q.G.; visualization, H.Z.; supervision, H.Z.; project administration, H.Z.; funding acquisition, H.Z. and Q.G. All authors have read and agreed to the published version of the manuscript.

Funding: This work was supported by the National Key Research and Development Program of China (2016YFB0500505) and the National Natural Science Foundation of China (NSFC) (Grant No. 61875004, No. 62075202).

Institutional Review Board Statement: Not applicable.

Informed Consent Statement: Not applicable.

Data Availability Statement: Not applicable.

Acknowledgments: We are very grateful to Zefu Xu and Chi Cheng for their dedication to the related discussions.

Conflicts of Interest: The authors declare no conflict of interest.

References

1. Zhao, H.; Wang, Z.; Jia, G.; Zhang, Y.; Xu, Z. Chromatic aberrations correction for imaging spectrometer based on acousto-optic tunable filter with two transducers. *Opt. Express.* **2017**, *25*, 23809–23825. [CrossRef] [PubMed]
2. Glenar, D.A.; Hillman, J.J.; Saif, B.; Bergstralh, J. Acousto-optic imaging spectropolarimetry for remote sensing. *Appl. Opt.* **1994**, *33*, 7412–7424. [CrossRef]
3. Zhang, C.; Wang, H.; Huang, J.; Gao, Q. The visible to the near infrared narrow band acousto-optic tunable filter and the hyperspectral microscopic imaging on biomedicine study. *J. Opt.* **2014**, *16*, 125303. [CrossRef]
4. Wachman, E.S.; Niu, W.H.; Farkas, D.L. Imaging acousto-optic tunable filter with 0.35-micrometer spatial resolution. *Appl. Opt.* **1996**, *35*, 5220–5226. [CrossRef]
5. Machikhin, A.; Batshev, V.; Pozhar, V.; Naumov, A.; Gorevoy, A. Acousto-optic tunable spectral filtration of stereoscopic images. *Opt. Lett.* **2018**, *43*, 1087–1090. [CrossRef] [PubMed]
6. Kastelik, J.C.; Champagne, J.; Dupont, S.; Yushkov, K.B. Wavelength characterization of an acousto-optic notch filter for unpolarized near-infrared light. *Appl. Opt.* **2018**, *57*, C36–C41. [CrossRef] [PubMed]
7. Gupta, N.; Suhre, D.R. Notch filtering using a multiple passband AOTF in the SWIR region. *Appl. Opt.* **2016**, *55*, 7855–7860. [CrossRef]
8. Yushkov, K.B.; Molchanov, V.Y.; Belousov, P.V.; Abrosimov, A.Y. Contrast enhancement in microscopy of human thyroid tumors by means of acousto-optic adaptive spatial filtering. *J. Biomed. Opt.* **2016**, *21*, 016003. [CrossRef]

9. Zhang, C.; Wang, H.; Zhang, Z.; Yuan, J.; Shi, L.; Sheng, Z.; Zhang, X. Narrowband double-filtering hyperspectral imaging based on a single AOTF. *Opt. Lett.* **2018**, *43*, 2126–2129. [CrossRef]
10. Zhang, X.; Liu, W.; Tong, H.; Liu, Y.; Wang, X.; Zhang, C.; Wang, Y.; Wang, H.; Sheng, Z.; Tan, Z.; et al. High-Resolution Hyperspectral Microscopic Imaging With Single Acousto-Optic Tunable Filter Based on Double Filtering. *IEEE Access.* **2020**, *8*, 11570–11576. [CrossRef]
11. Dyakonov, E.; Porokhovnichenko, D.; Ryu, J.; Balakshy, V. Implementation of the wide-angle acousto-optical interaction geometry in a mercury bromide single crystal. *Appl. Opt.* **2021**, *60*, 2348–2353. [CrossRef]
12. Hurtado-Aviles, E.A.; Torres, J.A.; Trejo-Valdez, M.; Urriolagoitia-Sosa, G.; Villalpando, I.; Torres-Torres, C. Acousto-plasmonic sensing assisted by nonlinear optical interactions in bimetallic Au-Pt nanoparticles. *Micromachines* **2017**, *8*, 321. [CrossRef]
13. Chang, I.C. Noncollinear acousto-optic filter with large angular aperture. *Appl. Phys. Lett.* **1974**, *25*, 370–372. [CrossRef]
14. Chang, I.C. Acousto-optic devices and applications. *IEEE Trans. Ultrason.* **1976**, *23*, 2–21. [CrossRef]
15. Yushkov, K.B.; Chizhikov, A.I.; Makarov, O.Y.; Molchanov, V.Y. Optimization of noncollinear AOTF design for laser beam shaping. *Appl. Opt.* **2020**, *59*, 8575–8581. [CrossRef] [PubMed]
16. Obydennov, D.V.; Yushkov, K.B.; Molchanov, V.Y. Ring-shaped optical trap based on an acousto-optic tunable spatial filter. *Opt. Lett.* **2021**, *46*, 4494–4497. [CrossRef] [PubMed]
17. Batshev, V.; Machikhin, A.; Gorevoy, A.; Martynov, G.; Khokhlov, D.; Boritko, S.; Pozhar, V.; Lomonov, V. Spectral Imaging Experiments with Various Optical Schemes Based on the Same AOTF. *Materials* **2021**, *14*, 2984. [CrossRef] [PubMed]
18. Batshev, V.I.; Gorevoy, A.V.; Pozhar, V.E.; Machikhin, A.S. Aberration analysis of AOTF-based stereoscopic spectral imager using optical design software. *J. Phys. Conf. Ser.* **2021**, *2127*, 012035. [CrossRef]
19. Gorevoy, A.; Machikhin, A.; Martynov, G.; Pozhar, V. Computational technique for field-of-view expansion in AOTF-based imagers. *Opt. Lett.* **2022**, *47*, 585–588. [CrossRef]
20. Bürmen, M.; Pernuš, F.; Likar, B. Spectral characterization of near-infrared acousto-optic tunable filter (AOTF) hyperspectral imaging systems using standard calibration materials. *Appl. Spectrosc.* **2011**, *65*, 393–401. [CrossRef]
21. Li, C.; Zhao, H.; Zhang, Y.; Zhou, P. Crystal geometry measurement of an acousto-optic tunable filter using the tested tuning curves. In Proceedings of the 2012 8th IEEE International Symposium on Instrumentation and Control Technology (ISICT) Proceedings, London, UK, 11–13 July 2012.
22. Vila-Francés, J.; Calpe-Maravilla, J.; Gómez-Chova, L.; Amorós-López, J. Improving the performance of acousto-optic tunable filters in imaging applications. *J. Electron. Imaging* **2010**, *19*, 043022. [CrossRef]
23. Machikhin, A.S.; Pozhar, V.E. Image aberrations in an acousto-optical tunable filter. *Opt. Quant. Electron.* **2010**, *55*, 1490–1496. [CrossRef]
24. Machikhin, A.; Batshev, V.; Pozhar, V. Aberration analysis of AOTF-based spectral imaging systems. *J. Opt. Soc. Am. A* **2017**, *34*, 1109–1113. [CrossRef]
25. Martynov, G.N.; Gorevoy, A.V.; Machikhin, A.S.; Pozhar, V.E.; Lehmann, P.; Osten, W.; Albertazzi Gonçalves, A. On inherent spatio-spectral image distortion in AOTF-based imagers. In Proceedings of the Optical Measurement Systems for Industrial Inspection XII, SPIE, Online, 21–26 June 2021; Volume 11782, pp. 268–277.
26. Liu, H.; Hou, X.; Hu, B.; Yu, T.; Zhang, Z.; Liu, X.; Liu, J.; Wang, X.; Zhong, J.; Tan, Z. Image blurring and spectral drift in imaging spectrometer system with an acousto-optic tunable filter and its application in UAV remote sensing. *Int. J. Remote Sens.* **2022**, *43*, 6957–6978. [CrossRef]
27. Zhao, H.; Shi, S.; Jiang, H.; Zhang, Y.; Xu, Z. Calibration of AOTF-based 3D measurement system using multiplane model based on phase fringe and BP neural network. *Opt. Express* **2017**, *25*, 10413–10433. [CrossRef]
28. Katrašnik, J.; Bürmen, M.; Pernuš, F. Spectral characterization and calibration of AOTF spectrometers and hyper-spectral imaging systems. *Chemometr. Intell. Lab.* **2010**, *101*, 23–29. [CrossRef]
29. Špiclin, Ž.; Katrašnik, J.; Bürmen, M.; Pernuš, F.; Likar, B. Geometrical calibration of an AOTF hyper-spectral imaging system. In Proceedings of the Design and Quality for Biomedical Technologies III, San Francisco, CA, USA, 23 February 2010.
30. Katrašnik, J.; Pernuš, F.; Likar, B. Radiometric calibration and noise estimation of acousto-optic tunable filter hyperspectral imaging systems. *Appl. Opt.* **2013**, *52*, 3526–3537. [CrossRef]
31. Healey, G.E.; Kondepudy, R. Radiometric CCD camera calibration and noise estimation. *IEEE. Trans. Pattern. Anal.* **1994**, *16*, 267–276. [CrossRef]
32. Tian, H.; Fowler, B.; Gamal, A.E. Analysis of temporal noise in CMOS photodiode active pixel sensor. *IEEE J. Solid-St. Circ.* **2001**, *36*, 92–101. [CrossRef]
33. Trutna, W.R.; Dolfi, D.W.; Flory, C.A. Anomalous sidelobes and birefringence apodization in acousto-optic tunable filters. *Opt. Lett.* **1993**, *18*, 28–30. [CrossRef]
34. Yano, T.; Watanabe, A. Acoustooptic TeO 2 tunable filter using far-off-axis anisotropic Bragg diffraction. *Appl. Opt.* **1976**, *15*, 2250–2258. [CrossRef] [PubMed]
35. Yushkov, K.B.; Dupont, S.; Kastelik, J.C.; Voloshinov, V.B. Polarization-independent imaging with an acousto-optic tandem system. *Opt. Lett.* **2010**, *35*, 1416–1418. [CrossRef] [PubMed]
36. Gorevoy, A.V.; Machikhin, A.S.; Martynov, G.N.; Pozhar, V.E. Spatiospectral transformation of noncollimated light beams diffracted by ultrasound in birefringent crystals. *Photonics Res.* **2021**, *9*, 687–693. [CrossRef]

37. Berny, J.G.; Bourgoin, J.P.; Ayrault, B. Dispersion des indices de refraction du Molybdate de Plomb ($PbMoO_4$) et de la paratellurite (TeO_2). *Opt. Commun.* **1972**, *6*, 383–387. [CrossRef]
38. Georgiev, G.; Glenar, D.A.; Hillman, J.J. Spectral characterization of acousto-optic filters used in imaging spectroscopy. *Appl. Opt.* **2002**, *41*, 209–217. [CrossRef]
39. Yu, K.; Zhao, H. Analysis on the Influence of Incident Light Angle on the Spatial Aberrations of Acousto-Optical Tunable Filter Imaging. *Materials* **2022**, *15*, 4464. [CrossRef] [PubMed]

materials

MDPI

Article

Optical Characterization of Thin Films by Surface Plasmon Resonance Spectroscopy Using an Acousto-Optic Tunable Filter

Ildus Sh. Khasanov [1,*], Boris A. Knyazev [2,*], Sergey A. Lobastov [1], Alexander V. Anisimov [1], Pavel A. Nikitin [1] and Oleg E. Kameshkov [2]

[1] Scientific and Technological Centre of Unique Instrumentation RAS, 117342 Moscow, Russia
[2] Budker Institute of Nuclear Physics SB RAS, 630090 Novosibirsk, Russia
* Correspondence: khasanov@ntcup.ru (I.S.K.); b.a.knyazev@gmail.com (B.A.K.)

Abstract: The paper presents the application of the acousto-optic tunable filter (AOTF) in surface plasmon resonance (SPR) spectroscopy to measure the optical thickness of thin dielectric coatings. The technique presented uses combined angular and spectral interrogation modes to obtain the reflection coefficient under the condition of SPR. Surface electromagnetic waves were excited in the Kretschmann geometry, with the AOTF serving as a monochromator and polarizer of light from a white broadband radiation source. The experiments highlighted the high sensitivity of the method and the lower amount of noise in the resonance curves compared with the laser light source. This optical technique can be implemented for nondestructive testing in the production of thin films in not only the visible, but also the infrared and terahertz ranges.

Keywords: acousto-optic; tunable filter; surface plasmon resonance; surface electromagnetic wave spectroscopy; thin film thickness; wavelength and angular interrogation; refractive index sensing

check for updates

Citation: Khasanov, I.S.; Knyazev, B.A.; Lobastov, S.A.; Anisimov, A.V.; Nikitin, P.A.; Kameshkov, O.E. Optical Characterization of Thin Films by Surface Plasmon Resonance Spectroscopy Using an Acousto-Optic Tunable Filter. *Materials* **2023**, *16*, 1820. https://doi.org/10.3390/ma16051820

Academic Editor: Polina P. Kuzhir

Received: 9 December 2022
Revised: 20 February 2023
Accepted: 21 February 2023
Published: 22 February 2023

1. Introduction

Recent advances in nanofabrication techniques have facilitated the development of new materials such as metasurfaces [1], gradient-index films [2], and so on. Such nanomaterials have sparked renewed interest in plasmonics [3], a field of research that aims to control light–matter interactions at the nanoscale at metal–dielectric interfaces. For the same reason, the methods for controlling the deposition and characterization of thin films for accurate measurement of their plasmonic performance have been receiving increased interest [4].

Surface plasmon resonance (SPR) is one of the most sensitive techniques for the optical characterization of thin layers whose thickness is much less than the wavelength of the probing light [5,6]. SPR is a well-established biosensing technique for environmental and medical applications for the detection and characterization of various analytes [7–10]. Like other reflectometry techniques, the SPR method consists of the measurement of the reflection coefficient R [6]. R depends on the wavelength, angle of incidence θ, film thickness, and optical constants of the multilayer structure under study. SPR was observed not only in the visible, but also in the infrared [11,12] and terahertz ranges [13,14]. Usually, the wavelength interrogation mode, i.e., SPR spectroscopy, is used in combination with optical fibers [15], thus the angular interrogation mode cannot be applied. Application of the acousto-optic tunable filter (AOTF) allows combining angular and spectral modes.

Since its development in 1969 [16], the AOTF has found applications in many optical research methods, such as spectroscopy [17], profilometry [18], microscopy [19], endoscopy [20], stereoscopy [21], colorimetry [22], holography [23,24], and so on, thanks to its high tuning speed, narrow spectral width, low distortion of the passing collimated beam [25], compactness, absence of moving mechanical parts, and electronic control. The AOTF is suitable for any research that requires spectral image processing [26]. The AOTF enables a dynamic increase in the imaging contrast of objects under study [27], attenuation of the laser radiation [28], and so on. The AOTF relies on the interaction of light with sound [29]; a piezoelectric transducer—when radiofrequency is applied—generates an acoustic wave, which creates in the birefringent medium periodic regions of compression and decompression with different refractive indices. The traveling light wave diffracts on these areas like on a diffraction grating, the period of which can be adjusted through the sound frequency for dynamic spectral selection. The diffraction efficiency of the AO interaction in new AO devices has improved significantly [30–32], and the number of applicable materials with optical anisotropy has increased [33,34]. Owing to the choice of new AO media (including liquid ones [35]), the spectral range has been extended from the ultraviolet and visible [36] to the infrared [37,38] and terahertz [39,40] ranges.

The narrow spectral width (which can theoretically be as low as 2.5 Å for $\lambda = 630$ nm [41]) from the emitted spectrum of the radiation source and the high degree of polarization is an ideal combination for surface plasmon resonance (SPR) spectroscopy, as this effect is observed in a monochromatic *p*-polarized radiation [42]. In this case, using a radiation with a short coherence length, as in the case of an AOTF-filtered white light source, may be preferable to the use of coherent laser radiation. Although the SPR phenomenon has a quantum nature as it occurs when light tunnels through a thin metal film, we can accurately calculate the reflection coefficient R within the transfer matrix formalism, similarly to its use in classical optics, to describe interference in thin films. However, it is essential to note that the thickness of the metallic film supporting SPR should not exceed half the wavelength and is usually much smaller than it. So, the coherence length of light does not affect the calculation of R because it does not exceed the thickness of the film. Therefore, SPR can be similarly observed with the coherent laser light or incoherent quasimonochromatic light. The latter is more favorable owing to the suppression of parasitic interference in lenses, as well as of spurious diffraction such as speckles from scattering on surface roughness, dust particles, and other minor inhomogeneities of the films. Therefore, the use of the white light filtered by the AOTF for the creation of monochromatic incoherent radiation makes it possible to obtain resonance curves containing less noise compared with the laser radiation.

The AOTF was first applied to SPR measurements in [43,44] in the fixed-angle wavelength interrogation mode. It has been shown that, in the Kretschmann configuration, it is possible to increase the sensitivity of the sensor by combining angular and spectral modulation via selection of the optimal wavelength and angle [45]. Recently, we have developed an algorithm [46] that allows using the full range of scanning angles and wavelengths to increase the accuracy of SPR measurements. Our goal is to show the effectiveness and prospects of our approach to popularize the application of the AOTF in SPR spectroscopy.

2. Materials and Methods

To experimentally demonstrate the AOTF operation in SPR spectroscopy in the combined spectral and angular interrogation mode, we use an AO cell made of paratellurite at the STC UI RAS. The main parameters of the AO cell are described in [36] (TeO_2 material, $\gamma = 7°$, diffraction mode e $\rightarrow$ o, crystal length of 25 mm, 10 mm $\times$ 8 mm light aperture, and spectral range of 450–900 and 900–1700 nm). The chromatic shift of the AO cell is compensated for with a heavy flint (TF-1) wedge and is equal to approximately 5′.

2.1. Optical Setup

The optical scheme of the experimental setup is shown in Figure 1.

To create a uniform collimated light without chromatism, we use a parabolic off-axis mirror with the fiber optic output of the white light source in the focus. The AO cell control driver enables wavelength selection. A collimated *p*-polarized radiation beam deflected by the AO cell is directed to the object under study; that is, a prism with a thin film coating on its base. The right-angle prism was made of LC-7 glass at the optical division of the STC UI RAS. LC-7 is a Russian National State Standard (GOST) light crown glass with a refractive index n of about 1.48 and dispersion coefficient $\nu_e = 66.17$. With the help of a beam-splitting cube, part of this radiation is deflected for analysis to the spectrometer. This enables determination of the wavelength without cell pre-calibration, as well as control of possible temperature deviations during the measurements [47]. We found that the AOTF characteristics remained stable during the experiment. The measurements are performed at a fixed angle of the rotating miniplatform, which is preselected such that the resonance dip is present in all images (as the resonance dip shifts monotonically towards lower angles with increasing wavelength). The angular interrogation is provided by the concentration of the beam by a focusing lens. The focal spot is at the base of the prism, i.e., on the thin film. The accuracy of the lens focusing is controlled by means of the camera via choice of the lens position corresponding to the minimum light spot visible from the back side of the prism. The focus spot does not exceed 0.5 mm. The cleanliness and homogeneity of the measured region of the thin film are also controlled by means of the HeNe laser, as a coherent radiation is more sensitive to roughness, defects, and contamination of the thin film surface. We are able to select a region of the thin film using micromechanical horizontal and vertical translators.

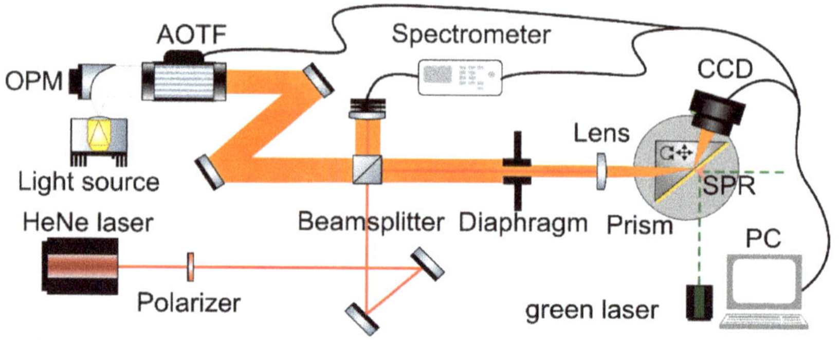

Figure 1. The optical scheme of the AOTF-based spectroscopy. The light source is the ISTEQ XWS-65 laser-pumped plasma broad white light radiation source collimated with the OPM (off-axis parabolic mirror) Thorlabs MPD399-M03 with a diameter of Ø3″ and reflected focal length = 9″; HeNe laser is a helium-neon laser LGN-301 with an emission wavelength of 632.8 nm; Beamsplitter is a non-polarizing 50/50 cube with a 25 mm facet; Diaphragm is an 8 mm aperture for beam collimation; Lens is a Thorlabs achromatic lens Ø1″ with a spectral range of 400–700 nm and focal distance of 50 mm; Polarizer is a visible range polarizer; Spectrometer is a tunable grating spectrometer Avesta ASP-150 with Hamamatsu S8378-1024Q CCD array and a spectral resolution of 0.06 nm; Prism is a right-angle prism (produced at STC UI RAS) with base dimensions of 22 mm and height of 11 mm, made of LC-7 glass with a metal (silver, Ag) film and dielectric (silicon dioxide, SO_2) coating and placed on a rotating miniplatform Standa 7R7 mounted on a translation stage Standa 7T173 and vertical translation stage Standa 7VT188-20; CCD is a GigE monochrome industrial camera Imaging Source DMK 23G445 with no IR filter, with a resolution of 1280 × 960 pixels and sensitivity of down to 0.015 lx; PC is a personal computer with the control software written in Python 3 and AutoIt 3.

2.2. Preparation of Samples

The object of the study was a thin film of silver (Ag) with a thickness of 55 nm. The coating was performed with a vacuum unit for ion-beam sputtering of optical coatings (the Aspira sputtering unit, OOO "FLAVT", Serpukhov, Russia).

At first, a set of silver-coated prisms was made, and then each prism was separately coated with a thin layer of silicon dioxide of 13 nm, 17 nm, and 30 nm thick. Initially, thicknesses of 10 nm, 20 nm, and 30 nm were planned, but this result could be achieved only after additional calibration of the deposition rate in the vacuum unit. The thickness of the deposited coating was under broadband optical control with time correction (at a known rate of substance deposition) (see also Appendix A).

2.3. Experiment

Using the PC, we sequentially tune the AOTF, starting with a wavelength of approximately 480 nm and ending at approximately 780 nm. As we approach the extremes of the wavelength range, the light intensity decreases. If necessary, the measuring range can be extended from 450 nm–900 nm to 900–1700 nm via proper cell driver tuning. The camera shoots at four exposures (0.1 ms, 0.5 ms, 2 ms, and 5 ms) to ensure a high dynamic range. A similar uncoated LC-7 glass prism is used as a reference of the incident radiation for calculation of the reflection coefficient R. As the observation takes place at angles exceeding the angle of total internal reflection, the reflection coefficient for the reference prism is close to 1. The reference prism allows us to take into account the deviations caused by the aberrations in the lenses, the insufficient uniformity of illumination, and the Fresnel reflections. To determine the angular scales in the images, we utilize a green laser to measure the angular position of the central laser spot from the reflection from the base. Then, using the rotating platform, we mark the positions of the light spot with increments of 30 angular minutes. We note that the resulting angular scale remains linear across all measurements, which indicates negligible aberrations in the lens and sufficient collimation of the beam. The measured angular scale refers to the angle of incidence of light onto the face of the prism. We recalculate it into angles of incidence onto the base of the prism, taking into account Snell's law and the known dispersion in the glass.

3. Results

The obtained experimental data in the form of a set of images from the camera and their corresponding spectra were processed with a program written in Python. The source code and images are available on Github. As the AO cell control driver was not pre-calibrated, the measurements were taken with a constant radiofrequency step of 1 MHz, starting at 62 MHz and up to 110 MHz (48 wavelength measurements). The measurement data were then recalculated and linearly interpolated to a uniform grid with wavelengths of 515 to 750 nm, as shown in Figure 2.

In Figure 2, we observe a spectrum in terms of the wavelength λ and angle θ of the resonance curves. This spectrum clearly repeats in shape the known dispersion curves ω (k), as the wavelength is inversely proportional to the frequency ω and the wave vector k, and k of the surface plasmons is proportional to the angle of incidence θ. Figure 2 is inspired by Figure 2 in [48], where SP dispersion curves for the flat and textured metal surfaces are shown, obtained by the modified prism coupling method using a photographic film.

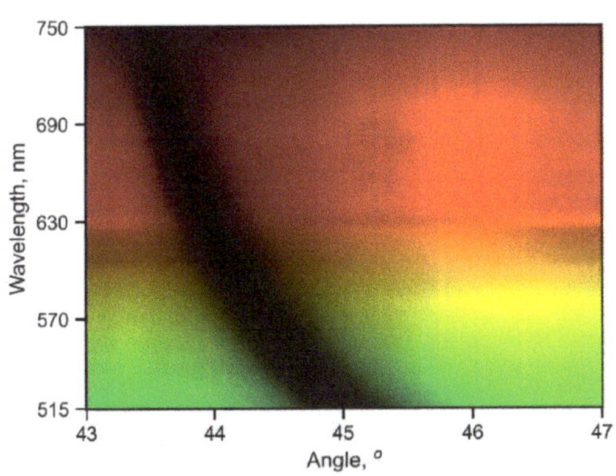

(a)

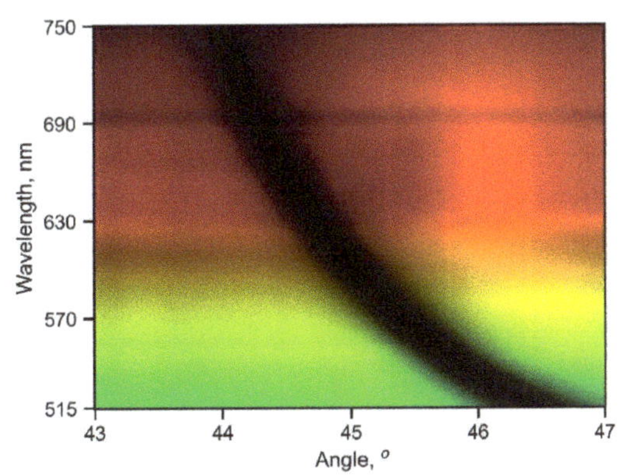

(b)

Figure 2. *Cont.*

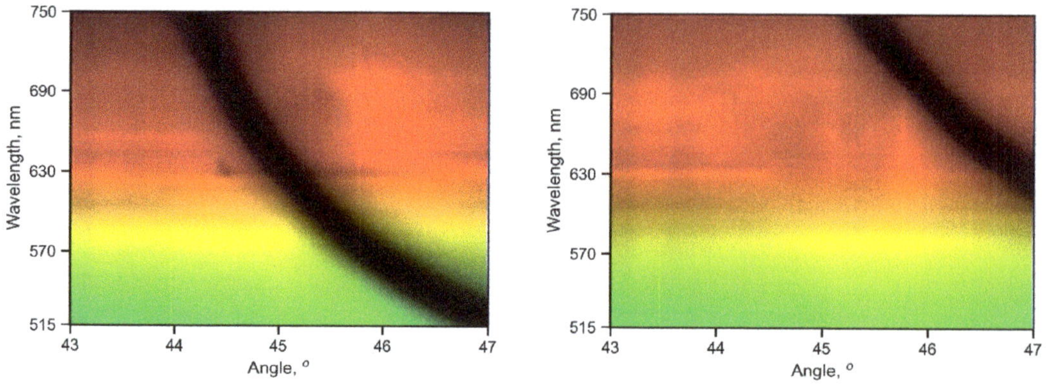

(c) (d)

Figure 2. Experimental spectral images of the resonance curves (dependence of the reflection coefficient R on the angle of incidence θ) for the LC-7 glass prism with the thin 55 nm silver film and various thicknesses of the dielectric coating (silicon dioxide). The color corresponds to the real color visible to the human eye. The thickness of the dielectric coating measured by the spectrophotometer is (**a**) 0 nm, Sample 1; (**b**) 13 nm, Sample 2; (**c**) 17 nm, Sample 3; and (**d**) 30 nm, Sample 4.

Having the known optical constants for silver (e.g., from refractiveindex.info [49]), we found that the experimental curves for the resonance angles in Figure 3a (value of the angle θ at the minimum of the reflection coefficient R) agree well with the theoretical values. We calculated the reflection coefficient R for the multilayer structure by the transfer matrix method [50,51]. However, the values of the reflection coefficient R in Figure 3b differ. The larger values for the reflection minimum can be explained by the fact that the calculation was performed for a monochromatic light, but the light after the acousto-optic cell has a wider spectral width, which leads to a decrease in the contrast of the resonance curve.

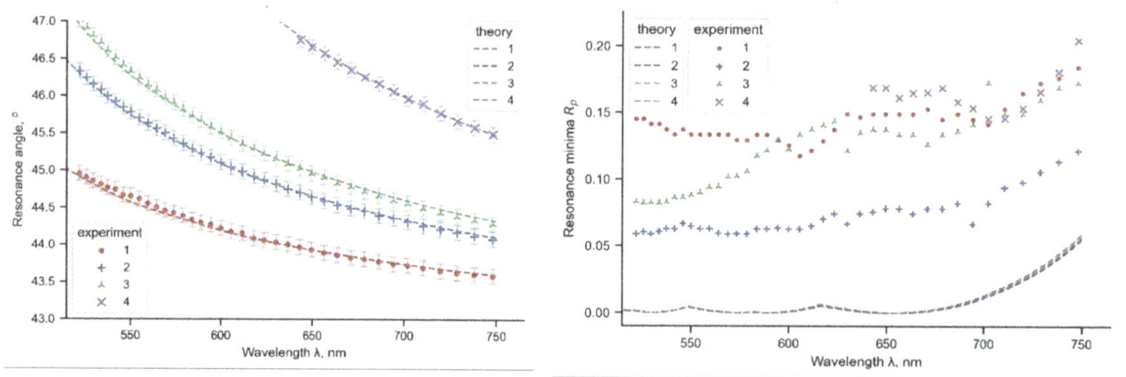

(**a**) (**b**)

Figure 3. Experimental (solid) and simulated (dashed) resonance curves for different samples: 1—without a dielectric coating; 2, 3, and 4—with sequentially increasing coating thickness (see Table 1). The wide semi-transparent line shows the confidence interval: (**a**) dependence of the resonance angle on the wavelength; (**b**) dependence of the minimum of the reflection coefficient on the wavelength.

Table 1. Measured thickness values in nm for the dielectric coating.

Method	Sample 1	Sample 2	Sample 3	Sample 4
Spectrophotometry	0	13.0 ± 3.0	17.1 ± 4.2	30.0 ± 5.2
SPR spectroscopy	0	8.7 ± 1.5	12.0 ± 1.4	25.9 ± 1.4

From Table 1, we see a difference in the dielectric coating thickness measurements. However, the measured values fall within the confidence intervals. We should note that the accuracy of the spectrophotometric control applied in the thin film production is not the best in its class and is given only for reference purposes. In our case, the SPR spectroscopy is a more precise and convenient technique because, at a well-defined resonance dip, a shift in the position of the resonance angles at multiple wavelengths is easier to interpret than a shift in the spectral maxima and minima of the gradual spectral curve obtained by spectrophotometry (compare Figures 3a and A1c). Besides that, SPR spectroscopy allows measuring the dielectric thickness directly on the sample itself, whereas spectrophotometry measurements are performed indirectly on a specially prepared control sample (see Appendix A).

4. Discussion

In the presented experimental setup, we used the AOTF produced by the Scientific and Technological Centre of Unique Instrumentation of the Russian Academy of Sciences. The AOTF combines the functions of a monochromator and a light polarizer; the AO crystal used is transparent in the visible and infrared ranges, and its optical design is suitable for imaging purposes, allowing parallel angular and spectral SPR measurements. This makes the AOTF an ideal tool for miniaturizing an SPR spectroscopy setup and allows using it for the development of an AOTF-based SPR videospectrometer.

The determination of angular coordinates in this experiment has a high error of 10′. However, the precision can be improved up to 1′, which is the limiting precision of the nonius in the rotating platform. This improvement would lead to a more accurate determination of the thickness of the thin film, up to 1 Å. To achieve this, it is necessary to reduce the influence of the error sources and increase the sensitivity of measurements.

Several sources of errors can be listed:

- chromatism in optical elements, that is, in the AOTF and lenses;
- misalignments in the setup;
- inaccuracy in the initial assumptions of the optical model.

Chromatism. The angular error introduced by the chromatic dispersion of light in the AOTF was compensated for to some extent by the use of the glass wedge. We can further improve the accuracy by calibrating the angular scale of the camera at different wavelengths instead of just one. Additionally, we can use additional calibration lasers of different wavelengths (e.g., the green KTP laser (λ = 532 nm) and red-diode-pumped semiconductor laser (λ = 671 nm)) to ensure coincidence of the resonant angle positions in both coherent and noncoherent light. The chromatism caused by the collecting lens in front of the prism can also be eliminated via application of a second off-axis parabolic mirror.

Misalignments. Another source of error in this experiment is the placement of the prism on the 3D-printed holder. This allows accurate positioning of the prism relative to the rotating platform. However, owing to the imperfection of the 3D printing (accuracy of about 0.4 mm), the holder plane is slightly inclined relative to the optical table, which results in a slightly inclined SPR line, with the bottom of the line deviating from the top by about 2′. This error can be eliminated via the introduction of a tilt platform.

Inaccurate optical model. Another factor not taken into account in these measurements is the roughness of the metal and dielectric films. The optical model used assumes the existence of sharp boundaries between the metal, dielectric, and air layers. However, if the surface is rough, then, instead of a sharp boundary between the layers, there will be a boundary with a gradient change in the refractive index. The roughness of the silver film can be estimated at 2–3 nm [52]. We can take this into account by adding gradient-index transition layers at the metal–dielectric and dielectric–air interfaces. The optical properties of these layers can be described with an effective medium approximation, e.g., the Maxwell–Garnett model [53].

To increase the sensitivity of measurements, it is better to recalibrate the AOTF for a shorter wavelength range, down to 400 nm, because, at shorter wavelengths, the resonance angle deviates by larger values. Additionally, to improve the determination of the reflection coefficient R, one could increase the light output via compact assembling of the optical setup. Currently, the light from the AOTF travels a distance of about 1 m to the prism, which results in a relatively low intensity compared with the brightness of the light exiting the AOTF.

Despite the experimental errors listed above, it is known that SPR spectroscopy is more sensitive than spectroreflectometry [54]. Besides, the use of the AOTF allows obtaining less noisy images, because there is no parasitic interference, which can be seen from the comparison of coherent and incoherent monochromatic light sources in Figure 4.

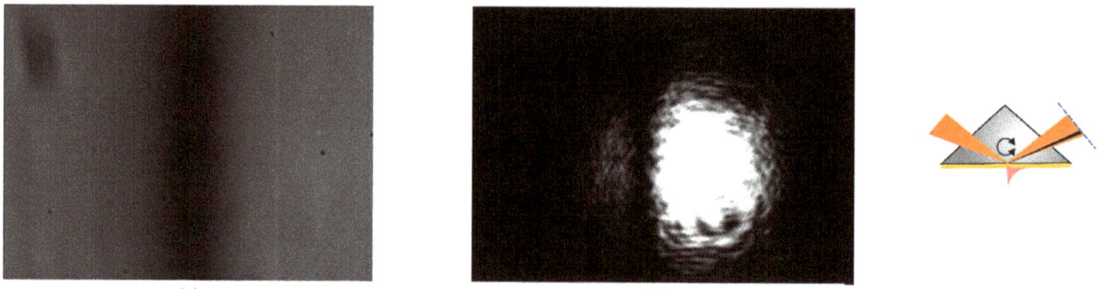

(a) (b) (c)

Figure 4. Comparison of images of the SPR curve with a monochrome camera: (**a**) for an AO filter with the central wavelength $\lambda \approx 633$ nm; (**b**) for a helium-neon laser, $\lambda = 632.8$ nm. The images are the angular distribution of the reflected radiation from the base of the prism; (**c**) SPR observation design (top view); the dashed line is the image plane.

However, the AOTF has the disadvantage of its light spectrum being broader than that of the laser. This leads to a slight decrease in the contrast of the overall picture. In addition, the focusing spot of the AO light on the surface of a thin film is an order of magnitude larger than the laser spot. Thus, the laser source has a higher lateral resolution in the case of surface microscopy [55]. However, for non-destructive testing, a wide spot can be an advantage, as it excludes the accidental influence of microscopic defects, such as pinholes, known as "starry sky", or settled dust particles.

Furthermore, noise can be seen on the resonance minimum plot R_{min} in Figure 3b. However, the data from Figure 3b were not used in the determination of the film thickness. Meanwhile, an analysis of Figure 3b shows that the SPR efficiency varies with different dielectric coating thicknesses, which contradicts our expectations. According to the calculations, the depth of the resonance curve (i.e., the value of R_{min}) shall not change, as seen for the dashed lines in Figure 3b. The depth of the SPR curve depends on the thickness of the metal film. If diffusion of the dielectric into the metal during the sputtering is assumed, we can see that the metal film gradually decreases in thickness, which leads to conditions closer to optimal SPR at the 9 nm coating (line 2 in Figure 3b). It is also noticeable that the corresponding resonance curve (the thick black line in Figure 2) has the most pronounced profile. However, as almost all experimental lines lie in the error interval, we cannot insist on the correctness of our hypothesis. Meanwhile, the results of the measurements of thicker coatings are not presented in this article; the downward trend in the SPR efficiency persists. Further investigation into the issue of metal diffusion into the dielectric is needed, but our approach to SPR spectroscopy can offer a promising way to examine the transient diffusion layers, which are important in understanding of the corrosion processes of thin metal films.

The high accuracy of the SPR technique in combination with spectroscopy will make it possible to study ultrathin films and, in particular, gradient-index films. The physical reason for the ability of each wavelength of light to provide information about the inhomogeneities of the film thickness, in addition to the dispersion of its optical constants, is the nonlinear interaction of surface electromagnetic waves with the substance of the film. The energy of the field is distributed exponentially with respect to the normal surface, allowing for subwavelength characterization of the film structure. This is important, for example, because graded-index thin films enable the achievement of high laser stability of mirrors [56], change in the mechanical properties of films, increase in the efficiency of scintillation detector BGO crystals [57], and so on. However, their manufacture requires precise non-destructive testing. The common X-ray diffraction analysis is difficult to use continuously in production and requires an ultra-smooth surface, and traditional spectrophotometry does not provide the necessary accuracy, as the strength of the interaction of light with ultrathin layers is too small [58]. SPR spectroscopy using compact AOTFs makes it possible to bring real-time optical control into existing thin film deposition facilities.

5. Conclusions

We obtained combined spectral and angular surface plasmon resonance curves and observed the expected difference in the SPR spectra for slightly different coating thicknesses. The image of the SPR curve obtained with the AOTF-filtered noncoherent light source is less noisy compared with the laser image. This promising result shows the potential of using the AOTF for SPR spectroscopy in the angular interrogation mode and for hyperspectral SPR imaging for sensorics and non-destructive testing applications, in not only the visible, but also the IR and THz ranges.

Author Contributions: Conceptualization, B.A.K.; methodology, I.S.K. and S.A.L.; software, I.S.K., O.E.K. and A.V.A.; validation, I.S.K.; formal analysis, I.S.K.; investigation, I.S.K.; sample preparation, S.A.L.; resources, I.S.K. and B.A.K.; data curation, I.S.K.; writing—original draft preparation, I.S.K. and P.A.N.; writing—review and editing, B.A.K.; visualization, I.S.K. and O.E.K.; supervision, B.A.K.; project administration, I.S.K.; funding acquisition, I.S.K. All authors have read and agreed to the published version of the manuscript.

Funding: Research funded by Council on grants of the President of the Russian Federation for state support of young Russian scientists (MK-3998.2021.1.2).

Institutional Review Board Statement: Not applicable.

Informed Consent Statement: Not applicable.

Data Availability Statement: The source code is publicly available at the author's Github repository: https://github.com/KhasanovISh/Public. The data presented in this study are openly available in FigShare at 10.6084/m9.figshare.22015910.

Acknowledgments: The authors thank Batshev V.I., Gorevoy A.V., and Sharikova M.O. for the help in setting up the acousto-optical equipment. This work was performed using the equipment of the Center for Collective Use of the Scientific and Technological Center of Unique Instrumentation of the Russian Academy of Sciences (STC UI RAS) [http://ckp.ntcup.ru].

Conflicts of Interest: The authors declare no conflict of interest. The funders played no role in the design of the study; collection, analyses, or interpretation of data; writing of the manuscript; or decision to publish the results.

Appendix A. Method of Determination of the Thickness of the Deposited Layer Using a Spectrophotometer

For measurements of the thin layer thickness, we used a control sample of high-purity fused silica glass KU-1 ($n_D = 1.4584$, $n_e = 1.4601$) with a diameter of 22 mm and thickness of 6 mm, with a pre-sprayed multi-layer dielectric coating (two layers: Ta_2O_5 with a thickness of 185 nm and SiO_2 with a thickness of 264 nm). This coating was applied in two steps: tantalum oxide deposition (working pressure of 2.8×10^{-2} Pa) and silicon dioxide deposition (working pressure of 1.75×10^{-2} Pa). The sputtering of coatings was carried out by ion-beam at an initial pressure in the chamber not worse than 8×10^{-4} Pa. After each step, the transmittance of the sample was measured with the Cary5000 spectrophotometer in the wavelength range of 400–1100 nm. Pre-calibration (normalization) of the signal to zero and 100% transmittance was performed. Only one control sample was used in all spectrophotometry measurements, so the dielectric coatings were sprayed sequentially one on another (see Figure A1b), which led to an increase in the measurement error of each successive layer (Table 1). From the data obtained, the thicknesses of the layers of tantalum oxide and silicon dioxide were determined with the help of a program written in Python. During the sputtering process, the desired thickness is obtained by fitting theoretical to experimental spectra. The estimated error of the thickness measurements was found via Monte Carlo simulation on synthetic data sets (M = 100) from the spectral transmittance points (N = 700) [59].

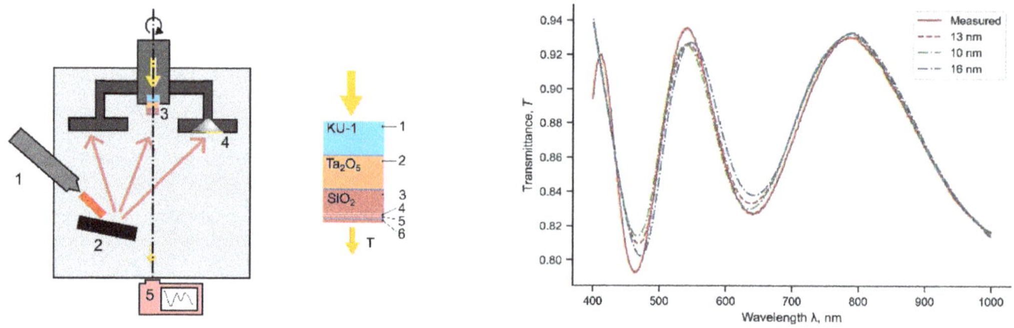

(a) (b) (c)

Figure A1. (**a**) The schematic of the thin-film sputtering chamber and optical control unit: ion gun (1) irradiates target (2); the dispersed substance is deposited on control sample (3) and SPR prism (4) fixed on the rotating carousel. The sputtering thickness is monitored with a spectrometer (5). (**b**) The final layout of the sprayed layers on the control sample: 1—KU-1 glass; 2—pre-sprayed Ta_2O_5 layer; 3—pre-sprayed SiO_2 layer; 4, 5, and 6—SiO_2 coatings. (**c**) An example of the measured transmission spectrum of the control sample for coating thickness (red solid line) and a model approximation for 13 nm coating thickness (dashed line), and model approximations within a confidence interval of 10–16 nm (dot-dashed lines).

The measurement error with Cary5000 was minor, about 0.1% for a sample after four measurements with a shift to take into account the uneven coating thickness. The sputtering error due to variations in the position of the parts on different tooling was 1%. This error was calculated from measurements taken on different holders for the same process. The error in measuring the transmittance spectrum during the sputtering process and in the control sample was influenced by the stability of the optical signal, which was disrupted by the vibration from the rotation of the carousel. The control sample and prism were located in different positions in the chamber (see Figure A1a), which could result in a slight difference in their coating thicknesses.

For calculation of the thickness of the sputtered layer, OptiLayer 7.68 software was used for the determination of the refractive index through the transmission of a thick (over 200 nm) layer of silica sputtered on a quartz substrate over a known layer of tantalum oxide. The refractive index dispersion for each dielectric used in the sputtering process must be measured beforehand. These values may vary slightly from one process to another, which complicates the film fabrication process. The main source of error in the determination of the silicon oxide thickness was the measurement of the refractive index used in the model for a sputtering process with a higher oxygen concentration. In our case, during the sputtering onto the SPR prism no supplemental oxygen flux was used in order to prevent oxidation of the metal film on the prism. This led to a change in the oxygen concentration. As a result, the model and measured spectra do not fit well (see Figure A1c).

References

1. Chen, H.-T.; Taylor, A.J.; Yu, N. A Review of Metasurfaces: Physics and Applications. *Rep. Prog. Phys.* **2016**, *79*, 076401. [CrossRef]
2. Shvartsburg, A.B.; Maradudin, A.A. *Waves in Gradient Metamaterials*; World Scientific: Singapore, 2013; ISBN 978-981-4436-95-3.
3. Luo, X.; Tsai, D.; Gu, M.; Hong, M. Extraordinary Optical Fields in Nanostructures: From Sub-Diffraction-Limited Optics to Sensing and Energy Conversion. *Chem. Soc. Rev.* **2019**, *48*, 2458–2494. [CrossRef]
4. Malureanu, R.; Lavrinenko, A. Ultra-Thin Films for Plasmonics: A Technology Overview. *Nanotechnol. Rev.* **2015**, *4*, 259–275. [CrossRef]
5. Abelès, F. Surface Electromagnetic Waves Ellipsometry. *Surf. Sci.* **1976**, *56*, 237–251. [CrossRef]
6. Arwin, H.; Poksinski, M.; Johansen, K. Enhancement in Ellipsometric Thin Film Sensitivity near Surface Plasmon Resonance Conditions. *Phys. Status Solidi (A)* **2008**, *205*, 817–820. [CrossRef]
7. Yesudasu, V.; Pradhan, H.S.; Pandya, R.J. Recent Progress in Surface Plasmon Resonance Based Sensors: A Comprehensive Review. *Heliyon* **2021**, *7*, e06321. [CrossRef] [PubMed]
8. Hamza, M.E.; Othman, M.A.; Swillam, M.A. Plasmonic Biosensors: Review. *Biology* **2022**, *11*, 621. [CrossRef] [PubMed]
9. Denizli, A. *Plasmonic Sensors and Their Applications*, 1st ed.; Wiley-VCH: Weinheim, Germany, 2021; ISBN 3-527-34847-6.
10. Shukla, N.; Chetri, P.; Boruah, R.; Gogoi, A.; Ahmed, G.A. Surface Plasmon Resonance Biosensors Based on Kretschmann Configuration: Basic Instrumentation and Applications. In *Recent Advances in Plasmonic Probes: Theory and Practice*; Lecture Notes in Nanoscale Science and Technology; Biswas, R., Mazumder, N., Eds.; Springer International Publishing: Cham, Switzerland, 2022; pp. 191–222. ISBN 978-3-030-99491-4.
11. Vinogradov, E.A.; Leskova, T.A. Polaritons in Thin Films on Metal Surfaces. *Phys. Rep.* **1990**, *194*, 273–280. [CrossRef]
12. Gu, G.; Lu, X.; Kemsri, T.; Zhang, Y. Surface Plasmonics and Its Applications in Infrared Sensing. In *Nanoplasmonics—Fundamentals and Applications*; Barbillon, G., Ed.; InTech: Rang-Du-Fliers, France, 2017; ISBN 978-953-51-3277-6.
13. Bogomolov, G.D.; Jeong, U.Y.; Zhizhin, G.N.; Nikitin, A.K.; Zavyalov, V.V.; Kazakevich, G.M.; Lee, B.C. Generation of Surface Electromagnetic Waves in Terahertz Spectral Range by Free-Electron Laser Radiation and Their Refractive Index Determination. *Nucl. Instrum. Methods Phys. Res. Sect. A Accel. Spectrometers Detect. Assoc. Equip.* **2005**, *543*, 96–101. [CrossRef]
14. Gerasimov, V.; Knyazev, B.; Nikitin, A.; Khasanov, I.; Lemzyakov, A.; Kameshkov, O.; Osintseva, N.; Kukotenko, V.; Pavelyev, V. Novel Experiments on THz Plasmonics Using Novosibirsk Free Electron Laser. In *Synchrotron and Free Electron Laser Radiation: Generation and Application (SFR-2022)*; Book of Abstracts; BINP: Novosibrsk, Russia, 2022; p. 40.
15. Filion-Côté, S.; Tabrizian, M.; Kirk, A.G. Real-Time Measurement of Complex Refractive Indices with Surface Plasmon Resonance. *Sens. Actuators B Chem.* **2017**, *245*, 747–752. [CrossRef]
16. Harris, S.E.; Wallace, R.W. Acousto-Optic Tunable Filter*. *J. Opt. Soc. Am. JOSA* **1969**, *59*, 744–747. [CrossRef]
17. Tran, C.D. Principles and Analytical Applications of Acousto-Optic Tunable Filters, an Overview. *Talanta* **1997**, *45*, 237–248. [CrossRef]
18. Tsepulin, V.G.; Tolstoguzov, V.L.; Perchik, A.V.; Karasik, V.E.; Arefiev, A.P. Thickness Distribution Measurement of Multilayer Film Structures by Spectral Reflectometry Methods. *Instrum. Metrol. Inf. Meas. Instrum. Syst.* **2016**, *3*, 3–12. [CrossRef]
19. Machikhin, A.S.; Pozhar, V.E.; Viskovatykh, A.V.; Burmak, L.I. Acousto-Optical Tunable Filter for Combined Wideband, Spectral, and Optical Coherence Microscopy. *Appl. Opt. AO* **2015**, *54*, 7508–7513. [CrossRef]
20. Machikhin, A.S.; Khokhlov, D.D.; Batshev, V.I.; Pozhar, V.E. An Acousto-Optic Endoscopic Module for Nondestructive Testing. *Bull. Russ. Acad. Sci. Phys.* **2018**, *82*, 1403–1405. [CrossRef]
21. Machikhin, A.; Batshev, V.; Pozhar, V.; Naumov, A.; Gorevoy, A. Acousto-Optic Tunable Spectral Filtration of Stereoscopic Images. *Opt. Lett. OL* **2018**, *43*, 1087–1090. [CrossRef]
22. Zajtsev, A.K.; Kludzin, V.V.; Kochin, L.B.; Polosin, L.L.; Sokolov, V.K. Characteristics of Acousto-Optic Tunable Filter for Colorimetry. In Proceedings of the 3rd International Conference on Optical Information Processing, Moscow, Russia, 28–31 May 1999; SPIE: Washington, DC, USA, 1999; pp. 54–59. [CrossRef]
23. Balakshy, V.I. Application of Acousto-Optic Interaction for Holographic Conversion of Light Fields. *Opt. Laser Technol.* **1996**, *28*, 109–117. [CrossRef]
24. Polschikova, O.; Machikhin, A.; Gorevoy, A.; Stoykova, E. Single-Shot Multiwavelength Digital Holography Using Bragg Diffraction of Light by Several Ultrasound Waves. *J. Opt. Soc. Am. A JOSAA* **2022**, *39*, A79–A85. [CrossRef] [PubMed]
25. Pozhar, V.; Machihin, A. Image Aberrations Caused by Light Diffraction via Ultrasonic Waves in Uniaxial Crystals. *Appl. Opt.* **2012**, *51*, 4513–4519. [CrossRef]
26. Batshev, V.; Machikhin, A.; Gorevoy, A.; Martynov, G.; Khokhlov, D.; Boritko, S.; Pozhar, V.; Lomonov, V. Spectral Imaging Experiments with Various Optical Schemes Based on the Same AOTF. *Materials* **2021**, *14*, 2984. [CrossRef]

27. Voloshinov, V.B.; Parygin, V.N.; Molchanov, V.Y. Tunable Acousto-Optic Filters and Their Applications in Laser Technology, Optical Communication, and Processing of Images. In Proceedings of the Laser Optics 2000: Control of Laser Beam Characteristics and Nonlinear Methods for Wavefront Control, Petersburg, Russia, 26–30 June 2000; SPIE: Washington, DC, USA, 2001; Volume 4353, pp. 17–22. [CrossRef]
28. Machikhin, A.S.; Sharikova, M.O.; Lyashenko, A.I.; Kozlov, A.B.; Pozhar, V.E.; Lomonov, V.A.; Stoikova, E. Attenuation of the Intensities of Spectral Components of a Multiwavelength Pulsed Laser System by Means of the Bragg Diffraction of Radiation by Several Acoustic Waves. *Quantum. Electron.* **2022**, *52*, 454. [CrossRef]
29. Gottlieb, M.S. Acousto-Optic Tunable Filters. In *Design and Fabrication of Acousto-Optic Devices*; CRC Press: Boca Raton, FL, USA, 1994; ISBN 978-1-00-321022-1.
30. Voloshinov, V.B.; Yushkov, K.B.; Linde, B.B.J. Improvement in Performance of a TeO2 Acousto-Optic Imaging Spectrometer. *J. Opt. A Pure Appl. Opt.* **2007**, *9*, 341. [CrossRef]
31. Antonov, S.N. Acousto-Optic Deflector with a High Diffraction Efficiency and Wide Angular Scanning Range. *Acoust. Phys.* **2018**, *64*, 432–436. [CrossRef]
32. Sopko, I.M.; Ignatyeva, D.O.; Knyazev, G.A.; Belotelov, V.I. Efficient Acousto-Optical Light Modulation at the Mid-Infrared Spectral Range by Planar Semiconductor Structures Supporting Guided Modes. *Phys. Rev. Appl.* **2020**, *13*, 034076. [CrossRef]
33. Mil'kov, M.G.; Volnyanskii, M.D.; Antonenko, A.M.; Voloshinov, V.B. Acoustic Properties of Biaxial Crystal of Double Lead Molybdate Pb_2MoO_5. *Acoust. Phys.* **2012**, *58*, 172–179. [CrossRef]
34. Kupreychik, M.I.; Balakshy, V.I. Peculiarities of Acousto-Optic Interaction in Biaxial Crystal of Alpha-Iodic Acid. *Appl. Opt. AO* **2018**, *57*, 5549–5555. [CrossRef]
35. Dürr, W. Acousto-Optic Interaction in Gases and Liquid Bases in the Far Infrared. *Int. J. Infrared Milli. Waves* **1986**, *7*, 1537–1558. [CrossRef]
36. Batshev, V.I.; Machikhin, A.S.; Kozlov, A.B.; Boritko, S.V.; Sharikova, M.O.; Karandin, A.V.; Pozhar, V.E.; Lomonov, V.A. Tunable Acousto-Optic Filter for the 450–900 and 900–1700 Nm Spectral Range. *J. Commun. Technol. Electron.* **2020**, *65*, 800–805. [CrossRef]
37. Voloshinov, V.B.; Mironov, O.V. Wide-aperture acoustooptical filter for the middle IR. *Opt. I Spektrosk.* **1990**, *68*, 452–457.
38. Voloshinov, V.B.; Gupta, N.; Kulakova, L.A.; Khorkin, V.S.; Melekh, B.T.; Knyazev, G.A. Investigation of Acousto-Optic Properties of Tellurium-Based Glasses for Infrared Applications. *J. Opt.* **2016**, *18*, 025402. [CrossRef]
39. Nikitin, P.; Voloshinov, V.; Knyazev, B.; Scheglov, M. Observation of Acousto-Optic Diffraction of Terahertz Radiation in Liquefied Sulfur Hexafluoride at Room Temperature. *IEEE Trans. Terahertz Sci. Technol.* **2019**, *10*, 44–50. [CrossRef]
40. Vogel, T.; Dodel, G. Acousto-Optic Modulation in the Far-Infrared. *Infrared Phys.* **1985**, *25*, 315–318. [CrossRef]
41. Voloshinov, V.B.; Mishin, D.D. Spectral Resolution Control of Acousto-Optical Cells Operating with Collimated and Divergent Beams. In Proceedings of the International Conference on Optical Information Processing, Saint Petersburg, Russia, 2–7 August 1993; SPIE: Washington, DC, USA, 1994; Volume 2051, pp. 378–385. [CrossRef]
42. Schasfoort, R.B.M. (Ed.) *Handbook of Surface Plasmon Resonance*; Royal Society of Chemistry: Cambridge, UK, 2017; pp. P001–P004, ISBN 978-1-78262-730-2.
43. Tian, Y.; Zhao, L.; Song, D.; Liu, X.; Cao, Y.; Peng, Z.; Liu, Z.; Zhang, H. Acousto-Optic Tunable Filter—Based Surface Plasmon Resonance Biosensor for Determination of Human Factor B. *Anal. Chim. Acta* **2004**, *511*, 97–104. [CrossRef]
44. Tian, Y.; Chen, Y.; Song, D.; Liu, X.; Bi, S.; Zhou, X.; Cao, Y.; Zhang, H. Acousto-Optic Tunable Filter-Surface Plasmon Resonance Immunosensor for Fibronectin. *Anal. Chim. Acta* **2005**, *551*, 98–104. [CrossRef]
45. Luo, Z.; Huang, Y. Sensitivity Enhancement of Surface Plasmon Resonance Sensor Based on Wavelength and Angular Combined Modulations. *Optik* **2018**, *168*, 271–277. [CrossRef]
46. Anisimov, A.V.; Khasanov, I.S. Algorithm for Optical Characterization of Dielectric Gradient Index Nanofilm by Surface Plasmon Resonance Spectroscopy. *J. Phys. Conf. Ser.* **2021**, *2091*, 012067. [CrossRef]
47. Nikitin, P.A.; Gerasimov, V.V.; Khasanov, I.S. Temperature Effects in an Acousto-Optic Modulator of Terahertz Radiation Based on Liquefied SF6 Gas. *Materials* **2021**, *14*, 5519. [CrossRef]
48. Barnes, W.L.; Dereux, A.; Ebbesen, T.W. Surface Plasmon Subwavelength Optics. *Nature* **2003**, *424*, 824–830. [CrossRef]
49. Johnson, P.B.; Christy, R.W. Optical Constants of the Noble Metals. *Phys. Rev. B* **1972**, *6*, 4370–4379. [CrossRef]
50. Byrnes, S.J. Multilayer Optical Calculations. *arXiv* **2020**, arXiv:1603.02720.
51. Mackay, T.G.; Lakhtakia, A. The Transfer-Matrix Method in Electromagnetics and Optics. *Synth. Lect. Electromagn.* **2020**, *1*, 1–126. [CrossRef]
52. McPeak, K.M.; Jayanti, S.V.; Kress, S.J.P.; Meyer, S.; Iotti, S.; Rossinelli, A.; Norris, D.J. Plasmonic Films Can Easily Be Better: Rules and Recipes. *ACS Photonics* **2015**, *2*, 326–333. [CrossRef]
53. Hunderi, O. Optics of Rough Surfaces, Discontinuous Films and Heterogeneous Materials. *Surf. Sci.* **1980**, *96*, 1–31. [CrossRef]
54. Schlesinger, Z.; Sievers, A.J. Broadband Surface Electromagnetic Wave Spectroscopy. *Surf. Sci.* **1981**, *102*, L29–L34. [CrossRef]
55. Johansen, K.; Arwin, H.; Lundström, I.; Liedberg, B. Imaging Surface Plasmon Resonance Sensor Based on Multiple Wavelengths: Sensitivity Considerations. *Rev. Sci. Instrum.* **2000**, *71*, 3530–3538. [CrossRef]
56. Liu, Y.; Chen, J.; Zhang, B.; Wang, G.; Zhou, Q.; Hu, H. Application of Graded-Index Thin Film in Laser Attack and Defense Equipment. *J. Phys. Conf. Ser.* **2020**, *1507*, 102037. [CrossRef]
57. Azarov, I.A.; Kuper, K.E.; Lemzyakov, A.G.; Porosev, V.V.; Shklyaev, A.A. Scintillator Surface Modification by Glancing Angle Deposition of Thin ZrO_2 Films. *J. Inst.* **2022**, *17*, T05013. [CrossRef]

58. Stenzel, O.; Ohlídal, M. *Optical Characterization of Thin Solid Films*, 1st ed.; Springer series in surface sciences 64; Springer International Publishing: Cham, Switzerland, 2018; ISBN 978-3-319-75324-9.
59. Harris, P.M.; Cox, M.G. On a Monte Carlo Method for Measurement Uncertainty Evaluation and Its Implementation. *Metrologia* **2014**, *51*, S176. [CrossRef]

materials

MDPI

Article

Acousto–Optic Modulation and Deflection of Terahertz Radiation

Pavel Alekseevich Nikitin [1,*], Vasily Valerievich Gerasimov [2,3] and Ildus Shevketovich Khasanov [1,*]

1 Scientific and Technological Centre of Unique Instrumentation RAS, 117342 Moscow, Russia
2 Department of Physics, Novosibirsk State University, 630090 Novosibirsk, Russia
3 Budker Institute of Nuclear Physics SB RAS, 630090 Novosibirsk, Russia
* Correspondence: nikitin.pavel.a@gmail.com (P.A.N.); khasanov@ntcup.ru (I.S.K.)

Abstract: It is known that one of the ways to increase the energy efficiency of acousto–optic devices is to use ultrasound beams with a higher power density. It has been established experimentally that the use of a partially electroded ultrasonic transducer significantlyincreases the energy efficiency of the acousto–optic modulator of terahertz radiation. In addition, the operation of an acousto–optic deflector of terahertz radiation with the use of a sectioned ultrasound transducer was theoretically investigated. It showed that a deflector of this kind enables one to achieve higher angular resolution.

Keywords: acousto–optic interaction; terahertz radiation; diffraction; liquefied inert gas

check for updates

Citation: Nikitin, P.A.; Gerasimov, V.V.; Khasanov, I.S. Acousto–Optic Modulation and Deflection of Terahertz Radiation. *Materials* **2022**, *15*, 8836. https://doi.org/10.3390/ma15248836

Academic Editors: Polina P. Kuzhir and Zine El Abiddine Fellah

Received: 16 September 2022
Accepted: 8 December 2022
Published: 10 December 2022

1. Introduction

The amount of transmitted information increases exponentially every year. The solution to this problem is the use of data compression algorithms as well as the use of signals with higher frequencies. Therefore, the communication development in the terahertz (THz) range is now relevant [1]. There are a number of sources and detectors of THz radiation [2], and a digital signal processing for high-speed THz communications has also made significant progress [3]. The main problem is the lack of devices for effective real-time control of THz radiation. Due to the specificity of this range, it was possible to create fairly simple metamaterials with a negative refractive index [4] as well as the adaptive metasurfaces for deflecting and modulating THz radiation [5,6]. Their optical properties are highly correlated with conductivity changes which can be caused by impact of the laser radiation or electrical signals. The best results have been achieved by graphene-based devices: the modulation depth is close to 100% and the operation speed is approximately kHz.

Another method for radiation control is based on a well-known acousto–optic (AO) interaction. AO devices are widely used for real-time optical processing of information [7]. This is achieved via control of the parameters of the ultrasound wave, which forms a phase diffraction grating in the medium, resulting in high operation speed of the commercial AO devices in the order of MHz. An important feature is that the intensity of the diffracted radiation resonantly depends on the wavelength of the radiation. Therefore, AO devices can be used for spectrometric measurements. For example, in 2020, an AO spectrometer was developed for the ExoMars space mission, which makes it possible to estimate H_2O content [8]. AO devices can also be used as a phase modulator in frequency-modulation heterodyne spectroscopy, which allows the analysis of both absorption and dispersion properties of optical resonances [9]. Finally, significant progress has been made in AO-modulated diffuse correlation spectroscopy to monitor blood flow in tissues [10].

To work with visible radiation, ultrasound frequencies of about 100 MHz are used [11]. As the ultrasound transducer is an open resonator, its thickness h is inversely proportional to the resonance frequency F_{res} [12]:

$$F_{\text{res}} = \frac{V_{\text{PZT}}}{2h}, \tag{1}$$

where V_{PZT} is the sound velocity in the transducer material. At $V_{PZT} \approx 5$ km/s and $F_{res} = 100$ MHz, the thickness h of the transducer is only 50 μm. Therefore, a transducer with a width d of a few millimeters can be considered as thin. As it is known, thin transducers have oscillations of piston type and generate a homogeneous acoustic field.

Meanwhile, when working with THz radiation, it is necessary to take into account the strong divergence, which is inversely proportional to the wavelength λ. Therefore, wide radiation beams with a diameter of about $D = 1$ cm are used [13]. Obviously, with an ultrasound transducer with a width $d = 1 \div 2$ mm, most of the radiation will not interact with the ultrasonic beam. On the other hand, it is not reasonable to use an ultrasound transducer with a width much greater than D, as it leads to a significant decrease in the acoustic power density. Therefore, the optimal width of the transducer should be comparable the diameter of the radiation beam D.

The energy efficiency $\xi_{norm} = \xi / P_a$ of the AO modulator is determined as diffraction efficiency ξ per 1 W of acoustic power P_a [14]:

$$\xi_{1D} = \frac{I_1}{I_0} \approx \frac{\pi^2}{2\lambda^2} \frac{M_2 P_a}{d} L \exp(-\alpha_s l), \tag{2}$$

where index 1D means that it corresponds to a simple 1D model; I_1 is intensity of the diffracted radiation, whereas I_0 corresponds to the transmitted radiation; M_2 is the coefficient of the AO figure of merit of the interaction medium; α_s is the sound power attenuation coefficient; l is the distance from the sound transducer to the THz beam; and L is the length of the ultrasound transducer.

There are several works related to AO in the THz range [15–17]. In these works, AO interaction was investigated in the following media: single crystals of germanium (Ge) and gallium arsenide (GaAs), TPX-plastics, and non-polar liquids (saturated hydrocarbons and their derivatives, in which one or more hydrogen atoms are replaced by other chemical elements). However, the AO diffraction efficiency in these media was only a few hundredth of a percent per 1 W of driving electrical power. It has been found in [18] that the best medium for AO diffraction of THz radiation is liquefied sulfur hexafluoride SF_6. It is characterized by an AO figure of merit M_2 two orders of magnitude greater than the abovementioned media. However, this medium is characterized by significant attenuation of ultrasound ($\alpha_s \approx 1.6$ dB/cm at $F = 300$ kHz and temperature $t = 21°$ C), which is proportional to the square of the ultrasound frequency F. Therefore, the frequency has to be about 300 kHz, which corresponds to a transducer thickness of 6 mm. The width ($\approx$10 mm) of such a transducer is comparable with its thickness ($\approx$6 mm), and it can no longer be considered "thin" plate.

Previously, we have performed a series of experiments on the AO diffraction of THz radiation in the liquefied SF_6 using ultrasound transducers of various widths d (from 6 to 14 mm) operating at a frequency of about $F \approx 300$ kHz [14]. As a result, the dependence of the energy efficiency ξ_{norm} on the width d of the transducer was determined. According to the formula (2), with reducing the width d of the transducer, the diffraction efficiency ξ should increase by the law $\xi \propto 1/d$. It was experimentally established that this law is implemented only for relatively wide ultrasound transducers with a width d greater than 12 mm. However, for narrower transducers, an unexpected result was obtained: the energy efficiency sharply decreased with a reduction in the width of the transducer at $d \leq 10$ mm (see Table 1).

Table 1. Characteristics of the AO modulator of the THz radiation with fully electroded ultrasound transducer obtained in [14].

	d (mm)	**14**	**12**	**10**	**8**	**6**
model	ξ_{norm} (%/W)	0.35	0.40	0.48	0.48	0.48
fitting	ξ_{norm} (%/W)	0.25	0.28	0.35	0.35	0.35
experiment	ξ_{norm} (%/W)	0.23	0.31	0.0076	0.11	0.003

For wide transducers (which can be approximately considered "thin"), the experimental data are consistent with the model, while for narrow transducers (which are "thick") the data differ from the model. A literature review showed that complex mechanical deformations occur in "thick" transducers, resulting in a complex structure of the ultrasonic field [19,20]. Therefore, we associate the unusual dependence of $\xi_{\mathrm{norm}}(d)$ in [14] with the inhomogeneity of the acoustic field caused by complex oscillations of the ultrasound transducer.

In [21], it is shown that in order to achieve greater uniformity of the transducer vibrations, only part of its radiating surfaces should be covered with electrodes. Thus, it is expected that this will lead to a significant improvement in performance of the THz AO modulator. Previously, we used ultrasound transducers that were completely covered with electrodes [14]. Now, we have applied the method of increasing the homogeneity of the acoustic field, proposed in [21]. In this paper, we present a more general model of the AO modulation as well as the results of experiments with transducers whose radiating surfaces were partially covered by electrodes.

In addition to improving the efficiency of AO modulation of THz radiation, another problem is the implementation of THz AO deflector. For effective AO deflection of THz radiation, the Bragg matching condition ($\eta = 0$) must be satisfied in a wide range of ultrasound frequency F (see Figure 1) [22]:

$$\mathbf{k_1} = \mathbf{k_0} + \mathbf{K} + \boldsymbol{\eta}, \quad (3)$$

where $\boldsymbol{\eta}$ is the mismatch vector, $\mathbf{K}$ is the wave vector of the ultrasound, and $\mathbf{k_0}$ and $\mathbf{k_1}$ are the wave vectors of the incident and diffracted radiation, respectively.

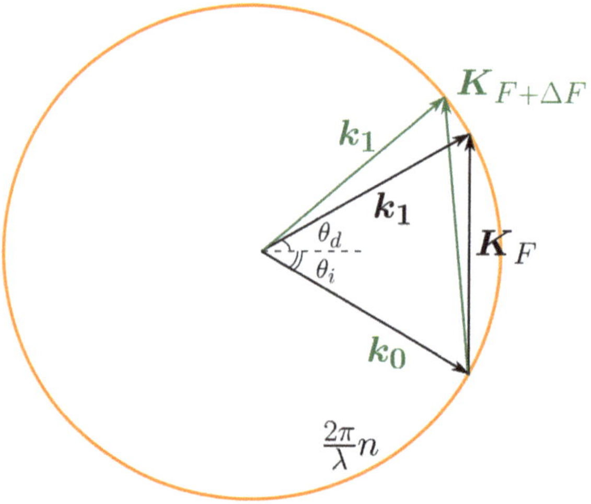

Figure 1. Wave vector diagram of AO diffraction on ultrasound with frequency F (black) and $F + \Delta F$ (green).

The mismatch vector $\boldsymbol{\eta}$ is perpendicular to the boundary of the sound column from the end of vector $\mathbf{K}$ to the radiation wave surface [23]. Therefore, as shown in Figure 1, the sound wave vector $\mathbf{K}$ must rotate with the frequency F so that its end touches the radiation wave surface. In [23,24], it was shown that a sectioned ultrasonic transducer can be used to solve this problem in the visible range. Earlier in [14], an ultrasound transducer completely covered with electrodes was used, which was only suitable for effective modulation, but not for broadband deflection of THz radiation. In this work, we theoretically investigate the operation of the THz AO deflector with a sectioned ultrasonic transducer with the aim of increasing the number of resolvable spots.

2. Theoretical Background

2.1. THz Acousto–Optic Modulator

The AO diffraction efficiency depends on the ratio between the width of the sound column and the diameter of the incident radiation beam (see Figure 2a). Let the width of the sound beam be equal to the width d of the region of the sound transducer covered by the electrode and the diameter D of the radiation beam be limited by the diaphragm (see Figure 2b).

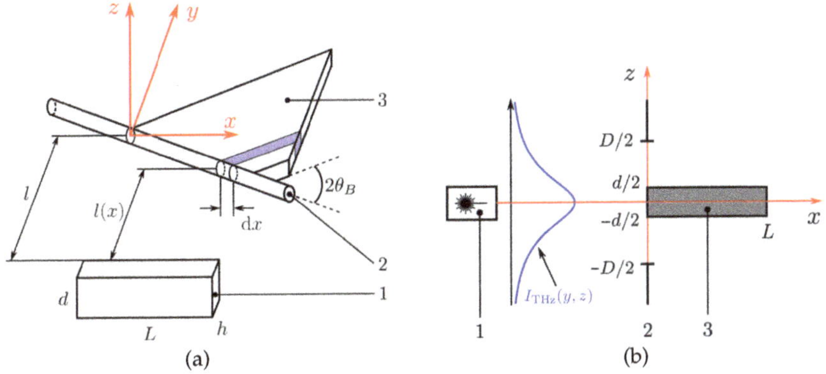

Figure 2. Schematic of AO interaction: (**a**) 1—ultrasound transducer, 2—transmitted radiation beam, 3—diffracted radiation beam; (**b**) 1—radiation source, 2—diaphragm, 3—region of AO interaction.

The intensity distribution over the cross section of the radiation beam for most sources has the form of the Gaussian function:

$$I_{\mathrm{THz}}(y,z) = I_s \frac{1}{\pi W^2} \exp\left(-\frac{y^2+z^2}{W^2}\right). \tag{4}$$

The integral intensity I_0 of the transmitted beam can be calculated by the following relation:

$$I_0 = \exp(-\alpha L) \int_{-\infty}^{+\infty} \mathrm{d}y \int_{-D/2}^{D/2} I_{\mathrm{THz}}(y,z)\mathrm{d}z = I_s \mathrm{erf}\left(\frac{D}{2W}\right) \exp(-\alpha L). \tag{5}$$

The integral intensity I_1 of the diffracted beam is limited by the smallest quantity $z_{\max}$ among d and D:

$$z_{\max} = \begin{cases} d & \text{for } d \le D; \\ D & \text{for } d > D. \end{cases} \tag{6}$$

To calculate the intensity of diffracted radiation, we divided the path of the radiation beam into small intervals $\mathrm{d}x$. The radiation on each is diffracted in accordance with Formula (2), with $\mathrm{d}x$ substituted for L. The oblique propagation of the radiation leads to a slight change in the length l in (2): $l - \tan\theta_{\mathrm{B}} x$. Finally, the radiation beam is not infinitely narrow, which must also be taken into account: $l(x) = l - \tan\theta_{\mathrm{B}} x - y$.

Thus, the integral intensity of the radiation diffracted on the ultrasonic field with plane wavefront can be calculated as follows:

$$I_1 = \frac{\pi^2}{2\lambda^2} \frac{M_2 P_{\mathrm{a}}}{d} \int_{-z_{\max}/2}^{z_{\max}/2} \mathrm{d}z \int_{-\infty}^{+\infty} \mathrm{d}y \int_0^L I_{\mathrm{THz}}(y,z) \exp[-\alpha_s (l - \tan\theta_{\mathrm{B}} x - y)]\mathrm{d}x. \tag{7}$$

The following expression for I_1 was obtained:

$$I_1 = I_s \frac{\pi^2}{2\lambda^2} \frac{M_2 P_{\mathrm{a}}}{d} \frac{\mathrm{erf}(z_{\max}/2W) \exp(-\alpha_s l + \alpha_s^2 W^2/4)[\exp(\alpha_s L \tan\theta_{\mathrm{B}}) - 1]}{\alpha_s \tan\theta_{\mathrm{B}}}. \tag{8}$$

Let us estimate the Bragg angle for AO diffraction of THz radiation in liquefied SF_6. Assume that the frequency of ultrasound is 300 kHz, the speed of sound—$V = 300$ m/s, and the radiation wavelength—$\lambda = 130$ μm. Under these conditions, the Bragg angle is only a few degrees, which corresponds to a quasi-orthogonal geometry of the AO interaction [25]:

$$\theta_B \approx \frac{\lambda F}{2V} \approx 4^\circ. \tag{9}$$

Since the Bragg angle θ_B is much less than unity, Relation (8) can be expanded into the Taylor series:

$$I_1 = I_s \frac{\pi^2}{2\lambda^2} \frac{M_2 P_a}{d} L \left(1 + \frac{\alpha_s L \theta_B}{2}\right) \mathrm{erf}\left(\frac{z_{\max}}{2W}\right) \exp\left(-\alpha_s l + \frac{\alpha_s^2 W^2}{4}\right). \tag{10}$$

The diffraction efficiency can now be calculated as the ratio of the integral intensity of the diffracted beam to the integral intensity of the transmitted beam:

$$\xi = \frac{I_1}{I_0} = \begin{cases} \xi_{1D}\left(1 + \frac{\alpha_s L \theta_B}{2}\right) \exp\left(\frac{\alpha_s^2 W^2}{4}\right) \mathrm{erf}\left(\frac{d}{2W}\right) / \,\mathrm{erf}\left(\frac{D}{2W}\right) & \text{for } d \le D; \\ \xi_{1D}\left(1 + \frac{\alpha_s L \theta_B}{2}\right) \exp\left(\frac{\alpha_s^2 W^2}{4}\right) & \text{for } d > D. \end{cases} \tag{11}$$

It follows from (11) that when the electrode width d is greater than the diameter D of the radiation beam incident on the AO cell, the integral efficiency ξ of AO diffraction decreases in inverse proportion to d. This can be explained by the decrease in the acoustic power density, which is inversely proportional to the area of the radiating surface of the ultrasound transducer. At the same time, it should be emphasized that when $d < D$, the AO diffraction efficiency does not depend on the sound beam width d. Note that according to the simple relation (2), the AO diffraction efficiency should be propotional to the acoustic power density. However, as can be seen from (11), this is not the case. This can be explained by the fact that the efficiency of AO diffraction increases, but locally, only in the region of AO interaction, and part of the radiation beam does not interact with the ultrasound. Let us consider an illustrative example. Suppose the initial transducer width d is equal to the diameter of the incident radiation beam. If we decrease d by a factor of two, the diffraction efficiency in the AO interaction region will increase by the same factor, but only half of the light beam will interact with the sound. Obviously, the influence of these factors is equal but opposite. Together, they balance each other, and the integral efficiency of AO diffraction will not change.

2.2. THz Acousto–Optic Deflector

The theory of AO beam steering using a sectioned ultrasonic transducer was developed in [26]. A general expression for the diffraction efficiency ξ was obtained for an even number m of the transducer sections. The simplest way to realize AO beam steering is to apply electrical signal to each section with a phase shift π. The operating point θ_i (see Figure 1) of the AO deflector in terms of the angle of incidence was chosen in [26] with the aim of maximizing the number of resolved light spots N. However, in that case there was a dip (of 50% of the maximum value) in the center of the frequency response $\xi(F)$. Our task is to choose the operating point θ_i for the diffraction efficiency to be independent on the deflection angle (when the dip in $\xi(F)$ can be neglected).

We started with calculating the frequency response for various angles of incidence θ_i in order to reveal the dynamics of the changes in $\xi(F)$. The following values were used: $\lambda = 130$ μm, $V = 300$ m/s, $n = 1.2$, $P_a = 1$ W, $M_2 = 15{,}000 \times 10^{-15}$ s^3/kg, $L = 8$ cm, $d = 1$ cm, $m = 8$. The results are shown in Figure 3.

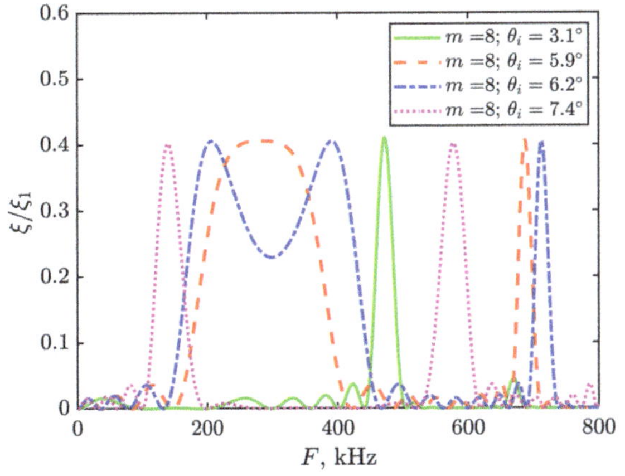

Figure 3. Frequency response of THz AO deflector having an ultrasound transducer with 8 sections for different operating points.

At an arbitrary angle of incidence, the AO diffraction occurs in a narrow frequency bandwidth ΔF (green curve in Figure 3), which limits the number of resolvable spots N. However, with a correct choice of operating point θ_i, a wide resonance occurs in a low frequency range, which is accompanied by the narrow peak shift to higher frequencies (red curve in Figure 3). A further increase in the angle of incidence leads to a slight shift in the central frequency F_d and to the dip presence on F_d (blue curve in Figure 3). Finally, the resonance curve splits into two substantially narrower ones (purple curve in Figure 3).

The analysis showed that the AO deflector mode is realized under the following conditions:

$$\theta_i \approx \sqrt{m\frac{\lambda}{nL}}; \qquad \theta_d = 0; \qquad F_d \approx V\sqrt{m\frac{n}{\lambda L}}, \tag{12}$$

whereas the number of resolved spots N depends on the quick action $\tau = V/D$ [27] and can be estimated as follows:

$$N = \Delta F\frac{D}{V} = 1.9\sqrt{\frac{n}{\lambda L}}D. \tag{13}$$

When using an ultrasonic transducer completely covered with an electrode, the number of resolved spots $N_1 = 1.8nVD/\lambda FL$ is inversely proportional to the AO interaction length L [28]. At the same time, according to (13), with a sectioned ultrasonic transducer, $N \propto 1/\sqrt{L}$. Therefore, an AO deflector based on the optically isotropic medium should be equipped with a sectioned ultrasonic transducer. According to our estimates for the parameters given above, for a THz radiation beam with a diameter of 1 cm, the number of resolved spots can be increased up to 3 times: from $N_1 = 2$ to $N \approx 6.5$.

3. Experimental Results and Discussion

For the experimental investigation of the AO modulation of the THz radiation in liquefied SF_6, a specialized cuvette was used. A set of ultrasound transducers was made of CTS-19 piezoceramics in the form of rectangular parallelepipeds with dimensions of 6 mm × 14 mm × 80 mm (6 mm being the thickness of the transducer, corresponding to a resonant frequency of about 300 kHz). The electrodes were placed on both 14 × 80 mm surfaces of the transducer. The length of the electrodes was $L = 80$ mm, while the width d was 6, 8, 10, and 12 mm (see Figure 4).

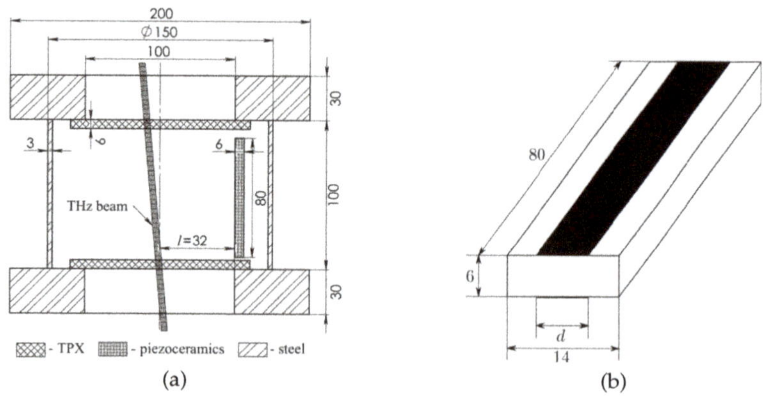

Figure 4. Schematic of THz AO modulator: (**a**) AO cell; (**b**) ultrasound transducer.

The Novosibirsk free-electron laser (FEL) *1* was used as a source of monochromatic THz radiation with the wavelength $\lambda = 130$ μm (see Figure 5). The THz beam was limited by iris diaphragm *2* with the hole diameter $D = 7$ mm. The radiation deflected by AO cell *3* was focused by lens *4* onto the radiation detector *5* (Golay cell) that has been oriented to achieve the maximum signal. The amplitude of electrical signal with frequency of about $F = 300$ kHz from RF generator *6* was modulated with a frequency 10 Hz. Therefore, the diffracted radiation had the same amplitude modulation increasing the signal-noise ratio, as the Golay cell has the maximum sensitivity at 10–15 Hz. The diffracted radiation intensity I_1 was proportional to the signal from lock-in detector *7*. Since the diffraction efficiency was about $\xi \approx 0.1\%$, the mechanical obturator and calibrated optical attenuator (not shown in Figure 5) at the input of the AO cell were employed for determination of the transmitted radiation intensity I_0. The experimental results are shown in Figure 6.

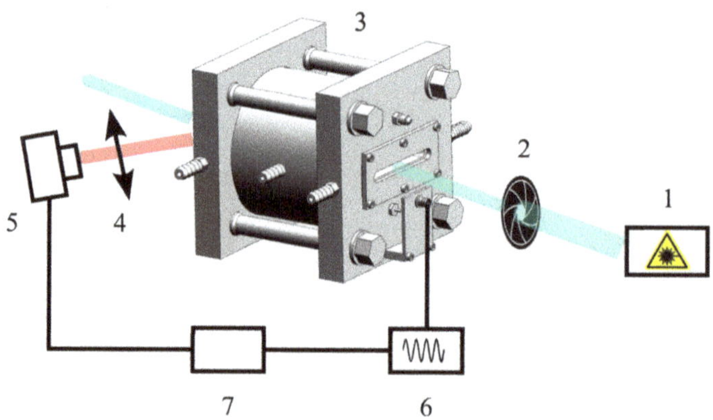

Figure 5. Schematic of experimental setup: 1—radiation source; 2—diaphragm; 3—AO deflector; 4—lens; 5—radiation detector; 6—RF generator; 7—lock-in detector.

For each ultrasound transducer, the optimal frequency F_{res} corresponding to the maximum value $\xi_{norm} = \xi / P_a$ of diffraction efficiency per 1 W of the driving electric power was determined. The data for the angular $\Delta\theta$ and frequency ΔF bandwidth, as well as F_{res} and ξ_{norm}, are summarized in Table 2. The experimental error in diffraction efficiency ξ_{norm} was mainly related to the FEL intensity instability and was determined from a 1-minute data sample.

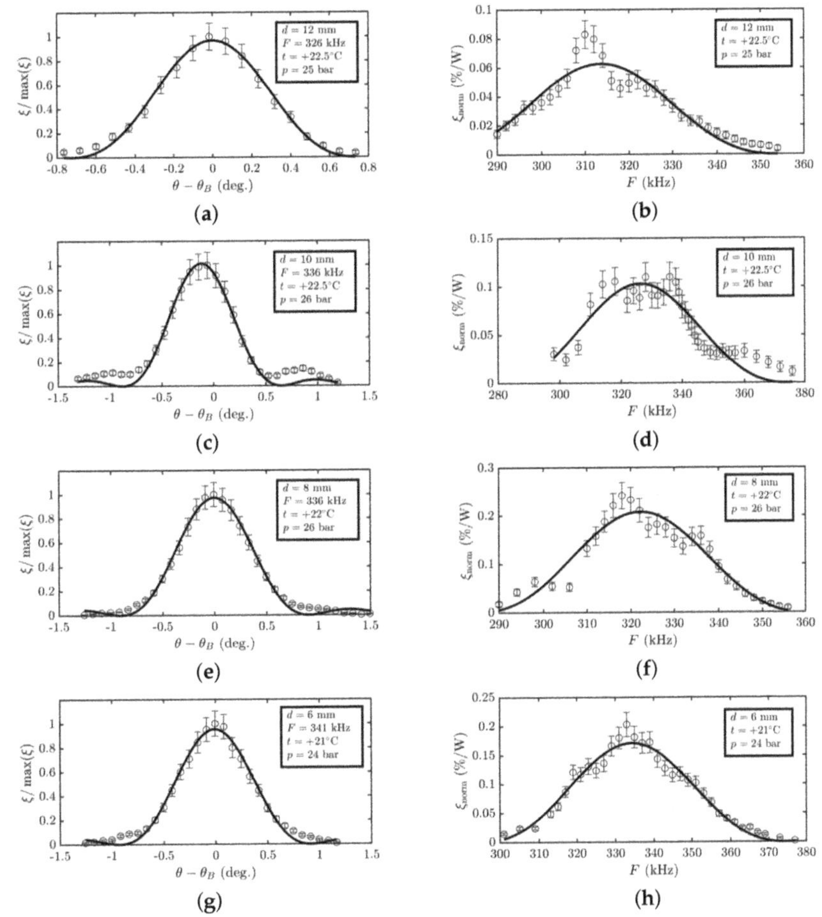

Figure 6. Experimental results: (**a**,**c**,**e**,**g**) AO diffraction efficiency $\xi/\max(\xi)$ vs. difference between angle θ of incidence of THz radiation on AO cell and Bragg angle θ_B; (**b**,**d**,**f**,**h**) frequency dependence of AO diffraction efficiency per 1 W of applied electric power.

Table 2. Characteristics of AO modulator of THz radiation revealed experimentally.

d (mm)	t (°C)	p (bar)	ξ_{norm} (%/W)	$\Delta\theta$ (deg)	ΔF (kHz)	F_{res} (kHz)
12	+22.5	25	0.08 ± 0.01	0.64 ± 0.03	35 ± 4	313.6 ± 1.4
10	+22.5	26	0.11 ± 0.01	0.68 ± 0.04	42 ± 5	326.3 ± 2.1
8	+22.0	26	0.21 ± 0.02	0.80 ± 0.02	33 ± 3	322.2 ± 1.3
6	+21.0	24	0.18 ± 0.02	0.82 ± 0.03	35 ± 2	334.4 ± 0.8

According to (1), the resonance frequency F_{res} of the sound transducer depends only on its thickness h. However, all the transducers had the same thickness $h = 6$ mm and different resonance frequencies F_{res}. This is related to the fact that the transducer width (14 mm) and the electrode width (from 6 to 12 mm) were of the same order as the transducer thickness (6 mm), which resulted in complex deformations. Therefore, a simple model (1) of a piston-type transducer allows only an estimation of the value of the resonance frequency F_{res}. A similar dependence $F_{res}(d)$ was revealed in [14] for fully electroded transducers with different widths d. The energy efficiency ξ_{norm} of the AO modulator is proportional to the AO figure of merit M_2 of liquefied SF_6, which depends on its pressure

p and temperature t. It should be noted that the experimental conditions were slightly different for each ultrasound transducer. Modelling was performed not for constant p and t, but for the experimental conditions. Therefore, the calculated dependence $\xi_{norm}(d)$ differs from $\xi \propto 1/d$ (11). The experimental data were fitted by the theoretical dependence reduced by a constant factor (see Figure 7).

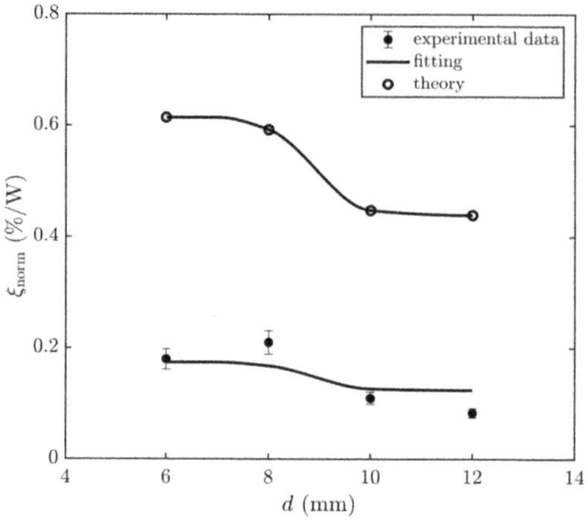

Figure 7. Dependence of energy efficiency of AO modulator on electrode width d of ultrasound transducer.

The experimentally determined diffraction efficiencies for all the ultrasound transducers were approximately three times lower than that predicted by the theory. So, one can admit the presence of a negative factor (for example, heat losses) in the set of transducers partially covered by electrodes. The physical mechanism can be revealed by modelling complex deformations of the ultrasound transducer. This problem can be solved, for example, in COMSOL multiphysics.

The fitted curve intersects only one of four error bar regions of the experimental data (see Figure 7). Nevertheless, there is a qualitative agreement of the model with all experimental data. At the same time, in [14] the model was valid only for $d = 14$ mm and $d = 12$ mm, whereas for $d = 10$ mm and $d = 8$ mm the difference between the model and experiment was several times, and for $d = 6$ mm this difference was of two orders of magnitude (see Table 1). Therefore, the results of [14] are in fact the first attempt at the experiment, while the current results are in good agreement with the more general theory.

In [14], the diffraction efficiency decreased with a diminishing of the transducer width d for $d < 12$ mm. In the current work, we experimentally found that the diffraction efficiency increased with a decrease in the electrode width d. For example, for $d = 8$ mm, the diffraction efficiency became two times higher relative to [14], and for $d = 6$ mm, this difference rose to 60 times. The main reason for this difference is the complex structure of the ultrasonic field, as the theory is only valid for an ultrasonic field with plane wavefront. Therefore, we can conclude that ultrasound transducers partially covered by electrodes generate a more homogeneous ultrasonic field and have a great potential in the THz acousto–optics. We also plan to equip the AO cell with the sectioned ultrasound transducer and to experimentally investigate the wide-angle AO deflection. For this purpose, the same electrical signal will be applied to neighboring sections with a phase shift π. This is the simplest design, and according to our estimations, it will enable one to increase the number of resolved spots of the AO deflector by several times.

4. Conclusions

The characteristics of the THz AO deflector with sectioned transducer have been estimated for the first time. It was shown that the use of a sectioned ultrasound transducer will increase the number of resolvable spots of the acousto–optic deflector by 3–4 times. It has been demonstrated that the ultrasound transducers partially covered by the electrodes can generate a more homogeneous acoustic field at 300 kHz compared with fully electroded transducers. The AO diffraction in liquefied SF_6 was investigated employing ultrasound transducers of this type. The diffraction efficiency for an electrode width of 6 mm was 0.18 (%/W), which was two times higher than that for a 12 mm electrode width. Moreover, with an electrode width of 6 mm and transducer width of 14 mm, the diffracted radiation intensity was increased by more than an order of magnitude, compared with fully electroded transducers with the width of 6 mm. This fact enables one to increase the acoustic power density and thus to achieve higher energy efficiency of the THz AO devices.

Author Contributions: Conceptualization, P.A.N.; methodology, P.A.N.; software, P.A.N.; investigation, P.A.N. and V.V.G.; resources, V.V.G. and I.S.K.; writing—original draft preparation, I.S.K.; writing—review and editing, V.V.G.; supervision, P.A.N.; project administration, P.A.N. All authors have read and agreed to the published version of the manuscript.

Funding: This research was funded by the Ministry of Science and Higher Education of the Russian Federation under State contract No. FFNS-2022-0009.

Institutional Review Board Statement: Not applicable.

Informed Consent Statement: Not applicable.

Data Availability Statement: The data presented in this study are available on request from the corresponding author. Because of the further research, the data are not publicly available.

Acknowledgments: The work was done at the shared research center SSTRC on the basis of the Novosibirsk FEL at BINP SB RAS with the use of equipment of the Shared Research Facilities of the Scientific and Technological Centre of Unique Instrumentation of the Russian Academy of Sciences (STC UI RAS).

Conflicts of Interest: The authors declare no conflict of interest.

Abbreviations

The following abbreviations are used in this manuscript:

AO	acousto-optic
THz	terahertz
FEL	free-electron laser

References

1. Rahaman, M.; Bandyopadhyay, A.; Pal, S.; Ray, K. Reviewing the scope of THz communication and a technology roadmap for implementation. *IETE Tech. Rev.* **2020**, *38*, 465–478. [CrossRef]
2. O'Sullivan, C.; Murphy, J. *Field Guide to Terahertz Sources, Detectors, and Optics*; SPIE: Bellingham, WA, USA, 2012; p. 144. [CrossRef]
3. Jianjun, Y.; Yi, W. Digital signal processing for high-speed THz communications. *Chin. J. Electron.* **2022**, *31*, 534–546. [CrossRef]
4. Smith, D.; Pendry, J.; Wiltshire, M. Metamaterials and negative refractive index. *Science* **2004**, *305*, 788–792. [CrossRef] [PubMed]
5. Ma, Z.; Geng, Z.; Fan, Z.; Liu, J.; Chen, H. Modulators for terahertz communication: The current state of the art. *Research* **2019**, *2019*, 1–22. [CrossRef] [PubMed]
6. Chen, B.; Wang, X.; Li, W.; Li, C.; Wang, Z.; Guo, H.; Wu, J.; Fan, K.; Zhang, C.; He, Y.; et al. Electrically addressable integrated intelligent terahertz metasurface. *Sci. Adv.* **2022**, *8*, eadd1296. [CrossRef]
7. Zhang, Y.; Fan, H.; Poon, T.C. Optical image processing using acousto-optic modulators as programmable volume holograms: A review. *Chin. Opt. Lett.* **2022**, *20*, 021101. [CrossRef]
8. Dobrolenskiy, Y.; Mantsevich, S.; Evdokimova, N.; Korablev, O.; Fedorova, A.; Kalinnikov, Y.; Vyazovetskiy, N.; Titov, A.; Stepanov, A.; Sapgir, A.; et al. Acousto-optic spectrometer ISEM for ExoMars-2020 space mission: ground measurements and calibrations. In Proceedings of the Fourteenth School on Acousto-Optics and Applications, Torun, Poland, 24–27 June 2019; Grulkowski, I., Linde, B.B.J., Duocastella, M., Eds.; SPIE: Bellingham, WA, USA, 2019; Volume 11210, p. 112100F. [CrossRef]
9. Baryshev, V.; Epikhin, V. Compact acousto-optic modulator operatingin the purely Raman-Nath diffraction regime as a phase modulator in FM spectroscopy. *Quantum Electron.* **2010**, *40*, 431–436. [CrossRef]

10. Robinson, M.; Carp, S.; Peruch, A.; Boas, D.; Franceschini, M.; Sakadžić, S. Characterization of continuous wave ultrasound for acousto-optic modulated diffuse correlation spectroscopy (AOM-DCS). *Biomed. Opt. Express* **2020**, *11*, 3071. [CrossRef]
11. Korablev, O.; Belyaev, D.; Dobrolenskiy, Y.; Trokhimovskiy, A.; Kalinnikov, Y. Acousto-optic tunable filter spectrometers in space missions. *Appl. Opt.* **2018**, *57*, C103–C119. [CrossRef]
12. Ishikawa, M.; Kurosawa, M.; Endoh, A.; Takeuchi, S. Lead zirconate titanate thick-film ultrasonic transducer for 1 to 20 MHz frequency bands fabricated by hydrothermal polycrystal Growth. *Jpn. J. Appl. Phys.* **2005**, *44*, 4342–4346. [CrossRef]
13. Kubarev, V.; Sozinov, G.; Scheglov, M.; Vodopyanov, A.; Sidorov, A.; Melnikov, A.; Veber, S. The radiation beamline of Novosibirsk free-electron laser facility operating in terahertz, far-infrared, and mid-infrared ranges. *IEEE Trans. Terahertz Sci. Technol.* **2020**, *10*, 634–646. [CrossRef]
14. Nikitin, P.; Gerasimov, V. Optimal design of an ultrasound transducer for efficient acousto-optic modulation of terahertz radiation. *Materials* **2022**, *15*, 1203. [CrossRef] [PubMed]
15. Durr, W.; Schmidt, W. Measurement of acousto-optic interaction in Germanium in the far infrared. *Int. J. Infrared Millim. Waves* **1985**, *6*, 1043–1049. [CrossRef]
16. Vogel, T.; Dodel, G. Acousto-optic modulation in the far-infrared. *Infrared Phys.* **1985**, *25*, 315–318. [CrossRef]
17. Nikitin, P.; Voloshinov, V.; Gerasimov, V.; Knyazev, B. Acousto-optic modulation and deflection of terahertz electromagnetic radiation in nonpolar liquids. *Tech. Phys. Lett.* **2017**, *43*, 635–637. [CrossRef]
18. Durr, W. Acousto-optic interaction in gases and liquid bases in the far infrared. *Int. J. Infrared Millim. Waves* **1986**, *7*, 1537–1558. [CrossRef]
19. Zhang, Y.; Du, J. Influence of boundary conditions on three-dimensional vibration characteristics of thick rectangular plates. *Sci. Prog.* **2020**, *103*, 0036850420969548. [CrossRef]
20. Khorsheed, S.; Yaseen, N. Effect of square shaped acousto-optic modulators on the Bragg diffraction. *J. Phys. Conf. Ser.* **2021**, *1897*, 012074. [CrossRef]
21. Imano, K. Use of energy trapping type piezoelectric transducer to suppress lateral vibration in the transducer. *IEICE Electron. Express* **2019**, *16*, 20190478. [CrossRef]
22. Kupreychik, M.; Balakshy, V. The spatial structure of acousto-optic phase matching in biaxial crystal of alpha-iodic acid. *Opt. Spectrosc.* **2017**, *123*, 463–470. [CrossRef]
23. Balakshy, V.; Kupreychik, M.; Mantsevich, S.; Molchanov, V. Acousto-optic cells with phased-array transducers and their application in systems of optical information processing. *Materials* **2021**, *14*, 451. [CrossRef] [PubMed]
24. Antonov, S.; Proklov, V.; Rezvov, Y.; Vainer, A. Two-element phased-array acousto-optic deflector at high diffraction efficiency: Scanning range broadening. *Univers. J. Phys. Appl.* **2014**, *8*, 90–95. [CrossRef]
25. Kupreychik, M.; Balakshy, V. Peculiarities of acousto-optic interaction in biaxial crystal of alpha-iodic acid. *Appl. Opt.* **2018**, *57*, 5549–5555. [CrossRef] [PubMed]
26. Gordon, E. A review of acoustooptical deflection and modulation devices. *Proc. IEEE* **1966**, *54*, 1391–1401. [CrossRef]
27. Reddy, G.; Saggau, P. Fast three-dimensional laser scanning scheme using acousto-optic deflectors. *J. Biomed. Opt.* **2005**, *10*, 064038. [CrossRef]
28. Kludzin, V.; Kulakov, S.; Molotok, V. Acousto-optic materials and acousto-optic cell applications. In Proceedings of the SPIE's 1996 International Symposium on Optical Science, Engineering, and Instrumentation, Denver, CO, USA, 5–9 August 1996; SPIE: Bellingham, WA, USA, 1996; Volume 2849. [CrossRef]

materials

Article

Temperature Effects in an Acousto-Optic Modulator of Terahertz Radiation Based on Liquefied SF_6 Gas

Pavel A. Nikitin [1,*], Vasily V. Gerasimov [2,3] and Ildus S. Khasanov [1]

[1] Scientific and Technological Centre of Unique Instrumentation RAS, 117342 Moscow, Russia; khasanov@ntcup.ru
[2] Department of Physics, Novosibirsk State University, 630090 Novosibirsk, Russia; v.v.gerasimov3@gmail.com
[3] Budker Institute of Nuclear Physics SB RAS, 630090 Novosibirsk, Russia
* Correspondence: nikitin.pavel.a@gmail.com

Abstract: The acousto-optic (AO) diffraction of terahertz (THz) radiation in liquefied sulfur hexafluoride (SF_6) was investigated in various temperature regimes. It was found that with the increase in the temperature from +10 to +23 °C, the efficiency of the AO diffraction became one order higher at the same amplitude of the driving electrical signal. At the same time, the efficiency of the AO diffraction per 1 W of the sound power as well as the angular bandwidth of the efficient AO interaction were temperature independent within the measurement error. Increase of the resonant sound frequency with decreasing temperature and strong narrowing of the sound frequency bandwidth of the efficient AO interaction were detected.

Keywords: acousto-optic interaction; terahertz radiation; diffraction; liquefied inert gas

Citation: Nikitin, P.A.; Gerasimov, V.V.; Khasanov, I.S. Temperature Effects in an Acousto-Optic Modulator of Terahertz Radiation Based on Liquefied SF_6 Gas. *Materials* **2021**, *14*, 5519. https://doi.org/10.3390/ma14195519

Academic Editors: Haim Abramovich and Antonio Politano

Received: 20 July 2021
Accepted: 17 September 2021
Published: 23 September 2021

1. Introduction

Acousto-optical (AO) devices have found wide application in optical information processing in real time and are actively used in fiber-optics, tomography, microscopy, and astronomy [1–4]. They have proven to be effective in the ultraviolet, visible, and infrared spectral ranges. The AO interaction is based on the photo-elastic effect, due to which the sound wave forms a phase diffraction grating in the medium. Birefringent single crystals usually serve as the interaction medium in the above mentioned ranges [5]. Most of the modern AO devices operate under normal conditions and provide efficiency close to 100%. The development of efficient AO filters and deflectors requires the use of birefringent crystals. This makes it possible to improve the characteristics of these devices by several orders of magnitude in comparison with analogs based on optically isotropic media [6,7]. Unfortunately, in the terahertz (THz) range, birefringent crystals are practically opaque, while optically isotropic crystals (for example, germanium) feature a lower value of the AO figure of merit [8]. Therefore, studies of the AO effect in the THz range have not been carried out for many years and were considered unpromising. However, the growing interest in THz radiation necessitated search for a suitable medium for AO interaction. It was found that inert gases (for example, Xe or Kr) can be such a medium since they are transparent to THz radiation and have good AO properties [9]. Unfortunately, ultrasonic attenuation is large in the gases even under high pressure. In the work [9] it was shown that the ultrasound attenuation in the liquid phase of CF6 is less than in the gaseous one. For this reason, experiments were carried out in work [9] with various inert gases in the liquid phase and the AO diffraction efficiency of about 80% was obtained. It was found that the best suitable medium was liquefied sulfur hexafluoride (SF_6), the density of which is about 1.5 times that of water, whereas its sound velocity is only about 300 m/s. Due to this rare combination of properties, liquefied SF_6 features an AO figure of merit one order higher than that of the best birefringent crystal (paratellurite, TeO_2) used in the visible range [10]. Liquefied SF_6 is an optically isotropic medium making it possible to manufacture only commercial AO modulators of powerful THz radiation. The prime cost of such modulators is low and is determined mainly by the manufacturing cost of a cuvette for liquefied gas. The advantages of AO modulators of THz radiation are: (1) low power consumption (several tens of watts); (2) high operating speed (about 10 μs); (3) spatial separation of the diffracted radiation beam from the transmitted one.

Research in work [9] was carried out at the temperature of +13 °C and +14 °C. It was found that the temperature increase of only 1 degree leads to the diffraction efficiency falling about three times at the same amplitude of the electrical signal. The author cannot provide an explanation of the effect discovered. The experiment we performed at a temperature of +24 °C gave only a qualitative confirmation of the model, which takes into account the effect of temperature on the physical properties of SF_6 [11]. We are confident that this discrepancy is due to the fact that the model does not take into account the structure of the sound field. If we assume that the ultrasound transducer acts like a piston, modeling of the structure of the sound beam can be performed even on low-power computers, using the expansion of the solution into Gaussian functions with known weight factors [12].

Note that because of the ultrasound attenuation in SF_6, it is necessary to use low frequencies of ultrasound. In works [9,11] it was about 300 kHz. The choice of the frequency of 300 kHz is due to the need to deflect the diffracted beam at an angle much larger than the divergence angle of this beam, which is about 1 degree for a beam diameter of 10 mm and a wavelength of 130 μm. At the ultrasound frequency of 300 kHz, the deflection angle is about 8 degrees, which is sufficient for its spatial separation from the transmitted radiation beam. The use of higher frequencies is not advisable because of the high attenuation of ultrasound, which is proportional to the square of the frequency. The resonant frequency F_0 is determined by the thickness h of the sound transducer and sound velocity V: $F_0 = 2h/V$. Accordingly, the thickness of the transducer ($h = 6$ mm at $F_0 = 300$ kHz) was comparable to its width, which was chosen close to the diameter of the radiation beam and was about 10 mm. For this reason, complex types of oscillations arise in the ultrasound transducer, and the distribution of the ultrasound amplitude on the surface of the transducer cannot be considered uniform. Modeling the oscillations of a rectangular transducer with its thickness close to its width is a difficult task, and due to the difference in the technology of manufacturing the transducers, the result of such modeling may differ significantly from the real state of affairs. We assume that the difference in the results of works [9,11] is associated with the use of different transducers. The motivation of the work is to obtain correct results. We are confident that it is necessary to carry out a series of experiments at different temperatures, but using the same ultrasound transducer. Therefore, in this work, we focused our efforts on the experimental study of temperature effects in AO modulators of THz radiation.

2. Materials and Methods

We used radiation of the Novosibirsk free electron laser 1 (FEL) with the wavelength $\lambda = 130$ μm, the polarization of which was controlled by wire polarizer 2, and the intensity was set by calibrated attenuators (see Figure 1). The radiation beam was incident on the center of the input optical window of the AO cell at the distance $l = 5$ cm from the sound transducer. For observation of the diffracted radiation, the AO cell 4 was tilted at the Bragg angle, and the signal (modulated at a frequency of 10 Hz) from generator of electrical signals 5 was applied to the ultrasound transducer via amplifier 6. At a distance of about 30 cm after the AO cell, lens 7 was located, focusing the radiation into the receiver, Golay cell 8. The signal from the receiver was isolated from the background noise using lock-in amplifier 9.

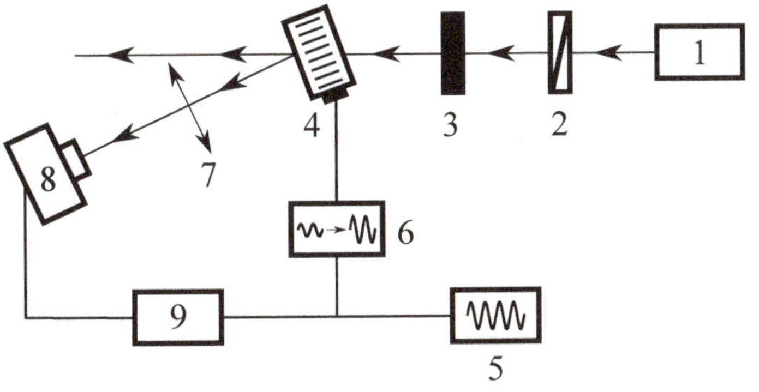

Figure 1. Schematic diagram of experimental setup: 1—FEL; 2—wire-grid polarizer; 3—set of calibrated radiation attenuators; 4—AO cell; 5—signal generator; 6—electrical amplifier; 7—lens; 8—Golay cell; 9—lock-in amplifier.

At the temperature of about +20 °C, SF_6 gas liquefies at a pressure of about 20 bar. Therefore, the AO cell body was built from high strength steel in the form of a cylinder (see Figure 2). A detailed description of the AO cell can be found in work [11]. We present only some of the modernizations of the AO cell during the preparation for this work. To generate a more uniform sound beam, we used a sound transducer ($d = 14$ mm wide and $L = 80$ mm long) made of piezoceramics PZT-19. To prevent the diffraction of THz radiation on the reflected sound wave, we disposed a duralumin plate (at a distance of about 10 cm) at an appropriate angle opposite the transducer. A closed loop water cooling system was used. Water came from a tank with a mixture of water with ice, then it passed between the double concentric walls of the AO cell, and returned back to the tank. Dry nitrogen blowing prevented water condensation on the outside of the optical windows.

Figure 2. Inside view of AO cell: 1—external wall; 2—internal wall; 3—optical window; 4—sound transducer holder; 5—sound transducer; 6—sound reflector; 7—SF_6 gas inlet/outlet; 8—cooling water inlet/outlet.

The optimal turn angle of the AO cell and the optimal frequency of ultrasound were determined as a result of sequential adjustment of the ultrasound frequency and the angle of incidence of THz radiation on the AO cell to achieve the maximum intensity I_1 of the diffracted radiation. In this mode, the dependence of the intensity I_1 on the square of the amplitude U of the driving electrical signal was investigated. The operating point was chosen on the linear section of the obtained dependence. The transducer resonant frequency varies with temperature. Therefore, at a given temperature, we determined the optimal frequency at which the AO diffraction efficiency reached its maximum value. Then the dependence of the intensity I_1 on the angle θ of the AO cell turn was measured. Further, the dependence of the intensity I_1 on the frequency F of the electrical signal was recorded. The described cycle of experiments was carried out for different temperature regimes.

To interpret and compare the results, as a measure of the efficiency of the AO diffraction, we used the ratio of the intensity I_1 of the diffracted radiation to the intensity I_0 of the radiation transmitted through the AO cell. As a result, the diffraction efficiency ξ and diffraction efficiency ξ_{norm} per 1 W of the electric power P_{el} were determined [13]:

$$\xi = \frac{I_1}{I_0}, \quad \xi_{\text{norm}} = \frac{\xi}{P_{\text{el}}}, \tag{1}$$

$$P_{\text{el}} = \frac{U^2}{2|Z|}\cos\phi = \frac{U^2}{2|Z|}\frac{\text{Re}(Z)}{|Z|} = \frac{1}{2}U^2\frac{\text{Re}(Z)}{|Z|^2}, \tag{2}$$

where $\cos\phi$ is the so-called power factor, equal to the ratio of the instantaneous real power used by an electrical load to the apparent power running through the circuit; $|Z|$ and $\text{Re}(Z)$ are the absolute value and real part, respectively, of the frequency dependent impedance measured with a vector network analyzer.

Since the electrical impedance Z of the ultrasound transducer depends on the frequency F, the amplitude U of the electrical signal changed together with the frequency. This can lead to asymmetry in the frequency dependence of the diffraction efficiency $\xi(F)$, while the dependence $\xi_{\text{norm}}(F)$ remains symmetric. Moreover, it is inappropriately to compare the diffraction efficiency ξ at different temperatures t and the same amplitude U of the electrical signal, because the impedance Z is a temperature dependent physical quantity [14]. Instead, it is necessary to use the diffraction efficiency per 1 W of electric power, i.e., $\xi_{\text{norm}}(F)$.

The experimental value of the diffraction efficiency ξ was compared with the one predicted by our model [11]:

$$\xi = \frac{\pi^2}{2\lambda^2}\frac{M_2 P_{\text{a}}}{d}L\exp(-\alpha_{\text{s}} l), \tag{3}$$

where M_2 is the AO figure of merit; P_{a} is the acoustic power, which is usually considered equal to the input electric power P_{el}; d and L are the width and length of the sound transducer; α_{s} is the sound power attenuation coefficient; l is the distance from the sound transducer at which the THz beam traveled.

All values of the parameters of the experimental setup required for calculations are given in Table 1.

Table 1. Parameters of experimental setup.

λ (μm)	**L (mm)**	**d (mm)**	**l (cm)**
130	80	14	5

As one can see from Equations (2) and (3), the diffraction efficiency ξ is proportional to the input electric power P_{el} and the square of the voltage amplitude U:

$$\xi = k_P P_{\text{el}} = k_{\text{U}} U^2, \tag{4}$$

where the factors k_P and k_{U} can be determined using the least square method (LSM).

The AO figure of merit M_2 and the refractive index n of the liquefied gas were calculated using the Lorentz–Lorenz equation [9,15]:

$$M_2 = \left[\frac{(n^2-1)(n^2+2)}{6n}\right]^2 \frac{4}{\rho V^3}, \qquad n = \sqrt{1 + \frac{2A\rho}{1 - 2A\rho/3}}, \tag{5}$$

where ρ is the density; $A = 1.64 \cdot 10^{-4}$ m^3/kg is a factor proportional to the mean polarizability of the gas molecules [11].

The density ρ of the liquefied gas was represented by a rational function [16]:

$$\rho = \frac{a_1 + a_2 t + a_3 t^2 + a_4 t^3}{1 + a_5 t}, \tag{6}$$

where the coefficients a_n depend on the pressure p; t is in (°C) and ρ is in (kg/m^3). Numerical values of a_n can be found in [16] for the temperature range from 0 to +50 °C and pressures from 12 to 200 bar. The values of a_n used in our work can be found in Table 2.

Table 2. Numerical values of coefficients a_n in dependence of density ρ on pressure p and temperature t.

p (bar)	a_1	a_2	a_3	$a_4 \cdot 10^3$	$a_5 \cdot 10^3$
20	1570.47	−25.6846	0	0	−11.8502
25	1577.29	33.0886	−0.210490	−4.73339	25.3988

The sound power attenuation coefficient α_s was calculated using the relation obtained in work [11] by the LSM for the temperature range from +9 to +30 °C:

$$\alpha_s = \left(\frac{F}{F_0}\right)^2 \cdot [2466 + 4.81 t^{1.875}] \cdot 10^{-4}, \tag{7}$$

where α_s is in (cm^{-1}), t is in (°C), and $F_0 = 300$ kHz.

Data on the sound velocity V in liquefied SF_6 gas for a wide range of temperatures (from −40 to +60 °C) and pressures (from 20 to 600 bar) are given in work [17] and are approximated by the LSM. We found a misprint in the values of the coefficients: the calculated data differed from the measured ones by more than 10%, while according to the authors, the difference should be less than 1%. We used the following approximating function for the sound velocity V:

$$V^2 = \frac{\sum_{j=0}^{3}\sum_{k=0}^{2} a_{jk}(p-p_0)^j(t-t_0)^k}{\sum_{l=0}^{3}\sum_{m=0}^{2} b_{lm}(p-p_0)^l(t-t_0)^m}, \tag{8}$$

where $p_0 = 10$ MPa and $t_0 = 250$ K; the experimental data $V(p,t)$ were taken from work [17]: p is in (MPa), t is in (K) and V is in (m/s).

The initial coefficient values for the LSM were taken from work [17]. The adjusted values of the coefficients a_{jk} and b_{lm} are followed:

$$a_{jk} = \begin{pmatrix} 258 \cdot 10^3 & 40.6 \cdot 10^3 & 1.83 \cdot 10^3 & 20.9 \\ -4.29 \cdot 10^3 & -382 & 8.58 & 0.549 \\ 20.1 & 0.715 & -0.104 & 1.53 \cdot 10^{-3} \end{pmatrix}, \tag{9}$$

$$b_{jk} = \begin{pmatrix} 1 & 0.130 & 3.73 \cdot 10^{-3} & -0.513 \cdot 10^{-6} \\ -2.07 \cdot 10^{-3} & 401 \cdot 10^{-6} & 73.5 \cdot 10^{-6} & 0.476 \cdot 10^{-6} \\ -14.2 \cdot 10^{-6} & 1.17 \cdot 10^{-6} & 0.415 \cdot 10^{-6} & 44.0 \cdot 10^{-12} \end{pmatrix}. \tag{10}$$

The Table 3 summarizes the physical properties of liquefied SF_6 under the conditions that can be realized in experiment.

Table 3. Properties of liquefied SF_6 gas.

t (°C)	p (bar)	ρ (kg/m^3)	n	V (m/s)	α_s (cm^{-1})	M_2 (10^{-15} s^3/kg)
+23	25	1369.4	1.2362	227	0.40	15,720
+10	20	1490.2	1.2585	283	0.30	9130

The AO interaction is resonant in nature and, at a given ultrasound frequency F, the AO diffraction is observed only when the radiation is incident on the AO cell at angles θ close to the Bragg angle $\theta_B = \lambda F/V$. Similarly, when the radiation is incident on the AO cell at the Bragg angle, the diffracted beam is observed only for a narrow ultrasound frequency band. The resonant frequency F_{res} is determined by the thickness d of the sound transducer: $F_{res} = V_{PZT}/2d$ [18]. Here V_{PZT} is the sound velocity in the material of the sound transducer. The dependence of the AO diffraction efficiency ξ on θ and F has the form of the squared sinc-function [19], and the two-sided interaction bandwidths (−3 dB criterion for the diffraction efficiency) can be calculated using the relations [20]:

$$\Delta\theta = \frac{0.9nV}{FL_{eff}}, \qquad \Delta F = \frac{1.8nV^2}{\lambda FL_{eff}}, \tag{11}$$

where L_{eff} is the effective length of the AO interaction region, which is usually equal to or slightly shorter than the length of the sound transducer L.

3. Results and Discussion

It was found that at the room temperature of +23 °C, the diffraction efficiency ξ was proportional to the input electric power P_{el}, while the angular dependence of ξ was well fitted by the squared sinc-function, as was predicted by the theory (see Figure 3a,b). However, the frequency dependence of ξ turned out to be asymmetric. We suppose that this fact was related to the changes in the voltage amplitude U with the frequency due to the resonance character of the sound transducer electric properties (Figure 3c). After normalization of the diffraction efficiency to 1 W of the input electric power, the dependence $\xi_{norm}(F)$ became symmetric and well-fitted by the squared sinc-function (Figure 3d). The same dependencies measured at the lower temperature $t = 10$ °C are shown in Figure 4. The experimental results for both temperature regimes are summarized in Table 4.

Table 4. Properties of AO modulator of THz radiation revealed in the experiment.

t (°C)	p (bar)	k_U (1/kV^2)	ξ_{norm} (%/W)	$\Delta\theta$ (deg)	ΔF (kHz)	F_{res} (kHz)
+23	25	9.6 ± 0.3	0.23 ± 0.02	0.77 ± 0.03	37 ± 3	294 ± 1
+10	20	1.3 ± 0.1	0.24 ± 0.02	0.80 ± 0.02	12 ± 3	311 ± 1

We also calculated properties of the AO modulator using relations given in the previous section (see Table 5).

Table 5. Properties of AO modulator of THz radiation predicted by the theory.

t (°C)	p (bar)	k_U (1/kV^2)	ξ_{norm} (%/W)	$\Delta\theta$ (deg)	ΔF (kHz)
+23	25	14.6	0.35	0.62	38
+10	20	1.8	0.33	0.74	56

The angular bandwidth $\Delta\theta$ of the AO modulator is 0.1° greater than that predicted by the theory. We attribute this to the divergence of radiation and ultrasound, which is not taken into account in the model. At the same time, the frequency bandwidth ΔF coincided with theoretical predictions at +23 °C within the error. At the temperature of +10 °C, the frequency resonance narrowed significantly and side lobes appeared. In our assessment, this fact may be associated with the complex acoustical modes of the transducer at this temperature.

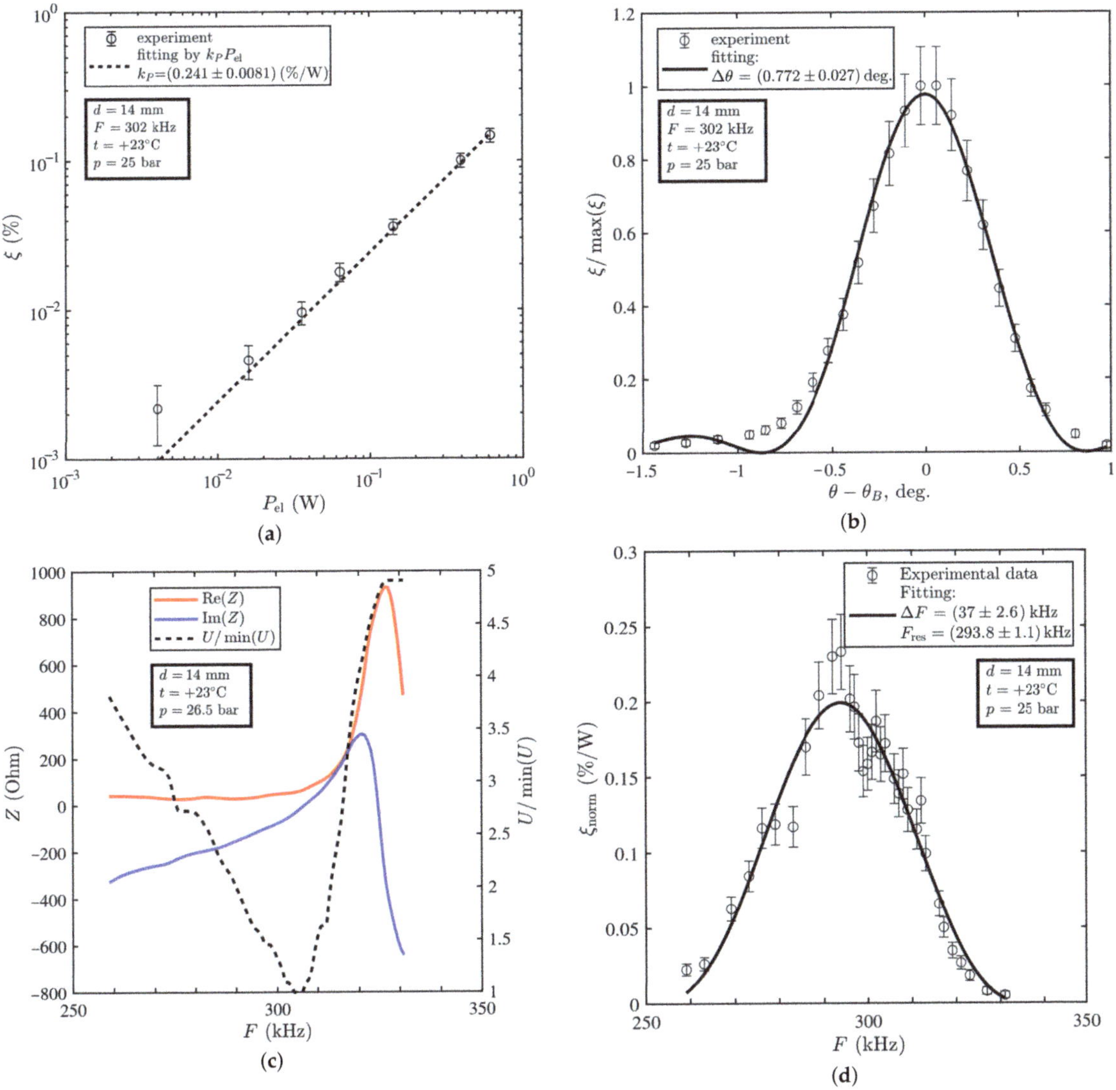

Figure 3. Experimental results obtained at +23 °C: (**a**) AO diffraction efficiency ξ vs. input electric power P_{el}; (**b**) AO diffraction efficiency $\xi/\max(\xi)$ vs. difference between angle θ of incidence of THz radiation on AO cell and Bragg angle θ_B; (**c**) frequency dependences of real and imaginary parts of complex impedance Z of sound transducer and amplitude U of electrical signal; (**d**) frequency dependence of AO diffraction efficiency per 1 W of input electric power.

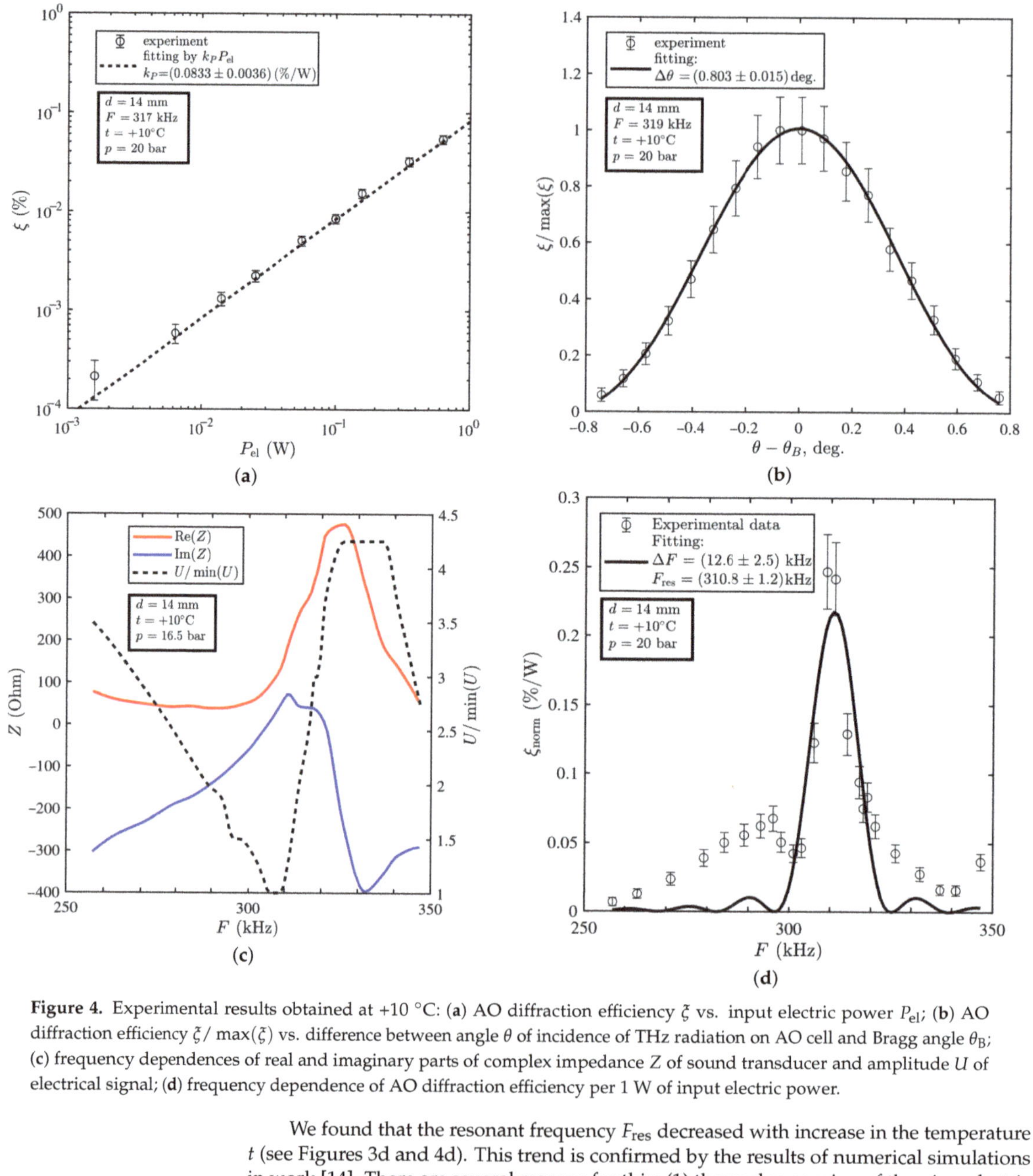

Figure 4. Experimental results obtained at +10 °C: (**a**) AO diffraction efficiency ξ vs. input electric power P_{el}; (**b**) AO diffraction efficiency $\xi/\max(\xi)$ vs. difference between angle θ of incidence of THz radiation on AO cell and Bragg angle θ_B; (**c**) frequency dependences of real and imaginary parts of complex impedance Z of sound transducer and amplitude U of electrical signal; (**d**) frequency dependence of AO diffraction efficiency per 1 W of input electric power.

We found that the resonant frequency F_{res} decreased with increase in the temperature t (see Figures 3d and 4d). This trend is confirmed by the results of numerical simulations in work [14]. There are several reasons for this: (1) thermal expansion of the piezoelectric material; (2) decrease in the sound velocity with increase in the temperature; (3) temperature dependence of the sound transducer impedance Z. The first and the second reasons directly affected the resonant frequency: $F_{res} = V_{PZT}/2d$. The last reason was indirect because the shift of the resonant frequency was affected by change in the electric power P_{el} (see Equation (2) and Figures 3c and 4c).

The diffraction efficiencies ξ at the same voltage amplitude U at +23 °C and +10 °C differed by one order. However, only this could not mean that at a temperature of +23 °C, the AO modulator performed an order of magnitude better than at +10 °C. As can be seen from Figures 3c and 4c, the electrical impedance Z of the sound transducer resonantly depended on the frequency F and varies with the temperature t. Therefore, it was necessary to compare the diffraction efficiency not at the same voltage amplitude U, but at the same electric power P_{el}. Unfortunately, we could not make a comparison with data of other researchers since in their works only the dependencies $\xi(U)$ of the diffraction efficiency on the voltage amplitude are given and there are no data on the frequency dependence of the electrical impedance $Z(F)$ of the ultrasound transducer.

The experimental value $\xi_{norm} \approx 0.23$ (%/W) of the normalized diffraction efficiency was the same for +23 °C and +10 °C within the error, as predicted by the theory. However, the value of the AO diffraction efficiency achieved in the experiment was less than the value $\xi_{norm} \approx 0.35$ (%/W), predicted by the theory. Therefore, we introduced a correction coefficient $\kappa \approx 0.7$ that was an attempt to fit the theory to the experimental results. This coefficient was not a universal constant and depended on the material of the ultrasound transducer and its dimensions, as well as on temperature, pressure and ultrasound frequency. There were also a number of factors that determined the value of this correction factor:

1. Relation (3) assumed that the radiation beam had a plane wavefront, although in fact there was diffraction divergence. However, as the THz radiation beam used in the experiments was about 10 mm wide [21], the divergence could be neglected ($\lambda/d = 130$ μm/10 mm $\approx 1°$).
2. The same applied to the sound beam. However, due to the low sound velocity ($V \approx 300$ m/s), the diffraction divergence of the sound beam was significant ($\lambda/d = V/Fd \approx 4°$) at the frequency $F = 300$ kHz. Because of strong attenuation ($\alpha_s \propto F^2$), it is not advisable to use ultrasound with a higher frequency for reducing the divergence [9].
3. Only part of the electric power was converted into the acoustic power, because of losses [22] and significant difference in the acoustic impedances of the ultrasound transducer material and the liquefied gas [23].
4. Since the width of the ultrasound transducer was only two times greater than its thickness, a number of acoustic modes with complex structures appeared [24], which was not taken into account in Equation (3). As a result, the radiation effectively interacted with only a part of the sound beam. Influence of the phase structure of the ultrasonic beam was also possible. Calculation of the structure of the sound field is very complicated and is beyond the scope of this work.

A few words should be said about the stability of measurements. The temperature was controlled by a thermocouple and the accuracy was 0.5 °C. When ultrasound with the power of 1 W was turned on for more than 30 s, the diffraction efficiency fell by about a factor of two and stabilized at this lower level. The thermocouple readings remained the same, since it was located at a distance of 5 cm from the ultrasound transducer. Nevertheless, we assumed that the decrease in the diffraction efficiency was associated with an increase in the ultrasound attenuation due to the heating of the SF_6 near the ultrasound transducer. At a power of about 0.1 W, no such effect was observed, which made it possible to measure the dependences of the diffraction efficiency on the ultrasound frequency and the angle of incidence of radiation. When we measured the dependence of the diffraction efficiency on the ultrasound power, it was necessary to turn on the ultrasound transducer only for a short time, in order to then correctly carry out averaging. The measurement error was about 10% due to the instability of the laser source.

4. Conclusions

In this study, we investigated the operation of an AO modulator based on liquefied SF_6 gas in two temperature regimes and took into account the effect of electrical impedance on its characteristics. The dependences of the diffraction efficiency on the voltage amplitude of the electrical signal, on the angle of incidence of THz radiation, and on the sound frequency were measured and analyzed in detail. The developed analytical model of the AO diffraction of THz radiation in liquefied SF_6 gas showed good agreement with the experimental results. The temperature was found to affect not only the optical and acoustic properties of the medium of AO interaction, but also the electrical impedance of the sound transducer. For the first time, the influence of the electrical properties of the sound transducer on the characteristics of AO THz radiation modulator based on liquefied SF_6 gas was taken into account. This made it possible to estimate the following characteristics of the monochromatic THz radiation AO modulator more correctly: its angular and frequency bandwidths and the diffraction efficiency per 1 W of the input electric power. Moreover, it enabled us to make a correct comparison of the operation of the AO modulator in different temperature regimes. The ultrasound attenuation of liquefied SF_6 gas enables the use of low-frequency ultrasound to observe the AO diffraction. As a result, the thickness of the piezoelectric plate must be comparable to its width, which leads to complex types of vibrations of the sound transducer. Therefore, in future work, we plan to investigate the effect of the ratio between the width and thickness of the sound transducer on the characteristics of the AO modulator based on liquefied SF_6 gas. We will also continue the search for optimal operating conditions for the AO modulator of THz radiation. Although, according to the theory, high temperatures are more preferable, they lead to decrease in the cavitation threshold and transparency of liquefied SF_6. Therefore, we are confident that the AO modulator will have the best performance at lower temperatures.

Author Contributions: conceptualization, P.A.N.; methodology, P.A.N.; software, I.S.K.; investigation, P.A.N. and V.V.G.; resources, V.V.G.; writing—original draft preparation, P.A.N. and I.S.K.; writing—review and editing, P.A.N. and V.V.G.; supervision, P.A.N.; project administration, P.A.N.; funding acquisition, P.A.N. All authors have read and agreed to the published version of the manuscript.

Funding: This research was funded by the Ministry of Science and Higher Education of the Russian Federation under State contract No. 0057-2019-0006.

Institutional Review Board Statement: Not applicable.

Informed Consent Statement: Not applicable.

Data Availability Statement: The data presented in this study are available on request from the corresponding author. The data are not publicly available due to the further research.

Acknowledgments: The work was done at the shared research center SSTRC on the basis of the Novosibirsk FEL at BINP SB RAS, using equipment of the Shared Research Facilities of the Scientific and Technological Centre of Unique Instrumentation of the Russian Academy of Sciences (STC UI RAS).

Conflicts of Interest: The authors declare no conflict of interest.

Abbreviations

The following abbreviations are used in this manuscript:

AO	acousto-optic
THz	terahertz
FEL	free-electron laser

References

1. Lu, J.; Shi, F.; Xu, J.; Meng, L.; Zhang, L.; Cheng, P.; Zhou, X.; Pang, F.; Zeng, X. Recent progress of dynamic mode manipulation via acousto-optic interactions in few-mode fiber lasers: Mechanism, device and applications. *Nanophotonics* **2020**, *10*, 983–1010. [CrossRef]
2. Gunther, J.; Andersson-Engels, S. Review of current methods of acousto-optical tomography for biomedical applications. *Front. Optoelectron.* **2017**, *10*, 211–238. [CrossRef]
3. Duocastella, M.; Surdo, S.; Zunino, A.; Diaspro, A.; Saggau, P. Acousto-optic systems for advanced microscopy. *J. Phys. Photonics* **2020**, *3*, 012004. [CrossRef]

4. Chanover, N.; Voelz, D.; Glenar, D.; Young, E. AOTF-based spectral imaging for balloon-borne platforms. *J. Astron. Instrum.* **2014**, *3*, 1440005. [CrossRef]
5. Valle, S.; Ward, J.; Pannell, C.; Johnson, N. Imaging AOTFs with low RF Power in deep-UV and mid-IR. *Phys. Procedia* **2015**, *70*, 707–711. [CrossRef]
6. Krauz, L.; Páta, P.; Bednář, J.; Klíma, M. Quasi-collinear IR AOTF based on mercurous halide single crystals for spatio-spectral hyperspectral imaging. *Opt. Express* **2021**, *29*, 12813–12832. [CrossRef] [PubMed]
7. Kastelik, J.C.; Dupont, S.; Yushkov, K.; Molchanov, V.; Gazalet, J. Double acousto-optic deflector system for increased scanning range of laser beams. *Ultrasonics* **2017**, *80*, 62–65. [CrossRef] [PubMed]
8. Khorkin, V.S.; Voloshinov, V.B.; Efimova, A.I.; Kulakova, L.A. Acousto-optic properties of germanium-, selenium-, silicon-, and tellurium-based alloys. *Opt. Spectrosc.* **2020**, *128*, 244–249. [CrossRef]
9. Durr, W. Acousto-optic interaction in gases and liquid bases in the far infrared. *Int. J. Infrared Millim. Waves* **1986**, *7*, 1537–1558. [CrossRef]
10. Kotov, V.M.; Averin, S.V. Acoustooptic diffraction of three-color radiation on a single acoustic wave. *Int. J. Opt.* **2019**, *2019*, 1–5. [CrossRef]
11. Nikitin, P.; Knyazev, B.; Voloshinov, V.; Scheglov, M. Observation of acousto-optic diffraction of terahertz radiation in liquefied sulfur hexafluoride at room temperature. *IEEE Trans. Terahertz Sci. Technol.* **2020**, *10*, 44–50. [CrossRef]
12. Cervenka, M.; Bednarik, M. On the structure of multi-Gaussian beam expansion coefficients. *Acta Acust. United Acust.* **2015**, *101*, 15–23. [CrossRef]
13. Morris, N. Volt-amperes, power, reactive VA and power factor consumption. In *Mastering Electronic and Electrical Calculations*; Palgrave: London, UK, 1996; pp. 213–228. [CrossRef]
14. Upadhye, V.; Agashe, S. Effect of temperature and pressure on the thickness mode resonant spectra of piezoelectric ceramic. *IOP Conf. Ser. Mater. Sci. Eng.* **2017**, *225*, 012125. [CrossRef]
15. Uchida, N. Elastooptic coefficient of liquids determined by ultrasonic light diffraction method. *Jpn. J. Appl. Phys.* **1968**, *7*, 1259–1266. [CrossRef]
16. Keramati, B.; Wolgemuth, C. Thermodynamic properties of liquid sulfur hexafluoride from 0 to 50 C to 200 bars. *J. Chem. Eng. Data* **1976**, *21*, 423–428. [CrossRef]
17. Vacek, V.; Zollweg, J. Speed of sound in compressed sulfur hexafluoride. *Fluid Phase Equilibria* **1993**, *88*, 219–226. [CrossRef]
18. Sunol, F.; Ochoa, D.; Sune, L.; Garcia, J. Design and characterization of immersion ultrasonic transducers for pulsed regime applications. *Instrum. Sci. Technol.* **2018**, *47*, 213–232. [CrossRef]
19. Balakshy, V.; Kupreychik, M.; Mantsevich, S.; Molchanov, V. Acousto-optic cells with phased-array transducers and their application in systems of optical information processing. *Materials* **2021**, *14*, 451. [CrossRef] [PubMed]
20. Shcherbakov, A.; Bliznetsov, A.; Castellanos, A.; Lucero, D. Acousto-optical spectrum analysis of ultra-high-frequency radio-wave analogue signals with an improved resolution exploiting the collinear acoustic wave heterodyning. *Optik* **2010**, *121*, 1497–1506. [CrossRef]
21. Kubarev, V.; Sozinov, G.; Scheglov, M.; Vodopyanov, A.; Sidorov, A.; Melnikov, A.; Veber, S. The radiation beamline of novosibirsk free-electron laser facility operating in terahertz, far-infrared, and mid-infrared ranges. *IEEE Trans. Terahertz Sci. Technol.* **2020**, *10*, 634–646. [CrossRef]
22. Wild, M.; Bring, M.; Halvorsen, E.; Hoff, L.; Hjelmervik, K. The challenge of distinguishing mechanical, electrical and piezoelectric losses. *J. Acoust. Soc. Am.* **2018**, *144*, 2128–2134. [CrossRef]
23. Rathod, V. A review of acoustic impedance matching techniques for piezoelectric sensors and transducers. *Sensors* **2020**, *20*, 4051. [CrossRef]
24. Ma, C.C.; Huang, C.H. The investigation of three-dimensional vibration for piezoelectric rectangular parallelepipeds using the AF-ESPI method. *IEEE Trans. Ultrason. Ferroelectr. Freq. Control* **2001**, *48*, 142–153. [CrossRef]

materials MDPI

Article

Optimal Design of an Ultrasound Transducer for Efficient Acousto-Optic Modulation of Terahertz Radiation

Pavel Alekseevich Nikitin [1,2,*] and Vasily Valerievich Gerasimov [3,4]

1 Scientific and Technological Centre of Unique Instrumentation RAS, 117342 Moscow, Russia
2 Moscow Power Engineering Institute, Krasnokazarmennaya Str. 14, 111250 Moscow, Russia
3 Department of Physics, Novosibirsk State University, 630090 Novosibirsk, Russia; v.v.gerasimov3@gmail.com
4 Budker Institute of Nuclear Physics SB RAS, 630090 Novosibirsk, Russia
* Correspondence: nikitin.pavel.a@gmail.com

Abstract: Acousto-optic (AO) interaction in the terahertz range was investigated with the use of monolithic ultrasound transducers of various widths. Sulfur hexafluoride (SF_6) liquefied at a temperature of about 23 °C and a pressure of 25 bar was used as a medium for AO interaction. The angular and frequency bandwidths of effective AO interaction, as well as the diffraction efficiency per 1 W of the driving electric power, were determined. For the first time, a correct comparison of the AO diffraction efficiency in SF_6 with the use of ultrasound transducers with different widths was carried out. In the experiments performed, the highest energy efficiency of the AO modulator was achieved with a transducer with a width of 12 mm.

Keywords: acousto-optic interaction; terahertz radiation; diffraction; liquefied inert gas

check for updates

Citation: Nikitin, P.A.; Gerasimov, V.V. Optimal Design of an Ultrasound Transducer for Efficient Acousto-Optic Modulation of Terahertz Radiation. *Materials* **2022**, *15*, 1203. https://doi.org/10.3390/ma15031203

Academic Editors: Alexander S. Machikhin and Vitold Pozhar

Received: 24 December 2021
Accepted: 2 February 2022
Published: 5 February 2022

Publisher's Note: MDPI stays neutral with regard to jurisdictional claims in published maps and institutional affiliations.

1. Introduction

Acousto-optic (AO) interaction is an effective tool for modulation of laser beam intensity with a time response of about 1 μs [1,2]. The main concept is to use an acoustic wave to form a phase diffraction grating in a medium. Due to the photo-elastic effect, the refractive index variation is proportional to the amplitude of the acoustic wave. Therefore, by coding the electrical signal applied to the ultrasound transducer, one can modulate the intensity of the diffracted radiation. The best medium for AO interaction in the terahertz (THz) range is liquefied sulfur hexafluoride (SF_6) [3,4]. In AO modulators, the Bragg regime was used and there were only two diffraction orders with wave vectors $\vec{k}_0$ and $\vec{k}_1$ (see Figure 1). The direction of $\vec{k}_1$ is determined by the relation $\vec{k}_1 = \vec{k}_0 + \vec{K}$, where $\vec{K}$ is the wave vector of the sound wave.

The deflection angle between the transmitted and diffracted radiation beams equals the doubled Bragg angle θ_B, which depends on the sound frequency F, radiation wavelength λ, and sound velocity V [5]:

$$\sin\theta_B = \frac{K}{2k} = \frac{\lambda F}{2V}. \quad (1)$$

The diffracted radiation beam has to be deflected with an angle much larger than the divergence angle of this beam, which is about 1 degree for a beam diameter of 10 mm and a wavelength of 130 μm. The sound velocity in liquefied SF_6 is about $V \approx 300$ m/s. Therefore, at the sound frequency $F = 300$ kHz, the deflection angle is about 8 degrees, which is sufficient for its spatial separation from the transmitted radiation beam. The use of higher frequencies is not advisable because of the high attenuation of ultrasound, which is proportional to the square of the frequency. The resonant frequency $F_{res} = V_{PZT}/2h$ of sound transducers is proportional to the sound velocity V_{PZT} in piezoceramics and is inversely proportional to its thickness h. The typical sound transducer thickness is about $h \approx 6$ mm for $F_0 = 300$ kHz [3,4]. The sound transducer width d should not be much greater than the diameter of the radiation beam, which is about 1 cm for the Novosibirsk free-electron laser (FEL) [6]. Hence, the sound transducer width d is close to its thickness h. As known, with $d/h \approx 1$, complex types of mechanical vibrations arise in the transducer [7,8]. As a result, the wavefront of the generated sound wave can no longer be considered as flat, which negatively affects the efficiency of the AO interaction. In addition, with a transducer the width of which exceeds the diameter of the radiation beam, part of the sound beam does not interact with the radiation beam, and the efficiency of the AO diffraction becomes lower. Therefore, there is an optimal width of the transducer at which the diffraction efficiency reaches its maximum.

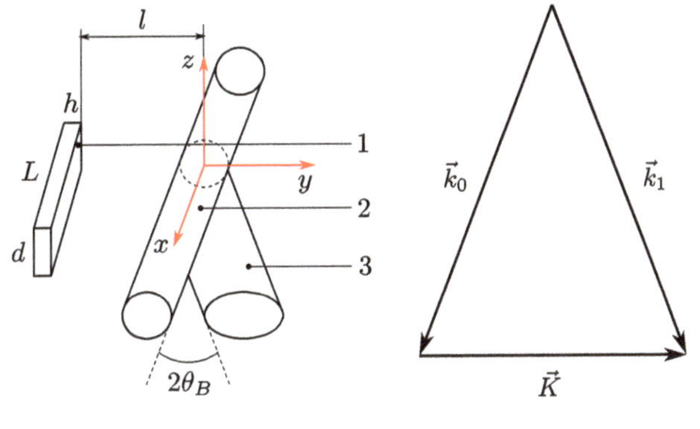

Figure 1. (**a**) Schematic of AO Bragg diffraction: 1—sound transducer; 2—transmitted radiation; 3—diffracted radiation; (**b**) wave vector diagram of AO interaction.

According to the simplest model, the real sound beam is replaced by a homogeneous sound column with a plane wave front [9]. In this case, due to a decrease in the acoustic power density, the AO diffraction efficiency is inversely proportional to the width of the ultrasound transducer. Despite the fact that this model seems to be crude, it describes well the results of most known works on acousto-optics [10–12]. This is because of the common use of high frequency ultrasound of about 100 MHz and above. As a result, the thickness of the ultrasound transducer is very small and is about $h = V_{PZT}/2F \approx 100$ µm, which is 2 orders of magnitude less than its width (about several millimeters) and length (about 1 cm). Therefore, such transducers behave like a piston. Since up to now there was no need to use thick ultrasound transducers, the simplest model was used and it was not necessary to take into account the influence of the transducer thickness on the characteristics of AO devices.

As mentioned above, for the effective AO modulation of THz radiation, it is necessary to use liquefied SF_6 as the medium for the AO interaction. As a result, the thickness of the transducer is comparable to its width. As far as we know, there are only a few works in which the AO diffraction of THz radiation in liquefied SF_6 [3,4,13] was investigated. The results obtained in these works cannot be compared correctly because of the lack of data about the active electrical power consumed by the sound transducer. The purpose of this work is to find the optimal width of the ultrasound transducer to increase the energy efficiency of the AO modulator of THz radiation based on liquefied SF_6.

2. Materials and Methods

2.1. Theoretical Background

In the Bragg regime of AO diffraction, only two diffraction maxima are observed: Zero and the first, with the intensities I_0 and I_1, respectively. As a measure of the efficiency of AO diffraction, the ratio I_1/I_0 is usually used. The power consumption required to achieve a given level of modulation depth is also important. Therefore, the diffraction efficiency ξ is normalized to the active electrical power P_{el} [14,15]:

$$\xi = \frac{I_1}{I_0}, \quad \xi_{norm} = \frac{\xi}{P_{el}}, \quad P_{el} = \frac{1}{2}U^2\frac{\mathrm{Re}(Z)}{|Z|^2}, \tag{2}$$

where ξ_{norm} is the diffraction efficiency per 1 W of the electrical power; U is the voltage amplitude on the sound transducer; $\mathrm{Re}(Z)$ and $|Z|$ are the real part and absolute value, respectively, of the frequency dependent impedance of the sound transducer.

To estimate the efficiency of AO diffraction, one can use the relation obtained within the simplest model of AO interaction [14] (which assumes the diffraction of a plane electromagnetic wave by an infinite sound column), introducing in it the additional exponential term considering the ultrasonic attenuation:

$$\xi_{1D} = \frac{\pi^2}{2\lambda^2}\frac{M_2 P_a}{d} L \exp(-\alpha_s l), \tag{3}$$

where M_2 is the AO figure of merit; P_a is the acoustic power, which is usually considered as equal to the RF driving power P_{el}; d and L are the width and length of the sound transducer; α_s is the sound power attenuation coefficient; l is the distance from the sound transducer at which the THz beam travels (in our experiment $l = 5$ cm). The dependence of physical properties of liquefied SF_6 on the temperature and pressure are summarized in work [4].

As one can see from Equations (2) and (3), the diffraction efficiency ξ is proportional to the RF driving power P_{el}, whereas the normalized diffraction efficiency ξ_{norm} does not depend on P_{el}:

$$\xi = k_P P_{el}, \tag{4}$$

where the factor k_P can be determined from experimental data by the least square method (LSM).

The diffraction efficiency ξ in a resonant manner depends on the angle of incidence θ of radiation on the AO cell, as well as on the ultrasound frequency F. For the interaction between plane electromagnetic and acoustic waves, this dependence has the form of $\mathrm{sinc}^2(x)$, whose argument is proportional to the deviation of θ or F from Bragg condition (1) [16]. The two-sided interaction bandwidths (-3 dB criterion for the normalized diffraction efficiency ξ_{norm}) can be calculated with the use of the known relations [17]:

$$\Delta\theta = \frac{0.9nV}{FL_{eff}}, \quad \Delta F = \frac{1.8nV^2}{\lambda F L_{eff}}, \tag{5}$$

where L_{eff} is the effective length of the AO interaction region, which is usually equal to or slightly shorter than the length of the sound transducer L. The relation for the angular bandwidth differs from the original one in [17] by the value n, since in [17] the angle is calculated for the interaction medium in the AO cell, while in this work the angle is calculated outside the cell.

Equation (3) was derived for the diffraction of plane electromagnetic wave. However, it is necessary to take into account the size d of the AO interaction region in z-direction and the size D of the optical window of the AO cell in z-direction (see Figure 2). If the optical window can be considered as infinitely wide in the direction of the y axis, the mathematical solution does not depend on y and we can use a one-dimensional distribution of the radiation beam intensity instead of a two-dimensional one.

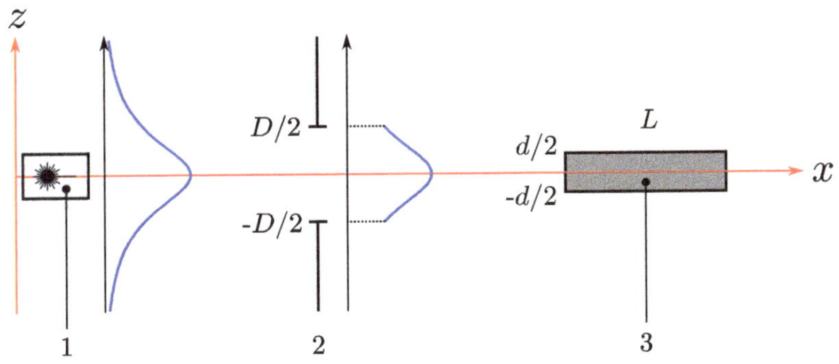

Figure 2. Schematic of acousto-optic Bragg diffraction: 1—THz laser; 2—optical window of AO cell; 3—region of AO interaction.

The intensity distribution of the THz beam is well fitted by the Gaussian function with amplitude A and width ω_z [6]. In the one-dimensional case, it can be written as follows:

$$I_{\mathrm{THz}}(z) = A \exp\left(-\frac{z^2}{\omega_z^2}\right). \tag{6}$$

It can be shown that the integral intensities of the transmitted and diffracted beams can be calculated by the following relations:

$$I_0 = \exp(-\alpha L) \int_{-D/2}^{D/2} I_{\mathrm{THz}}(z)\mathrm{d}z; \tag{7}$$

$$I_1 = \begin{cases} \xi_{1D} \exp(-\alpha L) \int_{-d/2}^{d/2} I_{\mathrm{THz}}(z)\mathrm{d}z & \text{for } d \leq D; \\ \xi_{1D} \exp(-\alpha L) \int_{-D/2}^{D/2} I_{\mathrm{THz}}(z)\mathrm{d}z & \text{for } d > D. \end{cases} \tag{8}$$

The derivation of relations (7) and (8) requires some explanation. The integral intensity I_0 of the transmitted radiation is limited only by the optical window size D. Therefore, the integration limits are $-D/2$ and $D/2$. The exponential term before the integral corresponds to the light absorption in the medium. The integral intensity I_1 of the diffracted radiation is proportional to the integral intensity of the THz beam in the region of the AO interaction. So, the limits in the integral are determined by the smallest of the optical window size D and the sound transducer width d.

Let us give an example of calculations for an experiment that uses a THz laser with a wide beam of radiation, such as the Novosibirsk free electron laser ($\omega_z = 1.44$ cm for $\lambda = 130$ μm and 8–11 stations) [6]. The minimum size of the optical window is limited by the enhancement of the effect of radiation diffraction at the edges of the optical window. Therefore, we assume $D = 1$ cm, as in the experiment in work [4]. As the variation of $I_{\mathrm{THz}}(z)$ for $|z| < 0.5$ cm is less than 10%, this function can be treated as constant. As a result, the integrals in (7) and (8) are proportional to the integration interval width. The diffraction efficiency can now be determined with the use of relation (2) in the following way:

$$\xi = \begin{cases} \xi_{1D} \dfrac{d}{D} & \text{for } d \leq D; \\ \xi_{1D} & \text{for } d > D. \end{cases} \tag{9}$$

The dependence $\xi(d)$ is a piecewise function, which is shown in Figure 3.

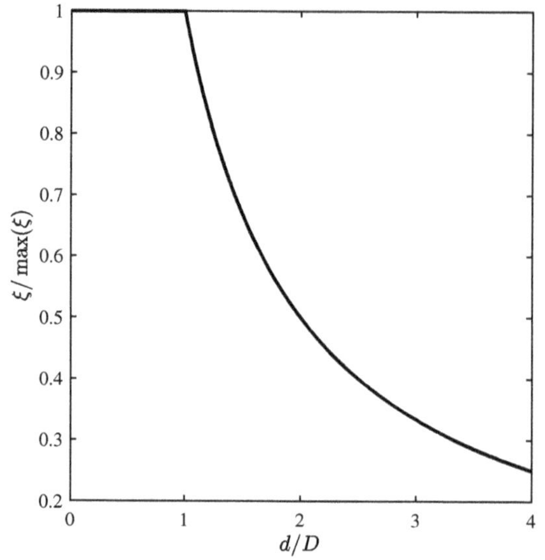

Figure 3. Theoretical dependence of diffraction efficiency on width of sound transducer.

It can be seen from (9) that when the radiation beams wider than the sound transducer are used, the diffraction efficiency decreases by a factor of d/D. It is related to the fact that part of the radiation passes away from the sound beam and does not interact with it. This factor is limiting and inhibits infinite increase in the efficiency of AO diffraction due to decrease in the width d of the sound transducer, as follows from (3).

2.2. Experimental Technique

Monochromatic coherent THz radiation of the Novosibirsk free electron laser 1 (FEL) with the wavelength $\lambda = 130$ µm (see Figure 4) was used in our experiments. Since an AO modulator based on a liquid medium is insensitive to radiation polarization, the intensity of the THz beam was set by wire polarizer 2. The radiation beam was incident on the center of the input optical window of AO cell 3 ($D = 10$ mm) at the distance $l = 5$ cm from the sound transducer. A detailed description of the AO cell is given in [13]. For observation of the diffracted radiation, the AO cell was turned through the Bragg angle, and a signal (modulated at a frequency of 10 Hz) from generator of electrical signals 4 was applied to the ultrasound transducer via amplifier 5. At a distance of about 30 cm after the AO cell, lens 6 was located, focusing the radiation into the receiver, Golay cell 7. The signal from the receiver was isolated from the background noise by means of lock-in amplifier 8.

The cycle of experiments can be broken into the following stages:

1. At the first stage, the setup was adjusted via change in the sound frequency F and the angle of incidence of THz radiation on the AO cell. The aim was to achieve the maximum intensity of the diffracted radiation.
2. Next, the dependence $I_1(U)$ of the intensity of the diffracted radiation on the amplitude of the electric signal was measured. In theory, this dependence is quadratic.
3. Further measurements were carried out at the maximum level of the voltage amplitude U, at which the $I_1(U)$ dependence is still quadratic.
4. The dependence $I_1(\theta)$ of the intensity of the diffracted radiation on the angle of incidence of the radiation was measured. The angle of incidence was changed via rotation of the AO cell.
5. Finally, the dependence $I_1(F)$ of the diffracted radiation intensity on the ultrasound frequency was measured at variation of the frequency of the electrical signal applied to the sound transducer.

6. Since the electrical impedance Z of the ultrasound transducer depends on the frequency, the amplitude U of the electrical signal was changed together with the frequency. Therefore, the dependencies $Z(F)$ and $U(F)$ were measured too.
7. The diffraction efficiency was normalized to 1 W of the electrical power for determination of the energy efficiency of the AO modulator.

The measurement error for the diffraction efficiency ξ was mainly caused by the FEL radiation intensity instability of about 10%. The sound frequency F was set with an accuracy of 0.001 kHz, while the angle θ with an accuracy of 0.5′. The temperature t and pressure p were measured with an accuracy of 0.5 °C and 0.5 bar, respectively. To avoid cavitation near the surface of the ultrasound transducer, we worked in a mode where the electrical power consumption was low (about 1 W) and the efficiency of AO diffraction was proportional to the square of the voltage on the transducer.

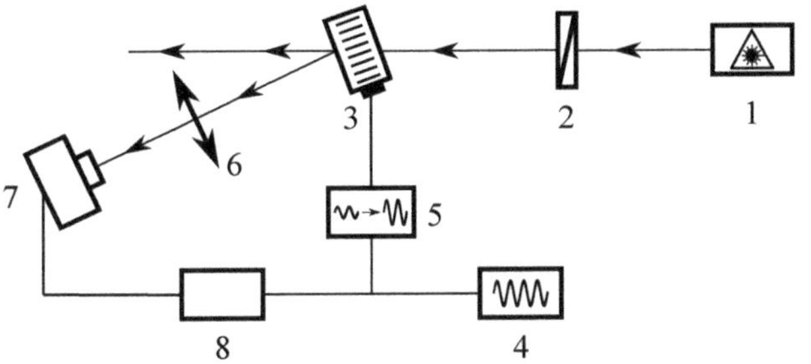

Figure 4. Schematic diagram of experimental setup: 1—FEL; 2—wire-grid polarizer; 3—AO cell; 4—signal generator; 5—electrical amplifier; 6—lens; 7—Golay cell; 8—lock-in amplifier.

3. Results and Discussion

The measured dependencies were approximated by the least squares method in accordance with the theoretical model and are shown in Figure 5. The error bars correspond to the experimental errors. As expected, they are of resonant nature, which made it possible to determine the working bandwidth of the angles of incidence of radiation and the working bandwidth of the ultrasound frequency. Table 1 shows the values of the parameters of the AO modulator based on liquefied SF_6 gas, obtained with various ultrasound transducers with a width d of 6 to 14 mm. The results for d = 14 mm was taken from our previous work [13]. For clarity, the results are presented in the form of graphs in Figure 6.

Table 1. Properties of AO modulator of THz radiation revealed in the experiment.

d (mm)	ξ_{norm} (%/W)	$\Delta\theta$ (deg)	ΔF (kHz)	F_{res} (kHz)
14	0.23 ± 0.02	0.77 ± 0.03	37.0 ± 2.6	293.8 ± 1.1
12	0.31 ± 0.03	0.71 ± 0.02	6.0 ± 0.4	322.5 ± 0.5
10	0.076 ± 0.007	0.60 ± 0.08	27.5 ± 5.8	289.0 ± 2.2
8	0.11 ± 0.01	0.75 ± 0.02	9.2 ± 1.3	319.9 ± 0.5
6	$(3.3 \pm 0.3) \times 10^{-3}$	1.53 ± 0.19	8.9 ± 2.1	369.6 ± 0.9

The measured dependencies (see Figure 5a,c,e,g,i) of the diffraction efficiency ξ_{norm} on the angle θ of incidence of radiation on the AO cell are fairly well described by a theoretical dependence of the form $\mathrm{sinc}^2(x)$. At the same time, some of them have pronounced side lobes, while others are of asymmetrical shape. This can be explained by the fact that the sound beam features a directional diagram, in which there are components propagating at an angle to the normal of the ultrasound transducer. In addition, the asymmetry of the frequency dependencies (see Figure 5b,d,f,h,j) of the AO diffraction efficiency indicates a complex structure of the acoustic modes of the transducer due to the fact that the width d of the ultrasound transducer is comparable to its thickness (6 mm). This is confirmed, among other things, by the dependence of the resonant frequency F_{res} of the ultrasound transducer on its width d (see Figure 6d).

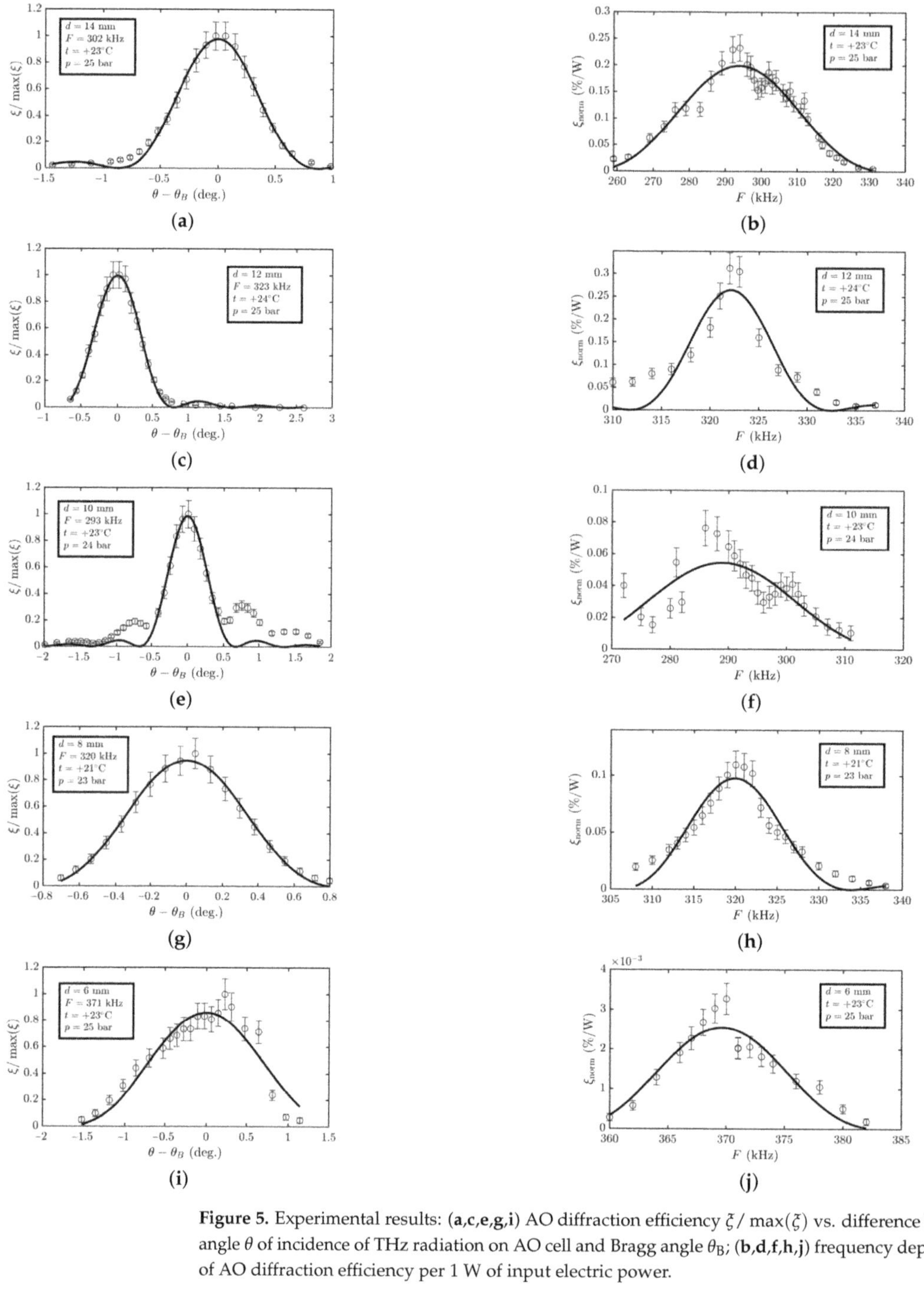

Figure 5. Experimental results: (**a**,**c**,**e**,**g**,**i**) AO diffraction efficiency $\xi/\max(\xi)$ vs. difference between angle θ of incidence of THz radiation on AO cell and Bragg angle θ_B; (**b**,**d**,**f**,**h**,**j**) frequency dependence of AO diffraction efficiency per 1 W of input electric power.

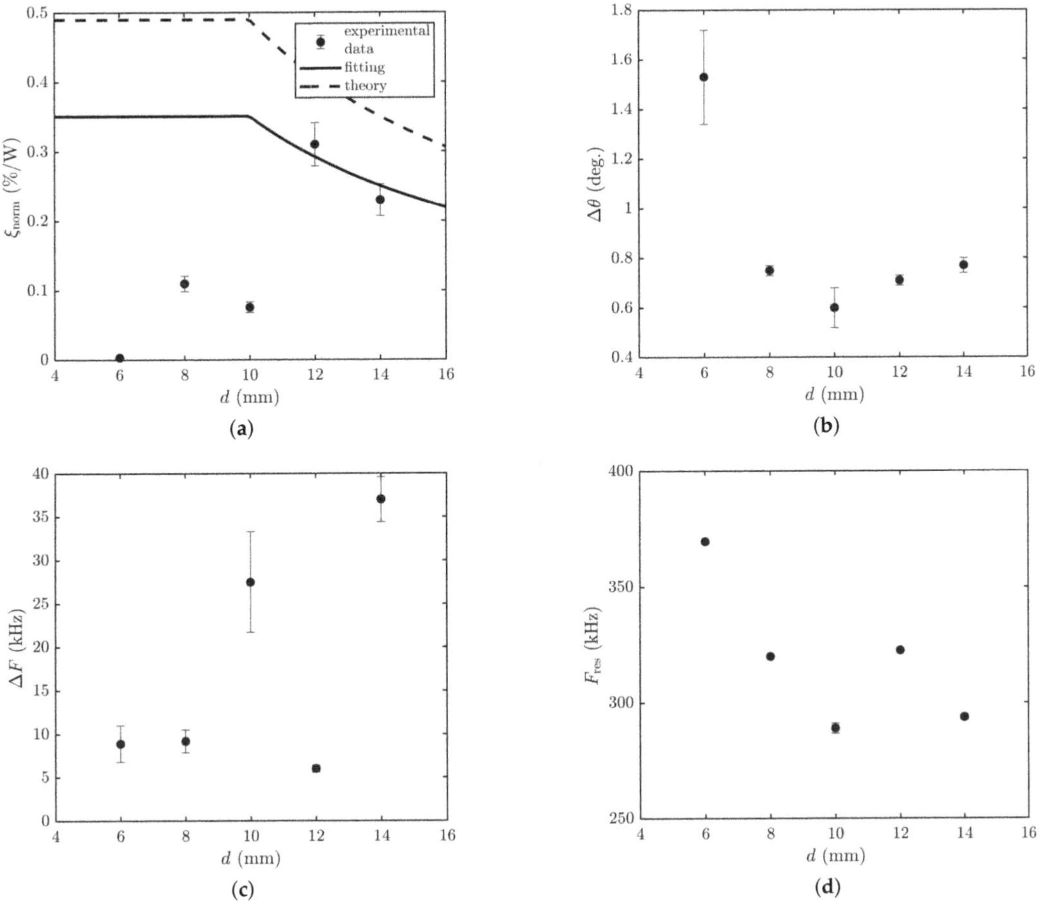

Figure 6. Dependence of parameters of AO modulator on width d of ultrasound transducer: (**a**) AO diffraction efficiency per 1 W of input electric power; (**b**) angular bandwidth; (**c**) frequency bandwidth; (**d**) operating frequency.

The angular $\Delta\theta$ and frequency ΔF interaction bandwidths were expected to not depend on the width of the ultrasound transducer and be determined only by the effective length of AO interaction L_{eff} (see Equation (5)). However, as can be seen from Figure 6b,c, this is not the case: With a decrease in the width d of the ultrasound transducer, the working bandwidth $\Delta\theta$ of the incidence angles of THz radiation increases, while the working frequency bandwidth ΔF of ultrasound, in contrast, decreases. These facts once again confirm that an ultrasound transducer with a width almost equal to its thickness behaves in a complex manner, and its behavior can no longer be described by the approximation of a piston-type transducer [18].

In accordance with Figure 3, the highest efficiency of AO diffraction would be achieved at $d < D$, i.e., at $d < 10$ mm, and at $d > D$; the AO diffraction efficiency would decrease according to the law $\xi \propto 1/d$. As can be seen from Figure 6a, with the ultrasound transducer width $d = 12$ mm and 14 mm, the model agrees well with the experimental results. However, for $d = 10$ mm and less, the experimental data differ significantly from the theoretical model. This can be explained by the influence of two factors. Firstly, the model assumes that the radiation is diffracted by a sound beam with a plane wavefront. However, at the frequency $F = 300$ kHz, the wavelength of sound in liquefied SF_6 is about 1 mm due to the low speed of sound of about 300 m/s. Therefore, at $d = 6$ mm, only 6 wavelengths fit on the ultrasound transducer, which leads to a strong diffraction divergence of the sound beam. Secondly, at $d = 6$ mm, the ultrasound transducer has a square cross-section, which leads to complex deformations when an alternating voltage is applied to the electrodes. As a result, the wavefront is also strongly distorted, which leads to a decrease in the AO diffraction efficiency.

4. Conclusions

Many parameters of AO devices designed for radiation in the ultraviolet, visible, and infrared spectral ranges, such as the angular $\Delta\theta$ and frequency ΔF bandwidths of effective interaction, as well as the operating frequency F_{res}, do not depend on the width d of the ultrasound transducer. However, our study has shown that this is not the case for AO modulators of THz radiation based on liquefied SF_6. It has been found found that with a decrease in the width d of the ultrasound transducer from 14 to 6 mm, the frequency bandwidth ΔF decreases. The angular bandwidth $\Delta\theta$ is approximately constant in the interval of d from 8 to 14 mm and increases by 2 times at $d = 6$ mm. At the same time, as the transducer width d decreases from 14 mm to 6 mm, the AO diffraction efficiency decreases by about two orders of magnitude. Such unusual characteristics, in our opinion, can be explained by the fact that the ultrasound transducer applied was "thick". This led to the complex deformations of the transducer and, as a consequence, to phase inhomogeneities of the ultrasonic field. Therefore, when fabricating THz radiation modulators based on liquefied SF_6 gas with an operating frequency of about 300 kHz, it is advisable to use an ultrasound transducer with a width d of about 12 mm. In this case, its characteristics are more predictable, and the diffraction efficiency ξ_{norm} reaches its maximum value.

Author Contributions: Conceptualization, P.A.N.; methodology, P.A.N.; software, P.A.N.; investigation, P.A.N. and V.V.G.; resources, V.V.G.; writing—original draft preparation, P.A.N. and V.V.G.; writing—review and editing, P.A.N. and V.V.G.; supervision, P.A.N.; project administration, P.A.N.; funding acquisition, P.A.N. All authors have read and agreed to the published version of the manuscript.

Funding: This research was financially supported by Russian Science Foundation grant № 20-72-00184.

Institutional Review Board Statement: Not applicable.

Informed Consent Statement: Not applicable.

Data Availability Statement: The data presented in this study are available on request from the corresponding author. Because of the further research, the data are not publicly available.

Acknowledgments: The work was done at the shared research center SSTRC on the basis of the Novosibirsk FEL at BINP SB RAS with the use of equipment of the Shared Research Facilities of the Scientific and Technological Centre of Unique Instrumentation of the Russian Academy of Sciences (STC UI RAS).

Conflicts of Interest: The authors declare no conflict of interest.

Abbreviations

The following abbreviations are used in this manuscript:

AO	acousto-optic
THz	terahertz
FEL	free-electron laser

References

1. Okada, A.; Yamazaki, R.; Fuwa, M.; Noguchi, A.; Yamaguchi, Y.; Kanno, A.; Yamamoto, N.; Hishida, Y.; Terai, H.; Tabuchi, Y.; et al. Superconducting acousto-optic phase modulator. *Opt. Express* **2021**, *29*, 14151–14162. [CrossRef] [PubMed]
2. Zeng, S.; Bi, K.; Xue, S.; Liu, Y.; Lv, X.; Luo, Q. Acousto-optic modulator system for femtosecond laser pulses. *Rev. Sci. Instruments* **2007**, *78*, 015103. [CrossRef] [PubMed]
3. Durr, W. Acousto-optic interaction in gases and liquid bases in the far infrared. *Int. J. Infrared Millim. Waves* **1986**, *7*, 1537–1558. [CrossRef]
4. Nikitin, P.; Knyazev, B.; Voloshinov, V.; Scheglov, M. Observation of acousto-optic diffraction of terahertz radiation in liquefied sulfur hexafluoride at room temperature. *IEEE Trans. Terahertz Sci. Technol.* **2020**, *10*, 44–50. [CrossRef]
5. Peled, I.; Kaminsky, R.; Kotler, Z. Acousto-optics bandwidth broadening in a Bragg cell based on arbitrary synthesized signal methods. *Appl. Opt.* **2015**, *54*, 5065–5073. [CrossRef] [PubMed]
6. Kubarev, V.; Sozinov, G.; Scheglov, M.; Vodopyanov, A.; Sidorov, A.; Melnikov, A.; Veber, S. The radiation beamline of novosibirsk free-electron laser facility operating in terahertz, far-infrared, and mid-infrared ranges. *IEEE Trans. Terahertz Sci. Technol.* **2020**, *10*, 634–646. [CrossRef]
7. Zhang, Y.; Du, J. Influence of boundary conditions on three-dimensional vibration characteristics of thick rectangular plates. *Sci. Prog.* **2020**, *103*. [CrossRef] [PubMed]
8. Diallo, O.; Clezio, E.; Delaunay, T.; Bavencoffe, M.; Feuillard, G. Electrical admittance of piezoelectric parallelepipeds: Application to tensorial characterization of piezoceramics. *AIP Adv.* **2014**, *4*, 017121. [CrossRef]
9. Proklov, V.; Rezvov, Y.; Podolsky, V.; Sivkova, O. Invariance of the transmission function of an acousto-optic device for a change in the drift angle of an acoustic beam. *Acoust. Phys.* **2019**, *65*, 385–390. [CrossRef]
10. Soos, J.; Rosemeier, R. High frequency GaAs bulk acousto-optic devices for modulators and frequency shifters at 1.3 μm and 1.5 μm in fiber-optics. In *High Bandwidth Analog Applications of Photonics II*; Neyer, B., Ed.; International Society for Optics and Photonics, SPIE: Bellingham, WA, USA, 1989; Volume 0987, pp. 119–127. [CrossRef]
11. Mantsevich, S.; Balakshy, V. Examination of optoelectronic feedback effect on collinear acousto-optic filtration. *J. Opt. Soc. Am. B* **2018**, *35*, 1030–1039. [CrossRef]
12. Antonov, S.; Kotelnikov, V. A review of physical principles and applications of acousto-optic deflectors on the basis paratellurite. *Phys. Astron. Int. J.* **2019**, *3*, 235–249.
13. Nikitin, P.; Gerasimov, V.; Khasanov, I. Temperature Effects in an Acousto-Optic Modulator of Terahertz Radiation Based on Liquefied SF6 Gas. *Materials* **2021**, *14*, 5519. [CrossRef] [PubMed]
14. Voloshinov, V.; Khorkin, V.; Kulakova, L.; Gupta, N. Optic, acoustic and acousto-optic properties of tellurium in close-to-axis regime of diffraction. *J. Phys. Commun.* **2017**, *1*, 025006. [CrossRef]
15. Morris, N. Volt-amperes, power, reactive VA and power factor consumption. In *Mastering Electronic and Electrical Calculations*; Palgrave: London, UK, 1996; pp. 213–228. [CrossRef]
16. Balakshy, V.; Kupreychik, M.; Mantsevich, S.; Molchanov, V. Acousto-optic cells with phased-array transducers and their application in systems of optical information processing. *Materials* **2021**, *14*, 451. [CrossRef] [PubMed]
17. Shcherbakov, A.; Bliznetsov, A.; Castellanos, A.; Lucero, D. Acousto-optical spectrum analysis of ultra-high-frequency radio-wave analogue signals with an improved resolution exploiting the collinear acoustic wave heterodyning. *Optik* **2010**, *121*, 1497–1506. [CrossRef]
18. Ding, D.S.; Shen, C.S.; Lu, H. A novel algorithm for the sound field of rectangular-shaped transducers. *Chin. Phys. Lett.* **2015**, *32*, 124304. [CrossRef]

materials

MDPI

Article

Acousto-Optic Cells with Phased-Array Transducers and Their Application in Systems of Optical Information Processing

Vladimir Balakshy [1,2,*], Maxim Kupreychik [1], Sergey Mantsevich [1] and Vladimir Molchanov [3]

[1] Faculty of Physics, Lomonosov Moscow State University, 119991 Moscow, Russia; info@physics.msu.ru (M.K.); np@ntcup.ru (S.M.)
[2] Scientific and Technological Centre of Unique Instrumentation of the Russian Academy of Sciences, 117342 Moscow, Russia
[3] Acousto-Optical Research Center at National University of Science and Technology MISIS, 119049 Moscow, Russia; aocenter@mail.ru
* Correspondence: balakshy@phys.msu.ru; Tel.: +7-906-075-0295

Abstract: This paper presents the results of theoretical and experimental studies of anisotropic acousto-optic interaction in a spatially periodical acoustic field created by a phased-array transducer with antiphase excitation of adjacent sections. In this case, contrary to the nonsectioned transducer, light diffraction is absent when the optical beam falls on the phased-array cell at the Bragg angle. However, the diffraction takes place at some other angles (called "optimal" here), which are situated on the opposite sides to the Bragg angle. Our calculations show that the diffraction efficiency can reach 100% at these optimal angles in spite of a noticeable acousto-optic phase mismatch. This kind of acousto-optic interaction possesses a number of interesting regularities which can be useful for designing acousto-optic devices of a new type. Our experiments were performed with a paratellurite (TeO_2) cell in which a shear acoustic mode was excited at a 9° angle to the crystal plane (001). The piezoelectric transducer had to nine antiphase sections. The efficiency of electric to acoustic power conversion was 99% at the maximum frequency response, and the ultrasound excitation band extended from 70 to 160 MHz. The experiments have confirmed basic results of the theoretical analysis.

Keywords: acousto-optics; phased-array piezoelectric transducers; acousto-optic materials; anisotropic Bragg diffraction; acousto-optic devices

check for updates

Citation: Balakshy, V.; Kupreychik, M.; Mantsevich, S.; Molchanov, V. Acousto-Optic Cells with Phased-Array Transducers and Their Application in Systems of Optical Information Processing. *Materials* **2021**, *14*, 451. https://doi.org/10.3390/ma14020451

Received: 30 November 2020
Accepted: 8 January 2021
Published: 18 January 2021

Publisher's Note: MDPI stays neutral with regard to jurisdictional claims in published maps and institutional affiliations.

1. Introduction

The basis of modern acousto-optic (AO) devices, applied in systems of optical information processing, is an AO cell that is usually made of a crystal with a certain cut. Ultrasonic waves are excited by piezoelectric plates attached to one of the cell faces. In these instruments, a type of light diffraction by ultrasound close to the Bragg regime is usually used [1–3]. This regime is realized at sufficiently high acoustic frequencies (usually over 100 MHz) and allows us to create devices with good performance and low light loss. However, as a disadvantage, this type of interaction has high selectivity, which means the AO effect magnitude has a strong dependence on the frequency of ultrasound f, the angle of light incidence θ_0 and the wavelength of optical radiation λ. An important characteristic of the effect is the Bragg angle θ_B. When the light beam falls at a Bragg angle (the phase-matching condition), the most effective light scattering into the diffraction order occurs, and a decrease in the diffraction efficiency by 3 dB determines the boundary of the AO interaction region.

In principle, the interaction region can be extended by reducing the length of AO interaction L (the width of the acoustic beam in the direction of light propagation), but this will result in a reduction of the diffraction efficiency and/or an increase in the acoustic power required for obtaining a necessary diffraction efficiency. Taking into consideration this negative situation, A. Korpel et al. [4] proposed the use of a step-sectional piezoelectric transducer for the excitation of an acoustic beam with a rotating radiation pattern. In their AO cell, the acoustic wavefront rotated with ultrasound frequency, adjusting to the optimal angle of light incidence.

However, such a cell was very difficult to manufacture, so planar structures were then proposed [5–9]. A planar structure is shown in Figure 1; in the first variant (Figure 1a), one piezoelectric plate (shown in green) is used, and separate sections are formed by partitioning the internal and external electrodes. In another case (Figure 1b), the internal electrode is first made solid, and then grooves are sawn at the final stage. The pictures show the AO cells with transducers containing four sections that are connected electrically in series, but in such a way that the direction of the electric field (shown by the arrows) in the adjacent sections is opposite. Due to this, acoustic beams from each section are excited with a phase shift of π, and a system of equivalent wavefronts (shown by two dashed lines) is formed, turned by an angle Ψ relative to the transducer plane. This angle is determined by the expression:

$$\Psi = \pm\frac{\Lambda}{2d} = \pm\frac{V}{2fd} \quad (1)$$

where $\Lambda = V/f$ is the acoustic wavelength, V is its velocity and d is the period of the transducer array.

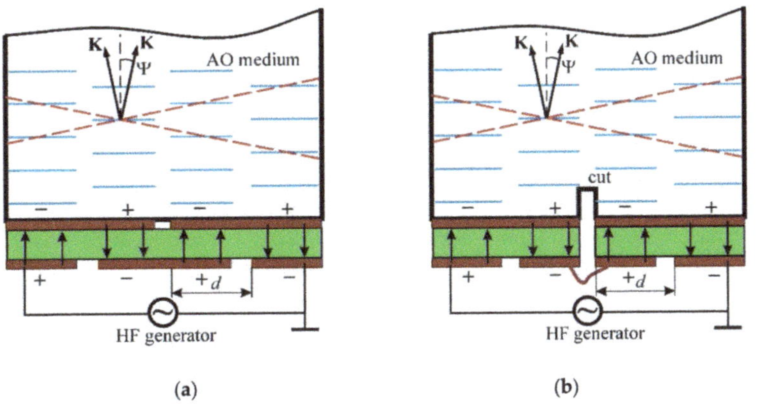

Figure 1. Two variants of flat phased-array transducers with antiphase excitation of adjacent sections: (**a**) Sectioning by partitioning the internal and external electrodes; (**b**) Sectioning by means of transducer cuts.

In comparison with a homogeneous (nonsectioned) transducer, here, the transfer function of the AO cell contains two main maxima, which are located symmetrically with respect to the Bragg angle θ_B, as shown in Figure 2 with the red curve. It follows from the figure that AO diffraction is absent when the light beam falls at the Bragg angle ($\theta_0 = \theta_B$). This is due to the fact that partial diffracted waves generated in neighboring acoustic beams are phase-shifted by π and therefore dampen each other during interference. However, there are other maxima situated equidistantly with the period Λ/d, which are inscribed into the dashed green curve showing the radiation pattern of a separate section of the transducer. The width of the maxima is equal to Λ/md, where m is the number of periods of the transducer array. Thus, we can conclude that in the case of the phased-array transducer, the concept of the Bragg angle as the angle of light incidence, at which the phase-matching condition is fulfilled and the maximum diffraction efficiency is observed, is incorrect. Here, we can talk about the optimal angles of light incidence $\theta_{opt} = \theta_B \pm \Lambda/2d = \theta_B \pm V/2fd$, which correspond to the maximal scattering of light, although the phase-matching condition is violated. It can be noticed that these optimal angles conform to the conventional Bragg angles when light falls on the equivalent wavefronts. Consequently, such a structure of the acoustic field can be considered as a superposition of two fields excited by two solid piezoelectric transducers rotated relative to each other by the angle Λ/d. Our calculations show that the diffraction efficiency here can reach 100%, despite a noticeable phase mismatch. However, this requires slightly higher acoustic power [10,11].

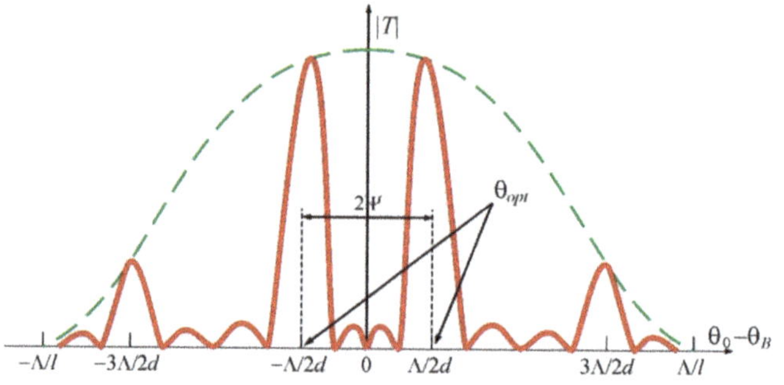

Figure 2. Radiation patterns of four-sectioned (red line) and one-sectioned (green line) transducers.

As follows from Equation (1), when the ultrasound frequency decreases, the main lobes of the transducer radiation pattern diverge. Therefore, if one chooses the true angle of light incidence, the equivalent wavefront will rotate, adjusting to the changing Bragg angle. This adjustment will not be complete since the Bragg angle depends on the frequency linearly, but the angle of rotation Ψ is in accordance with the hyperbolic law. The best correction of the Bragg angle is obtained when $d\theta_{opt}/df = 0$. This condition is satisfied at the frequency:

$$f^* = V\sqrt{\frac{n}{\lambda d}} \tag{2}$$

where n is the refractive index of the AO medium and λ is the optical wavelength in vacuum. Thus, by selecting an appropriate period d of the phased-array transducer, we can set the operating point f^* in any desired frequency range.

The interest in phased-array transducers weakened noticeably after anisotropic AO diffraction entered the practice of acousto-optics [12]. The main advantage of anisotropic diffraction is that it consists of a significantly more complicated frequency dependence on the Bragg angles in comparison with the AO interaction in an isotropic medium [1–3]. This feature makes it possible to choose optimal interaction geometry for each separate AO device. For example, for AO deflectors, the optimal area is that where $d\theta_B/df \to 0$, whereas for AO video-filters, it is the area where $d\theta_B/df \to \infty$.

The use of the phased-array transducers in combination with anisotropic diffraction gives more complicated types of Bragg curves, which open up new opportunities for improvement of AO device characteristics [13,14]. Therefore, the aim of this work is to study the peculiarities of anisotropic AO interactions in an acoustic field created by the phased-array transducers with antiphase excitation of adjacent sections.

2. Acousto-Optic Effect in the Field of Phased-Array Transducer

Figure 3 illustrates the problem in the most general case of an anisotropic medium in which acoustic beams propagate with a walk-off angle α. We assume that the inclined phase grating created by the acoustic wave occupies the area of space between the infinite planes $x = 0$ and $x = L$. The wave vector of ultrasound is inclined at an angle α. The width of each acoustic column is l. Therefore, the second column occupies the space between the planes $x = l + a$ and $x = 2l + a$, where a is the gap between the acoustic beams, etc. Thus, the period of the transducer structure is $d = l + a = l(1 + \xi)$. The initial acoustic phase in the first column is equal to $\Phi = 0$, and in the following column, is $\Phi = \pi$; at that time the phase shift between the adjacent beams is equal to $\Delta\Phi = \pi$. The total number of beams is m.

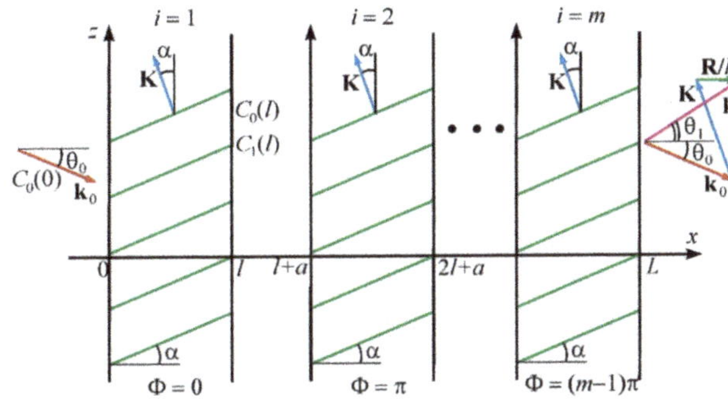

Figure 3. Statement of the problem of acousto-optic (AO) interaction in the field of the phased-array transducer.

The regime of anisotropic diffraction with Bragg scattering of light into two diffraction orders, zero and +1st or zero and −1st, is considered here for this structure. In this case, the calculation has to take into account two optical plane waves. The first wave is falling; it is characterized by the wave vector $\mathbf{k}_0$ and the incidence angle θ_0. The second one is a diffracted wave with the wave vector $\mathbf{k}_{\pm1}$ and the diffraction angle $\theta_{\pm1}$. The interaction of these waves is described by the following system of equations [15]:

$$\begin{cases} \frac{dC_0}{dX} = \pm\frac{\Gamma}{2}C_{\pm1}\exp[\pm j(R_\pm X - \Phi)] \\ \frac{dC_{\pm1}}{dX} = \mp\frac{\Gamma}{2}C_0\exp[\mp j(R_\pm X - \Phi)] \end{cases} \quad (3)$$

where, for the convenience of numerical calculations, dimensionless values are introduced: the normalized amplitudes of the incident C_0 and diffracted $C_{\pm1}$ waves, the coordinate $X = x/l$, the Raman–Nath parameter

$$\Gamma = \frac{2\pi l\Delta n}{\lambda\cos\theta_0} \quad (4)$$

and dimensionless phase mismatch

$$R_\pm = \frac{2\pi}{\lambda}l\left[n_0\cos\theta_0 \mp \frac{\lambda f}{V}\sin\alpha - \sqrt{n_{\pm1}^2 - \left(n_0\sin\theta_0 \pm \frac{\lambda f}{V}\cos\alpha\right)^2}\right] \quad (5)$$

In these formulas, n_0 and $n_{\pm1}$ are the refractive indices for incident and diffracted light, and Δn is the amplitude of the refractive index change under the action of the acoustic wave. A wave vector diagram of the AO interaction is shown in Figure 3 on the right; it corresponds to the vector relationship [1]:

$$\mathbf{k}_0 + \mathbf{K} + \frac{\mathbf{R}_\pm}{l} = \mathbf{k}_{\pm1} \quad (6)$$

Here, we have taken into consideration that the phase mismatch vector $\mathbf{R}_\pm$ is perpendicular to the boundaries of the acoustic beams.

When the light waves propagate in the periodic acoustic field, the optical energy is redistributed between them. Our task is to find the amplitudes of the waves at the output of the structure: $C_0(L)$ and $C_{\pm1}(L)$. This can be fulfilled by recording a recurrence relation connecting the input and output fields for the i-th acoustic beam. The boundary conditions at the input have the following form:

$$\begin{cases} E_0(x) = E_i C_0^{(i)} \exp[\mathrm{j}(\omega_o t - k_0 \cos\theta_0 x - k_0 \sin\theta_0 z)] \\ E_{\pm1}(x) = E_i C_{\pm1}^{(i)} \exp[\mathrm{j}(\omega_{\pm1} t - k_{\pm1} \cos\theta_{\pm1} x - k_{\pm1} \sin\theta_{\pm1} z)] \end{cases} \tag{7}$$

where E_i is the amplitude of the incident optical wave, ω_0 and $\omega_{\pm1}$ are the frequencies of incident and diffracted light, $\omega_{\pm1} = \omega_0 \pm \Omega$, and $\Omega = 2\pi f$ is the cyclic frequency of the ultrasound. Solving Equation (3) with the boundary conditions of Equation (7), we obtain the expressions for the amplitudes at the input of the (i + 1)-st acoustic column:

$$C_0^{(i+1)} = \left[C_0^{(i)}\left(A \mp \mathrm{j}\frac{R_\pm}{2}B\right) \pm C_{\pm1}^{(i)}\frac{\Gamma}{2}B\exp(\mp\mathrm{j}\Phi)\right]\exp\left(\pm\mathrm{j}\frac{R_\pm}{2}\right) \tag{8}$$

$$C_{\pm1}^{(i+1)} = \exp\left(\mp\mathrm{j}\frac{R_\pm}{2}\right)\left[C_{\pm1}^{(i)}\left(A \pm \mathrm{j}\frac{R_\pm}{2}B\right) \mp C_0^{(i)}\frac{\Gamma}{2}B\exp(\pm\mathrm{j}\Phi)\right]\exp[\pm\mathrm{j}R_\pm(1+\xi)] \tag{9}$$

The phase shift, introduced by the area of empty space between the acoustic columns, is then taken into account [16]:

$$A \equiv \cos\frac{\sqrt{\Gamma^2 + R_\pm^2}}{2},\ B \equiv \sin\mathrm{c}\frac{\sqrt{\Gamma^2 + R_\pm^2}}{2\pi} \tag{10}$$

However, the equal phase shift $\exp[-\mathrm{j}k_0 l \cos\theta_0(1+\xi)]$ is omitted. In Equations (3)–(10), the upper sign corresponds to light scattering into the +1st order, while the lower sign corresponds to the -1st order.

In acousto-optics, the angles of incidence and diffraction are usually calculated from the front of the acoustic wave. In accordance with this, we introduce angles $\phi_0 = \theta_0 - \alpha$ and $\phi_{\pm1} = \theta_{\pm1} - \alpha$. The condition of phase matching $R_\pm = 0$ determines the Bragg angle ϕ_B:

$$C_0(l+a) = C_0(l),\ C_{\pm1}(l+a) = C_{\pm1}(l)\exp[\pm\mathrm{j}R_\pm(1+\xi)] \tag{11}$$

Equation (11) indicates that the acoustic walk-off does not affect the AO phase matching. However, the parameters Γ and $R_\pm$ depend on the walk-off angle α, as well as the optimal angles θ_{opt}. This results in a change in the AO interaction range and, consequently, in the diffraction characteristics [17].

3. Computation Results

This section presents the results of our calculations for an acoustic field created by a phased-array transducer in a paratellurite (TeO_2) crystal. We used the original computer program developed on the MATLAB platform. This program takes into account the optical activity of the material as, for the considered diffraction variants, the optical beams propagate close to the optical axis of the crystal, and the optical activity can have a noticeable effect on diffraction characteristics. The calculations are performed for the crystallographic plane $(1\bar{1}0)$ when a shear acoustic wave propagates at an angle of $\chi = 4°$ to the plane (001). For this acoustic mode, the velocity is $V = 0.632 \times 10^5$ cm/s, and the walk-off angle is $\alpha = 35°$.

Figure 4a demonstrates the frequency dependences of the Bragg angles for optical wavelength $\lambda = 0.633$ μm. The four curves correspond to different polarizations of incident light (ordinary "*o*" or extraordinary "*e*") and to the scattering of light into the +1st or -1st orders of diffraction. For example, branch *+1e* characterizes the diffraction of the *e*-wave into the +1st order. Points D and T indicate areas optimal for AO deflectors and video-filters, respectively [1–3].

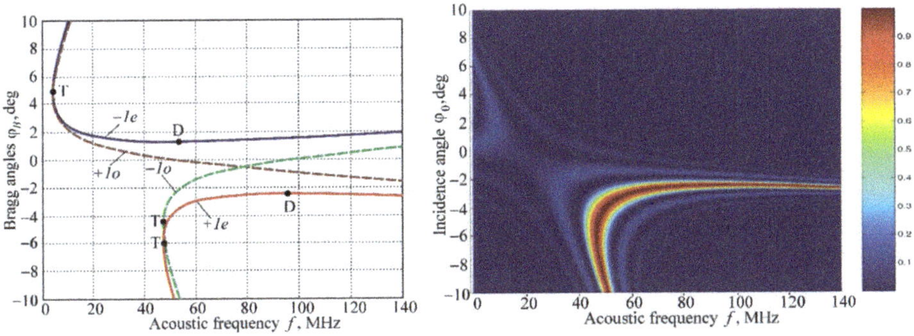

(**a**) (**b**)

Figure 4. AO interaction in a paratellurite crystal: (**a**) Frequency dependencies of the Bragg angles in the case of light scattering in the +1st and −1st diffraction orders at different polarizations of incident optical radiation; (**b**) The area of AO interaction when the optical beam with extraordinary polarization diffracts into +1st order.

Figure 4b displays the AO interaction area separately for branch *+1e* in coordinates $\phi_0 - f$. The calculations are carried out for the case of a homogeneous (nonsectioned) transducer with a width of l = 1 mm. The diffraction efficiency $\varsigma = |C_{+1}|^2$ is illustrated by the color scheme from zero (dark blue) to one (dark red). The values $\zeta = 1$ correspond to the Bragg angles. The fine structure in the picture is caused by the lateral maxima of the function $\sin c(\bullet)$ in Equation (9). The horizontal sections of the pattern determine frequency characteristics of the AO interaction $\varsigma(f)$ for specified incidence angles of light, and the vertical sections determine angular characteristics $\varsigma(\phi_0)$. The shape of the area reproduced curve *+1e* is shown in Figure 4a, and its width is defined primarily by the size of the transducer l.

In the case of a phased-array transducer with antiphase excitation of adjacent sections, the situation changes cardinally. Figure 5 shows this change when the period of structure d decreases. In these calculations, the number of sections is chosen equal to m = 4, and the normalized gap between sections is $\xi = 1$.

We can see from Figure 5 that, for large periods of the transducer structure, a common area of AO interaction splits into two symmetrical domains (Figure 5a). Then, an additional domain appears from the side of low ultrasound frequencies (Figure 5b). Thereafter, the left domains merge and form a common area with a very complex shape (Figure 5c). Finally, this area splits, and three nonoverlapping domains form (Figure 5d). It should be noted that such unusual dynamics of AO interaction areas differ significantly from the variant of a homogeneous (nonsectioned) acoustic field. This can be considered as an advantage of the phased-array transducers because the optimal geometry of interaction in the most convenient frequency range can be obtained when designing AO devices.

Of particular interest is the case of Figure 5c, which shows a completely unique AO interaction structure characterized by low angular and frequency selectivity. To evaluate this result, one has to take into account that the high selectivity of the Bragg diffraction in many AO devices is an interfering factor that leads to a decrease in the frequency and angular ranges of AO interaction, and, consequently, to a deterioration of characteristics of AO devices in resolution and operation speed. Figure 5c proves that by using the same AO cell, both collimated light beams and image-carrying beams can be processed in a wide frequency range.

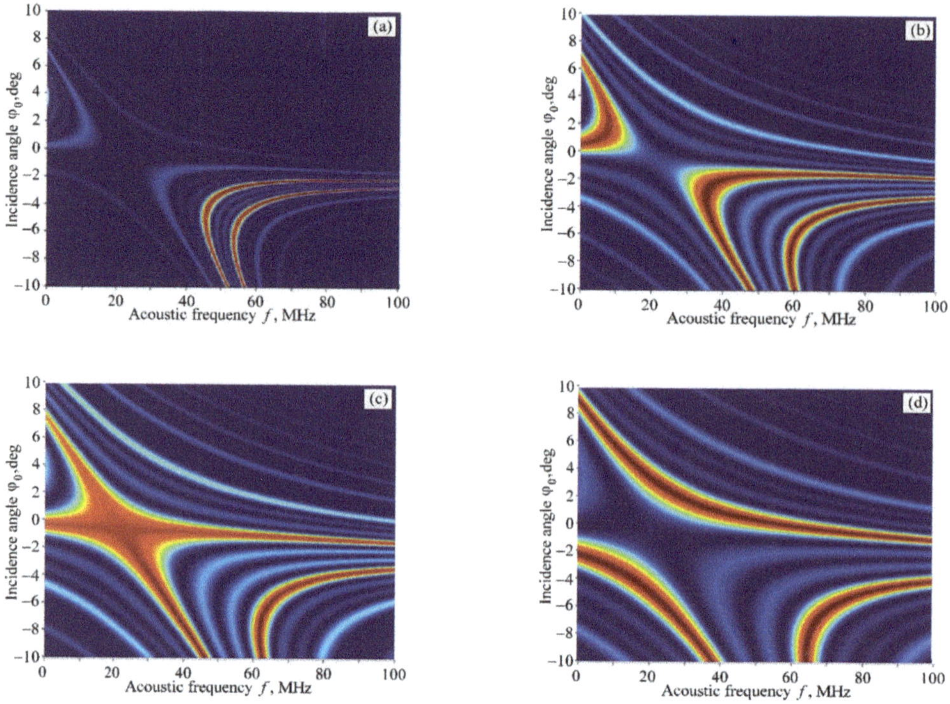

Figure 5. AO interaction areas for branch *+1e* in the case of $m = 4$ and different width of individual sections: (**a**) $l = 0.4$ mm; (**b**) $l = 0.15$ mm; (**c**) $l = 0.11$ mm; (**d**) $l = 0.07$ mm.

Two special points on the Bragg curves, D and T, are pointed out in Figure 4a. They determine the optimal AO interaction geometry for deflectors and video-filters [1–3]. In the case of anisotropic diffraction in uniaxial crystals, these points never coincide. This means that different AO interaction geometries must be applied when creating deflectors and filters. Our research has shown that this problem can be solved by using phased-array transducers, as Figure 5c demonstrates.

Figure 6 displays the characteristics of this unusual AO geometry. In Figure 6a, the operating frequency f_0, which corresponds to the coinciding points D and T, is presented in the dependence of the crystal cut angle χ and the period of the transducer structure d. The divergence angle between the incident and diffracted beams $\delta\phi$ (important characteristic for video-filters) increases with the acoustic frequency according to the formula $\delta\phi \approx \lambda f_0/nV$. Unfortunately, this positive dependence is accompanied by a decreasing AO figure of merit M (green curve in Figure 6a), which falls from a maximum value $M = 1200 \times 10^{-18}$ s^3/g at $\chi = 0°$ to $M = 42 \times 10^{-18}$ s^3/g at $\chi = 10°$ [1].

Figure 6b shows detailed characteristics related to the case $f_0 = 21$ MHz. The frequency dependences are constructed for different angles of light incidence: $\phi_0 = -0.5°$ (blue line), $\phi_0 = -1°$ (red line), and $\phi_0 = -1.4°$ (green line). It is seen that the frequency range $\Delta f_1 = 65$ MHz is confirmed to the angular range $\Delta\phi_0 = 0.4°$ (between green and red curves), while the range $\Delta f_2 = 42$ MHz can be realized in the angular range $\Delta\phi_0 = 0.9°$ (between the green and blue curves). At the same time, the frequency $f_0 = 21$ MHz corresponds to the angular range of the AO interaction $\Delta\phi_0 = 4.2°$, which is 3.1 times greater than in the case of the unsectioned transducer.

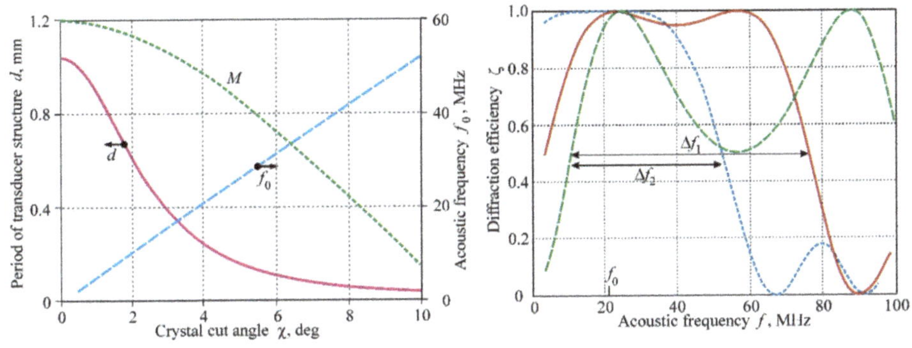

(a) (b)

Figure 6. Characteristics of low-selective area shown in Figure 5c: (**a**) Operating frequency f_0, period of transducer structure d and figure of merit M as functions of crystal cut angle χ; (**b**) Frequency characteristics of AO interaction at the operating frequency $f_0 = 21$ MHz ($\chi = 4°$).

The calculation of AO interaction areas for the variant of ordinary polarization of incident light (branch *+1o* in Figure 4a) gives a significantly different result. The phased-array transducer also leads to a splitting of the interaction area into two domains. However, the change of these domains when varying the period d of the transducer structure is greatly different in comparison with the variant discussed above: the region with extremely low angular and frequency selectivity does not appear at all, as seen in Figure 7.

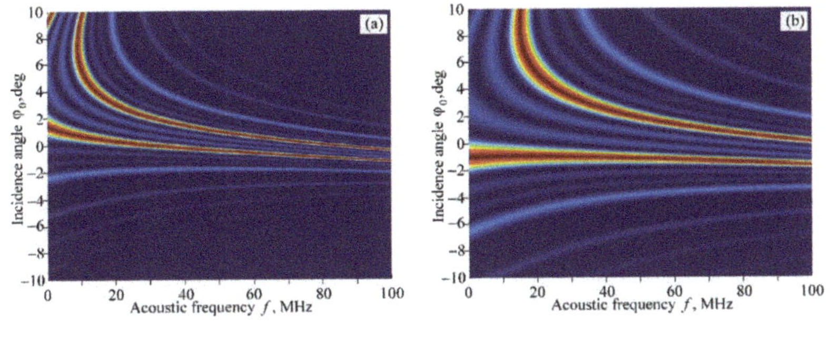

Figure 7. AO interaction areas for branch *+1o* in the case of $m = 4$ and different width of individual sections: (**a**) $l = 0.15$ mm; (**b**) $l = 0.07$ mm.

The comparison of the AO interaction areas for the branches *+1e* and *+1o* allows us to note another interesting and practically important feature, which is fundamentally impossible in the case of AO diffraction in the homogeneous acoustic field. Figure 8 demonstrates the frequency dependences of the optimal angles of light incidence ϕ_{opt}, combined for branches *+1e* and *+1o*. The graphs show that the curves, belonging to different diffraction orders *+1e* and *+1o*, intersect at the point M at the frequency $f = 67$ MHz (Figure 8a) or $f = 134$ MHz (Figure 8b). This means that, at these frequencies, the optical waves with ordinary and extraordinary polarizations have to diffract into the same diffraction maximum of the +1st order. Such AO interaction geometry makes it possible to modulate the intensity of unpolarized light beams without optical losses.

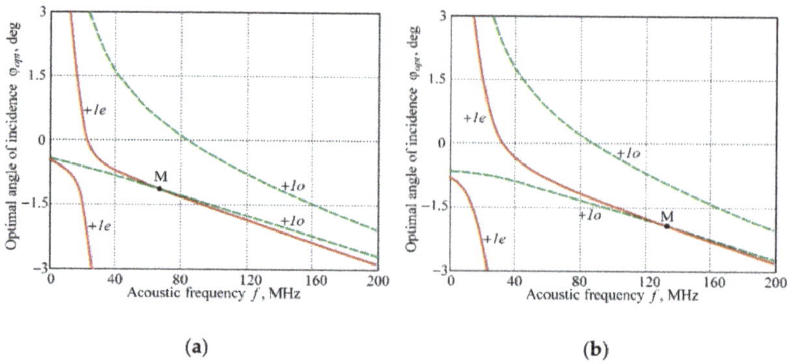

Figure 8. Frequency dependences of optimal angles for branches +1e and +1o: calculation for (**a**) d = 0.22 mm, and (**b**) d = 0.2 mm.

In this regard, it should be noted that the problem of AO modulation of unpolarized light has existed since the early 1960s and has not yet received an effective solution. The easiest solution is to search for cuts of crystals with close values of AO figure of merit M for both optical eigenmodes. A suitable material is lead molybdate ($PbMoO_4$) [1]. However, this crystal is characterized by a relatively small AO figure of merit value $M = 36 \times 10^{-18}$ s^3/g and, most importantly, has bad thermophysical properties. Modulators based on the paratellurite crystal with a longitudinal acoustic wave along the Z-axis are commercially available, but this cut of the crystal is characterized by a noticeable difference in the figure of merit for the ordinary and extraordinary waves: $M_o = 30 \times 10^{-18}$ s^3/g and $M_e = 22 \times 10^{-18}$ s^3/g. Besides, we may also mention an exotic scheme of an AO modulator with two multidirectional acoustic beams having different frequencies, which was studied in [18].

Contrary to that, we propose a significantly simpler variant of the modulator based on a paratellurite crystal with a phased-array transducer. In this device, the figure of merit coefficients for both optical eigenmodes are the same (due to the anisotropic type of diffraction) and equal to $M = 800 \times 10^{-18}$ s^3/g (for $\chi = 4°$). The position of the operating point M in Figure 8 is determined by the period of the transducer structure d. Choosing the period of the phased array transducer, one can change the operating point in a wide frequency range. Similar variants of the modulator are possible for other cuts of the paratellurite crystal.

A similar idea can be implemented for creating AO deflectors intended for the scanning of unpolarized optical beams. Figure 9a illustrates this situation for the variant $\lambda = 1.5$ μm, $\chi = 2°$, and $m = 8$ and $d = 0.24$ mm. Here, the dashed curves demonstrate frequency dependences of the Bragg angles for a nonsectioned transducer, whereas the AO interaction domains for the phased-array transducer are shown by red and blue colors. The yellow color represents the overlap area. Letter D shows the position of the deflector operating point at $f = 49$ MHz and $\phi_0 = 3°$.

Frequency characteristics of this deflector for the incidence angle $\phi_0 = 3°$ are presented in Figure 9b. The two curves relate to different polarizations of the incident light. It can be seen that the common frequency band is $\Delta f = 35$ MHz. This means that such a deflector can provide optical beam scanning in an angular range of $\Delta\phi = (\lambda/nV)\Delta f = 2.1°$ with spatial resolution $N = (w/V)\Delta f = 570$ resolvable points when the optical aperture of the AO cell is $w = 1$ cm.

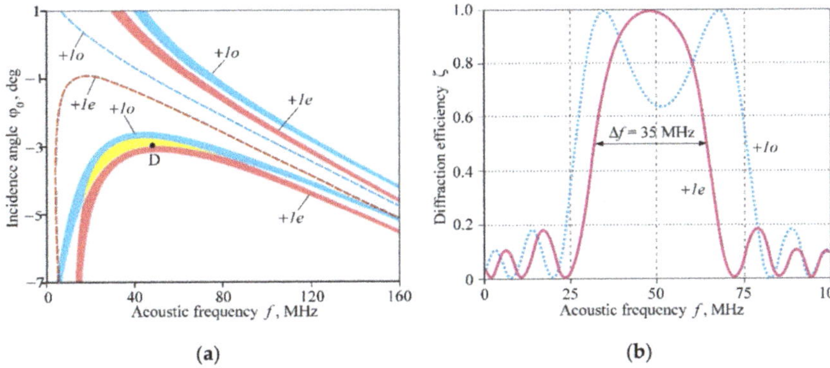

Figure 9. AO deflector of nonpolarized light: (**a**) Combined areas of AO interaction for *+1e* (red color) and *+1o* (blue color) diffraction branches, the overlap area is shown by yellow color; (**b**) Frequency characteristics of the AO deflector in the case of unpolarized light.

4. Experimental Results

Experimental studies were carried out with an AO cell made of a paratellurite crystal in the variant of Figure 1b. An acoustic wave in the form of a slow shear mode was excited in the crystallographic plane $(1\bar{1}0)$ at an angle $\chi = 9°$ to the direction [110]. For this crystal cut, the sound velocity was equal to $V = 0.69 \times 10^5$ cm/s, and the walk-off angle was $\alpha = 52.5°$. A phased-array transducer was made of an *X*-cut lithium niobate crystal with an electromechanical coupling coefficient $k = 68\%$. The transducer had $m = 9$ sections with a width of $l = 2$ mm each and a relative gap between the sections of $\xi = 0.2$ (Figure 10a). Thus, the total length of the transducer in the direction of light propagation was $L = 2.12$ cm. To effectively excite ultrasound, an RF generator was matched with the transducer with the help of reactive elements—an inductance coil and a capacitor. The efficiency of electric to acoustic power conversion in the maximum of the frequency characteristic was 99%, and the ultrasound excitation band extended from 70 to 160 MHz.

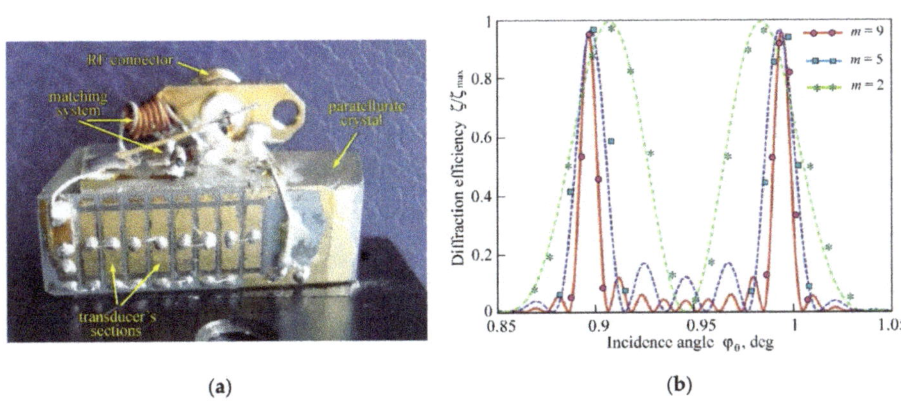

Figure 10. Experimental results: (**a**) AO cell with a nine-section transducer; (**b**) Angular characteristics of the AO diffraction for different number of connected sections: $m = 9$ (red), 5 (blue) and 2 (green).

Figure 10 shows the angular characteristics of AO interaction (normalized diffraction efficiency ζ/ζ_{max} as a function of the incidence angle ϕ_0) obtained at the acoustic frequency $f = 105$ MHz. The measurements were fulfilled at the scattering of the ordinary polarized optical radiation with a wavelength of $\lambda = 0.53$ μm into the +1st diffraction order.

The red curve relates to the case where all nine sections are connected in series. In good agreement with calculations, the angular characteristic consists of two main maxima located symmetrically on different sides of the Bragg angle. Decreasing the number of sections results in a broadening of the maxima and a reduction in their intensity. In our experiment, for the convenience of comparing the results, the ultrasound power was adjusted inversely proportional to the number of sections.

5. Conclusions

In this paper, we present the results of a study of anisotropic Bragg diffraction of light in a spatially periodical acoustic field created by a sectioned piezoelectric transducer with antiphase excitation of adjacent sections. The problem of AO interaction in this periodical acoustic field is solved, taking into account the strong optical and acoustic anisotropy of the interaction medium that is typical for many crystals used in modern acousto-optics. Numerical calculations are performed for an AO cell made of paratellurite crystal with the shear acoustic mode propagating at an angle $\chi = 4^\circ$ to the plane (001) of the crystal. It is shown that the AO interaction in such a structure is absent when the optical beam falls at the Bragg angle. However, angles of incidence exist, called "optimal", at which the diffraction efficiency can reach 100%, despite the violation of the phase-matching condition. The areas of interaction for different periods of the transducer structure are then calculated. A number of unusual regularities of AO scattering of light are established, which can be useful in the development of a new type of AO devices, such as modulators, deflectors and filters [1–3], as well as dual-channel monochromators for stereoscopic research [19,20]. In particular, the possibility of implementing AO modulators and deflectors for controlling nonpolarized light is shown, which could provide significantly better characteristics than the currently known devices. Preliminary experimental studies are performed with a paratellurite cell, with section numbers varying from nine to two. Results of the experiment are in a good agreement with the theoretical analysis.

Author Contributions: Conceptualization, V.B.; investigation, M.K., S.M. and V.M.; methodology, V.M.; resources, S.M. and V.M.; software, M.K.; supervision, V.B.; writing—original draft, M.K.; writing—review and editing, V.B. All authors have read and agreed to the published version of the manuscript.

Funding: This research was funded in part by the Russian Science Foundation (RSF), grant number 19-19-00606.

Institutional Review Board Statement: Not applicable.

Informed Consent Statement: Not applicable.

Data Availability Statement: Data sharing is not applicable to this article.

Conflicts of Interest: The authors declare no conflict of interest.

References

1. Balakshy, V.I.; Parygin, V.N.; Chirkov, L.E. *Physical Principles of Acousto-Optics*; Radio & Svyaz: Moscow, Russia, 1985.
2. Das, P.K.; DeCusatis, C.M. *Acousto-Optic Signal. Processing: Fundamentals and Applications*; Artech House: New York, NY, USA, 1991.
3. Xu, J.; Stroud, R. *Acousto-Optic Devices*; Wiley: New York, NY, USA, 1992.
4. Korpel, A.; Adler, R.; Desmares, P.; Watson, W. A television display using acoustic deflection and modulation of coherent light. *Proc. IEEE* **1966**, *54*, 1429–1437. [CrossRef]
5. Coquin, G.A.; Griffin, J.P.; Anderson, L.K. Wide-band acousto-optic deflectors using acoustic beam steering. *IEEE Trans. Son. Ultrason.* **1970**, *17*, 34–40. [CrossRef]
6. Alphonse, G.A. Broadband acoustooptic deflectors: New results. *Appl. Opt.* **1975**, *14*, 201–207. [CrossRef] [PubMed]
7. Kim, S.; Gao, L.; Wagner, K.; Weverka, R.; McLeod, R. Acousto-optic tunable filter using phased-array transducer with linearized RF to optical frequency mapping. *Proc. SPIE* **2005**, *5953*, 59530. [CrossRef]
8. Antonov, S.N.; Vainer, A.V.; Proklov, V.V.; Rezvov, Y.G. Extension of the angular scanning range of the acousto-optic deflector with a two-element phased-array piezoelectric transducer. *Techn. Phys.* **2013**, *58*, 1346–1351. [CrossRef]
9. Muromets, A.V.; Voloshinov, V.B.; Kononin, I.A. Transmission characteristics of acousto-optic tunable filters using sectioned transducers. *Appl. Acoust.* **2016**, *112*, 221–225. [CrossRef]

10. Balakshy, V.I.; Linde, B.B.J.; Vostrikova, A.N. Acousto-optic interaction in a non-homogeneous acoustic field excited by a wedge-shaped transducer. *Ultrasonics* **2008**, *48*, 351–356. [CrossRef] [PubMed]
11. Mantsevich, S.N.; Balakshy, V.I. Acousto-optic interaction in an inhomogeneous acoustic field. *Opt. Spectr.* **2015**, *118*, 617–622. [CrossRef]
12. Dixon, R.W. Acoustic diffraction of light in anisotropic media. *IEEE J. Quant. El.* **1967**, *3*, 85–93. [CrossRef]
13. Voloshin, A.; Balakshy, V.; Mantsevich, S. Unpolarized Light Diffraction in an Acoustic Field Created by a Phased Array Transducer. In Proceedings of the IEEE International Ultrasonics Symposium (IUS), Tours, France, 18–21 September 2016; pp. 767–770.
14. Balakshy, V.I.; Kupreychik, M.I.; Mantsevich, S.N. Light diffraction on ultrasound in a spatially periodic acoustic field. *Opt. Spectr.* **2019**, *127*, 712–718. [CrossRef]
15. Balakshy, V.I.; Voloshin, A.S.; Molchanov, V.Y. Anisotropic light diffraction in crystals with a large acoustic-energy walk-off. *Opt. Spectr.* **2014**, *117*, 801–806. [CrossRef]
16. Pozhar, V.E.; Pustovoit, V.I. Consecutive collinear diffraction of light in several acoustooptic cells. *Sov. J. Quant. Electron.* **1985**, *15*, 1438–1439. [CrossRef]
17. Balakshy, V.I.; Voloshin, A.S.; Molchanov, V.Y. Influence of acoustic energy walk-off on acousto-optic diffraction characteristics. *Ultrasonics* **2015**, *59*, 1–7. [CrossRef] [PubMed]
18. Antonov, S.N.; Mirgorodsky, V.I. Anisotropic acoustooptic modulator of nonpolarized light based on diffraction by a slow acoustic wave in a paratellurite crystal. *Techn. Phys.* **2004**, *49*, 83–86. [CrossRef]
19. Machikhin, A.; Batshev, V.; Pozhar, V.; Mazur, M.; Naumov, A.; Gorevoy, A. Development of dual-channel AOTF-based system for 3D imaging spectroscopy. *Proc. SPIE* **2019**, *11208*, 1120857. [CrossRef]
20. Machikhin, A.; Batshev, V.; Pozhar, V.; Naumov, A. Single-volume dual-channel acousto-optical tunable filter. *Opt. Expr.* **2020**, *28*, 1150–1157. [CrossRef] [PubMed]

materials

MDPI

Article

Analysis of Acousto-Optic Figure of Merit in KGW and KYW Crystals

Konstantin B. Yushkov *, Natalya F. Naumenko and Vladimir Ya. Molchanov

Acousto-Optical Research Center, University of Science and Technology MISIS, 119049 Moscow, Russia
* Correspondence: konstantin.yushkov@misis.ru

Abstract: Monoclinic potassium rare-earth crystals are known as efficient materials for solid-state lasers and acousto-optic modulators. A number of specific configurations for acousto-optic devices based on those crystals have recently been proposed, but the acousto-optic effect of those crystals has only been analyzed fragmentarily for some interaction directions. In this work, we numerically searched for the global maxima of an acousto-optic figure of merit for isotropic diffraction in $KGd(WO_4)_2$ and $KY(WO_4)_2$ crystals. It was demonstrated that the global maxima of the acousto-optic figure of merit in those crystals occur in the slow optical mode propagating along the crystal's twofold symmetry axis and in the acoustic wave propagating orthogonally, both for quasi-longitudinal and quasi-shear acoustic modes. The proposed calculation method can be readily used for the optimization of the acousto-optic interaction geometry in crystals with arbitrary symmetry.

Keywords: acousto-optics; diffraction; figure of merit; monoclinic crystal; potassium rare-earth tungstate

check for updates

Citation: Yushkov, K.B.; Naumenko, N.F.; Molchanov, V.Y. Analysis of Acousto-Optic Figure of Merit in KYW and KGW Crystals. *Materials* **2022**, *15*, 8183. https://doi.org/10.3390/ma15228183

Academic Editors: Alexander S. Machikhin and Vitold Pozhar

Received: 25 October 2022
Accepted: 15 November 2022
Published: 17 November 2022

Publisher's Note: MDPI stays neutral with regard to jurisdictional claims in published maps and institutional affiliations.

1. Introduction

In the recent decade, there has been a growth of interest in the acousto-optic (AO) properties of biaxial crystals [1–24]. The main factors motivating the research in this area are (1) the good optical and acousto-optic properties of some biaxial crystals used in nonlinear optics and laser physics, and (2) a larger number of material constants in biaxial crystals as compared to uniaxial ones, which provides more useful configurations of AO interactions. Moreover, biaxial crystals are of special interest in AO research since they can provide unique configurations of the Bragg diffraction for laser beam deflection and spatial filtering [11,19,24].

In particular, potassium rare-earth tungstate crystals, also known as the KREW family, have recently been recognized as efficient crystal materials for applications in acousto-optics [2–4]. Those crystals belong to the monoclinic crystal system, which ensures a large number of independent elastic and photoelastic constants. Potassium gadolinium tungstate (KGW), $KGd(WO_4)_2$, and potassium yttrium tungstate (KYW), $KY(WO_4)_2$, are the most prominent crystals of the KREW family. Previous studies included measurements and computations of the AO figure of merit M_2 only for certain geometries of an AO interaction, with light propagating near the twofold symmetry axis of the crystal and ultrasound propagating on the symmetry plane [3,4,7,14,22]. A fair AO figure of merit in KREW crystals has been combined with a high laser-induced damage threshold [23,25]. This allowed for the design and fabrication of several experimental configurations of AO modulators (AOMs) and Q-switches for infrared (IR)-pulsed lasers [14,15,17,23]. Specific combinations of the acoustic and photoelastic properties of those crystals has prompted the creation of novel designs for AO devices, including spatial light modulators [18] as well as two-coordinate monolithic deflectors and polarization switches [22].

In this work, we explore the spatial distributions of an AO figure of merit M_2 in monoclinic KREW crystals with an arbitrary orientation of light and ultrasound under the assumption of a small-angle orthogonal diffraction. A procedure that can be used to search for the global maxima of the M_2 for isotropic AO diffraction in crystals with arbitrary symmetry is proposed and validated. Numerical results for KGW and KYW crystals are obtained and analyzed.

2. Materials and Methods

2.1. KREW Crystals' Physical Properties

KYW and KGW are monoclinic crystals of the 2/m point group. The orientation of the dielectric axes relative to the crystal's unit cell is shown in Figure 1. We use the axes setting for monoclinic crystals, with the Z axis being parallel to the twofold symmetry axis of the crystal (**b** crystallographic axis and N_p dielectric axis); X and Y are the dielectric axes N_m and N_g, respectively.

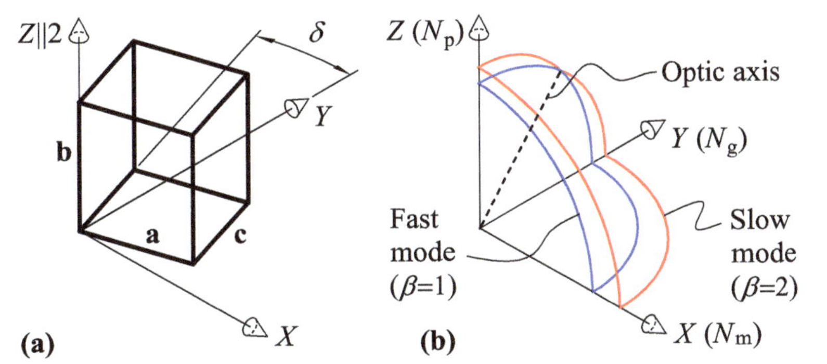

Figure 1. Dielectric properties of the KYW and KGW monoclinic crystals: (**a**) crystal's unit cell; (**b**) refractive index surface cut-out. Z is the twofold symmetry axis; XY is the symmetry plane; optic axis (dashed line) is on the ZY plane.

The acoustic and photoelastic constants of KREW crystals have been measured by Mazur et al. [2,6]. The stiffness and photoelastic tensors, c_{ijkl} and p_{ijkl}, are written in Voight's notation. In monoclinic crystals, there are 13 independent stiffness coefficients c_{qr}:

$$c_{qr} = \begin{pmatrix} c_{11} & c_{12} & c_{13} & 0 & 0 & c_{16} \\ c_{12} & c_{22} & c_{23} & 0 & 0 & c_{26} \\ c_{13} & c_{23} & c_{33} & 0 & 0 & c_{36} \\ 0 & 0 & 0 & c_{44} & c_{45} & 0 \\ 0 & 0 & 0 & c_{45} & c_{55} & 0 \\ c_{16} & c_{26} & c_{36} & 0 & 0 & c_{66} \end{pmatrix} \quad (1)$$

and 20 independent photoelastic coefficients p_{qr}:

$$p_{qr} = \begin{pmatrix} p_{11} & p_{12} & p_{13} & 0 & 0 & p_{16} \\ p_{21} & p_{22} & p_{23} & 0 & 0 & p_{26} \\ p_{31} & p_{31} & p_{33} & 0 & 0 & p_{36} \\ 0 & 0 & 0 & p_{44} & p_{45} & 0 \\ 0 & 0 & 0 & p_{54} & p_{55} & 0 \\ p_{61} & p_{62} & p_{63} & 0 & 0 & p_{66} \end{pmatrix}. \quad (2)$$

The values of the stiffness, photoelastic, and dielectric constants for the KGW crystal are listed in Appendix A.1, and those for the KYW crystal are in Appendix A.2.

2.2. Acousto-Optic Figure of Merit

The efficiency of the AO interaction in crystals depends on both acoustic and optical wave propagation directions with respect to the crystal axes. Any direction of the acoustic wave vector **s** is associated with three bulk acoustic wave (BAW) eigenmodes having orthogonal polarization vectors **u** and different phase velocities v. The exceptions are the acoustic axes, i.e., the directions corresponding to the degeneracy of two acoustic modes [21]. Any propagation direction of light **m** in a birefringent crystal is associated with two orthogonal eigenmodes that have polarization vectors **d** and refractive indices n. There are two types of AO diffraction: isotropic diffraction, which corresponds to the same polarization of incident and diffracted beams, and anisotropic diffraction, which corresponds to the orthogonal polarization of the beams.

Furthermore, we assume the isotropic diffraction type and remain within a small-angle approximation, i.e., that the optical wave normal vector **m** and the polarization vector **d** are the same for incident and diffracted waves. This approximation is valid for most of the configurations of isotropic diffraction, except for those where light propagates near the optic axis of a biaxial crystal. In the neighborhood of conical optic axes, the polarizations of eigenwaves **d** rapidly change with the wave normal direction **m**, and isotropic diffraction may be mixed with the anisotropic one [24]. In the other cases, i.e., those far from singular points of the refractive index surface and those in the typical range of ultrasound frequencies of the Bragg diffraction—from a few tens to a few hundred MHz—the double Bragg angle is on the order of 1°, and the difference between polarization vectors **d** in the directions of incident and diffracted light is small. Thus, the small-angle approximation allows one to eliminate the ultrasound frequency from the problem parameters. In addition, the error of the experimental measurement of the photoelastic constants is more than 10% [6], therefore the error of $M_2 \propto p_{\text{eff}}^2$ can be estimated to be at 20 %.

Figure 2 illustrates the orientation of the acoustic and optical wave vectors and polarizations. Standard procedures for calculating the optical and acoustic eigenmodes in crystals are described elsewhere [26,27]. Hereinafter, we sort the acoustic and optical modes of crystals from the fastest to the slowest, i.e., $v^{(1)} > v^{(2)} \geqslant v^{(3)}$ and $n^{(1)} \leqslant n^{(2)}$. The fastest acoustic mode $\mathbf{u}^{(1)}$ is a quasi-longitudinal (QL) wave, and the other modes, $\mathbf{u}^{(2)}$ and $\mathbf{u}^{(3)}$, are quasi-shear (QS) waves. The polarization angle of acoustic waves is defined as $\gamma = \angle(\mathbf{s}, \mathbf{u}^{(1)})$.

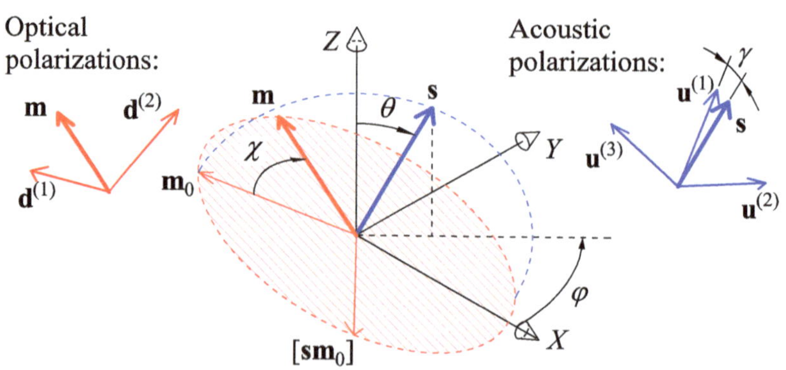

Figure 2. Spatial diagrams for M_2 calculations: a selection of acoustic and optical wave propagation directions and polarizations of optical and acoustic waves. Dashed plane is orthogonal to the acoustic direction unit vector **s**. The locus of optical wave normal vectors **m** for the given acoustic wave vector direction **s** is shown with hatching.

The direction of the acoustic wave vector is a function of two Euler angles, $\varphi \in [0, 360°]$ and $\theta \in [0, 90°]$, and is relative to the axes Z and X:

$$\mathbf{s}(\varphi, \theta) = \{\sin\theta\cos\varphi, \sin\theta\sin\varphi, \cos\theta\}. \tag{3}$$

Small-angle approximation is equivalent to orthogonal diffraction geometry, $(\mathbf{ms}) = 0$, so one can use the parametrization of $\mathbf{m}$ as follows:

$$\mathbf{m}(\mathbf{s}, \chi) = \mathbf{m}_0 \cos\chi + [\mathbf{m}_0\mathbf{s}] \sin\chi, \tag{4}$$

where the initial direction $\mathbf{m}_0$ is orthogonal to the line of nodes:

$$\mathbf{m}_0(\varphi, \theta) = \{-\cos\theta\cos\varphi, -\cos\theta\sin\varphi, \sin\theta\}. \tag{5}$$

For a certain interaction geometry, the effective photoelastic constant is expressed as follows [28]:

$$p_{\text{eff}}^{(\alpha,\beta)}(\mathbf{s}, \mathbf{m}) = d_i^{(\beta)} d_j^{(\beta)} p_{ijkl} s_k u_l^{(\alpha)} \tag{6}$$

where $\alpha = 1, 2, 3$ is the acoustic mode number, $\beta = 1, 2$ is the optical mode number, and summation over repeated lower indices is assumed. In Equation (7), d_i and d_j are the components of the optical polarization eigenvector $\mathbf{d}^{(\beta)}(\mathbf{m})$ corresponding to the β-th optical mode, and $\mathbf{u}(\mathbf{s})$ is the acoustic displacement vector corresponding to the α-th acoustic mode.

$$M_2^{(\alpha,\beta)}(\mathbf{s}, \mathbf{m}) = \frac{{n^{(\beta)}}^6 {p_{\text{eff}}^{(\alpha,\beta)}}^2}{\rho {v^{(\alpha)}}^3} \tag{7}$$

Numerical maximum search was used to find the optimal propagation of the optical beam that maximizes M_2 for a given BAW propagation direction. The result of this search is the optical direction, expressed through angle χ:

$$\chi_{\max}^{(\alpha,\beta)}(\mathbf{s}) = \underset{\chi \in [0, 360°]}{\arg\max}\, M_2^{(\alpha,\beta)}(\mathbf{s}, \mathbf{m}(\mathbf{s}, \chi)) \tag{8}$$

The final result is the maximum figure of merit for a chosen acoustic direction $\mathbf{s}$ and combination of modes (α, β).

$$M_{2\max}^{(\alpha,\beta)}(\varphi, \theta) = M_2^{(\alpha,\beta)}(\mathbf{s}, \mathbf{m}(\mathbf{s}, \chi_{\max}^{(\alpha,\beta)})) \tag{9}$$

In total, there are 6 combinations of optical and acoustic modes. Each combination was processed independently.

2.3. Software

The original software for computing $M_{2\max}$ was developed in the Fortran programming language. The initial data are the set of material constants (dielectric permittivity tensor ε_{ii}, stiffness c_{pq} and photoelastic p_{qr} matrices) defined in the crystals' dielectric axes. The software is based on numerical techniques of linear algebra instead of analytical solutions. Therefore, it is universal and can be applied to crystals of any symmetry.

Standard methods for calculating the optical and acoustic eigenmodes in crystals were used for every acoustic wave normal vector $\mathbf{s}$ and optical wave normal vector $\mathbf{m}$ [26,27]. Then, Equations (6)–(9) were used for each possible pair of mode indices α and β. In the computations, the grid steps were 2° for φ and 1° for θ and χ.

3. Results

The results of the numerical simulations for KGW and KYW are $M_{2\max}^{(\alpha,\beta)}$ and $\chi_{\max}^{(\alpha,\beta)}$, as functions of the Euler angles φ and θ for three BAW modes ($\alpha = 1, 2, 3$) and two optical modes ($\beta = 1, 2$) of the crystals. Full numerical data on $\chi_{\max}$ and $M_{2\max}$ can be found in the related dataset in [29]. All diagrams are plotted in stereographic projection showing the upper hemisphere, $\theta \in [0, 90°]$.

The data for the KGW crystal are plotted in Figures 3 and 4, including the acoustic properties of the crystal: BAW velocities $v^{(\alpha)}$ and normalized Gaussian curvatures $K^{(\alpha)}$ of the slowness surfaces. The Gaussian curvature is the product of two dimensionless BAW diffraction coefficients, which characterize the anisotropy of acoustic beam diffraction in crystals [30]. This coefficient tends to infinity in the neighborhood of conical acoustic axes. Similar results for the KYW crystal are plotted in Figures 5 and 6.

Table 1 summarizes diffraction geometry parameters that maximize the AO figure of merit M_2 in KGW and KYW crystals. The table includes the direction angles φ and θ of the BAW, the optical rotation angle χ that defines the diffraction plane, and polarizations of the interacting beams. The data for the QS modes include the global maxima over both QS modes.

Table 1. Maxima of the AO figure of merit in KGW and KYW crystals.

Parameter	KGW Crystal		KYW Crystal	
BAW mode, α	1	3	1	3
Polarization type and angle, γ	QL (0.7°)	QS (6.5°)	QL (1.0°)	QS (4.7°)
Velocity, v (10^3 m/s)	4.33	2.22	4.73	2.38
Azimuthal angle, φ	−62°	−16°	−58°	−12°
Polar angle, θ	90°	90°	90°	90°
Optical mode, β	2	2	2	2
Rotation angle, χ	0	0	0	0
Figure of merit, M_2 (10^{-15} s^3/g)	15.8	27.8	13.8	24.5

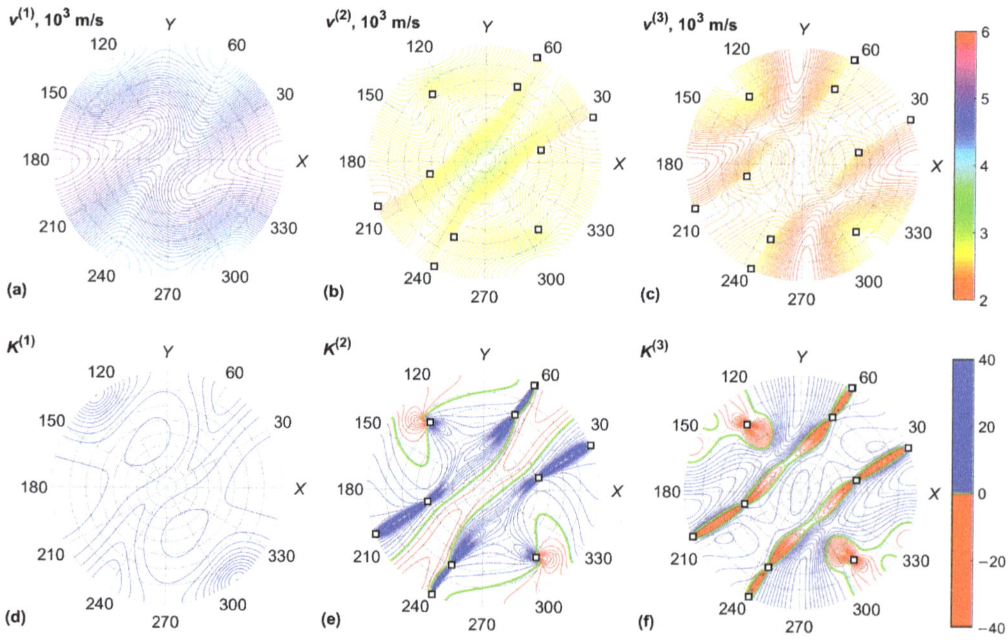

Figure 3. KGW crystal, acoustic properties: (**a**) QL BAW mode velocity; (**b**) fast QS BAW mode velocity; (**c**) slow QS BAW mode velocity; (**d**) QL BAW diffraction coefficient; (**e**) fast QS BAW diffraction coefficient; (**f**) slow QS BAW diffraction coefficient. Conical acoustic axes of the crystal are marked with squares.

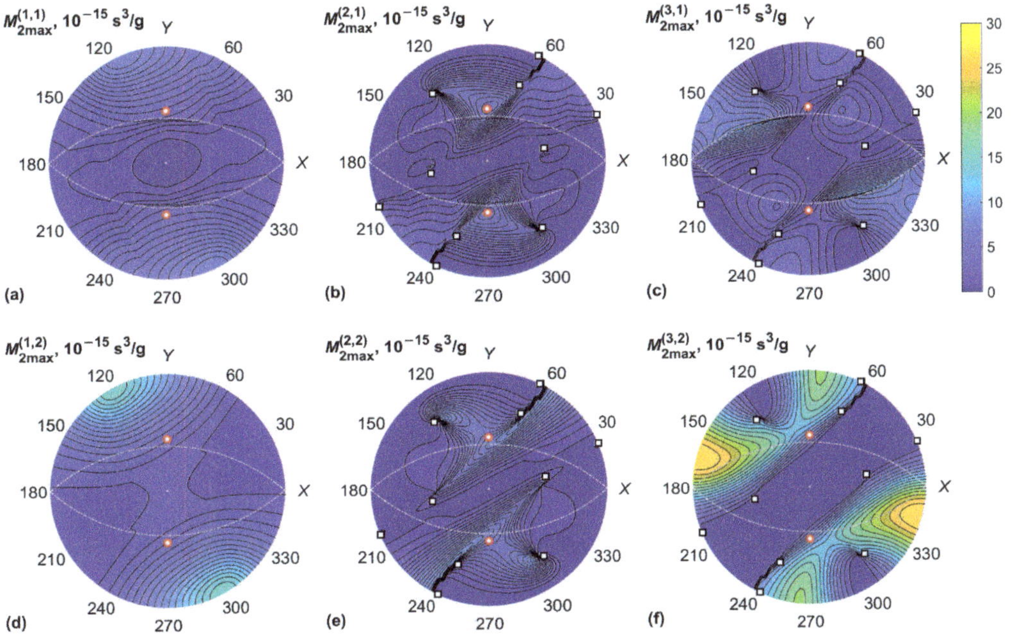

Figure 4. KGW crystal, maximum AO figure of merit, $M_{2\max}^{(\alpha,\beta)}$ vs. BAW direction (φ, θ): (**a**) QL BAW mode ($\alpha = 1$), fast optical mode ($\beta = 1$); (**b**) fast QS BAW mode ($\alpha = 2$), fast optical mode ($\beta = 1$); (**c**) slow QS BAW mode ($\alpha = 3$), fast optical mode ($\beta = 1$); (**d**) QL BAW mode ($\alpha = 1$), slow optical mode ($\beta = 2$); (**e**) fast QS BAW mode ($\alpha = 2$), slow optical mode ($\beta = 2$); (**f**) slow QS BAW mode ($\alpha = 3$), slow optical mode ($\beta = 2$). Conical acoustic axes of the crystal are marked with squares; optic axes are marked with red circles; directions orthogonal to optic axes are shown with dash-dotted lines.

General features of M_2 data presented in Figures 4 and 6 are the following: Firstly, the maxima of the figure of merit are associated with the minima of BAW velocity since $M_2 \propto v^{-3}$. Secondly, rapid changes in M_2 for the QS BAW modes took place near the acoustic axes of the crystal and the directions of high acoustic anisotropy between them. The reason for this is the fast rotation of BAW polarization vectors $\mathbf{u}^{(2)}$ and $\mathbf{u}^{(3)}$ near the directions of high anisotropy [30]. Rapid changes in $M_{2\max}$ were also observed for some BAW directions orthogonal to the optic axes of the crystal since the optical wave normal vector $\mathbf{m}$ in this case crossed the optic axis associated with polarization singularity. However, we note that the small-angle approximation has a limited validity when light propagates near an optic axis because of the high optical anisotropy and singularity of the polarization field affecting the AO phase matching [24]. For this reason, the AO diffraction of light propagating near the optic axis of a biaxial crystal cannot be used for the design of AO modulators. On the other hand, it enables the design of unique types of AO devices for the deflection of light and processing vector beams [1,5,11,19]. In this case, the AO figure of merit should be calculated with respect to the actual frequency of ultrasound and the polarizations of optical eigenmodes.

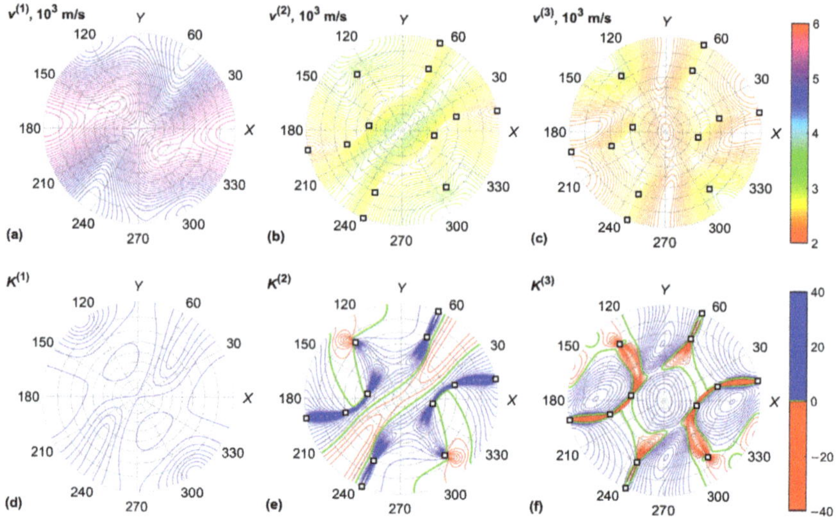

Figure 5. KYW crystal, acoustic properties: (**a**) QL BAW mode velocity; (**b**) fast QS BAW mode velocity; (**c**) slow QS BAW mode velocity; (**d**) QL BAW diffraction coefficient; (**e**) fast QS BAW diffraction coefficient; (**f**) slow QS BAW diffraction coefficient. Conical acoustic axes of the crystal are marked with squares.

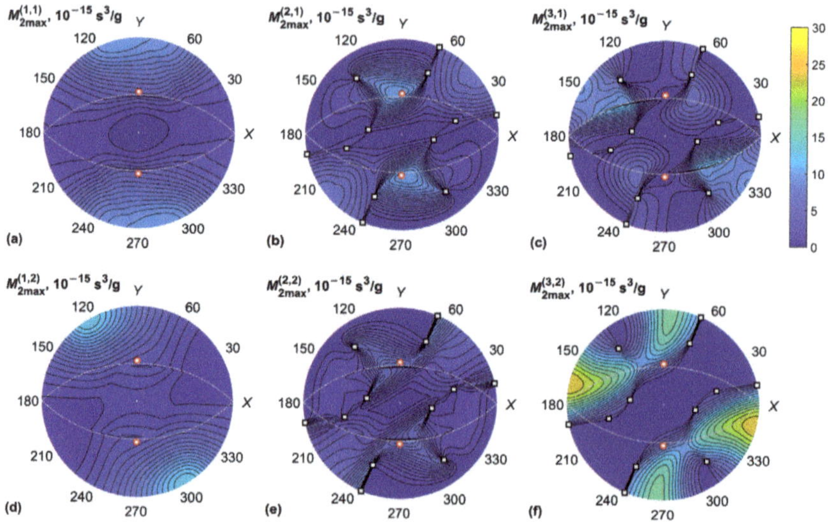

Figure 6. KYW crystal, maximum AO figure of merit: (**a**) QL BAW mode, fast optical mode; (**b**) fast QS BAW mode, fast optical mode; (**c**) slow QS BAW mode, fast optical mode; (**d**) QL BAW mode, slow optical mode; (**e**) fast QS BAW mode, slow optical mode; (**f**) slow QS BAW mode, slow optical mode. Conical acoustic axes of the crystal are marked with squares; optic axes are marked with red circles; directions orthogonal to optic axes are shown with dash-dotted lines.

4. Discussion

According to the plots in Figures 4 and 6, the global maxima of the figure of merit M_2 occurred for the QS BAW mode propagating on the *XY* plane and which has a displacement orthogonal to the Z axis. The peaks of M_2 were higher for the slow optical mode ($\beta = 2$). The global maxima were achieved at $\chi = 0$, which corresponds to the optical wave normal vector **m** parallel to the Z axis. To prove this, we selected the BAW propagation in the XY plane, i.e., $\theta = 90°$, and φ is a variable; we then calculated the figure of merit $M_2(\chi)$ analytically.

Analytical expressions for p_{eff} can be readily derived for the case of $\chi = 0$, i.e., an optical wave propagating along the Z axis [22]:

$$p_{\text{ql}}^{(i)} = p_{i1} \cos\varphi \cos(\varphi + \gamma) + p_{i2} \sin\varphi \sin(\varphi + \gamma) + p_{i6} \sin(2\varphi + \gamma); \tag{10}$$

$$p_{\text{qs}}^{(i)} = -p_{i1} \cos\varphi \sin(\varphi + \gamma) + p_{i2} \sin\varphi \cos(\varphi + \gamma) + p_{i6} \cos(2\varphi + \gamma), \tag{11}$$

where $i = 1$ for the fast optical mode with the polarization vector $\mathbf{d}^{(1)} = \{1, 0, 0\}$, and $i = 2$ for the slow optical mode with $\mathbf{d}^{(2)} = \{0, 1, 0\}$. Equations (10) and (11) describe the interaction geometry with the optical beam propagating along the Z axis, i.e., at $\chi = 0$. These equations are also valid at $i = 3$ for the fast optical mode propagating orthogonally to **s** on the XY plane, i.e., at $\chi = \pm 90°$. This mode has a polarization vector of $\mathbf{d}^{(1)} = \{0, 0, 1\}$. The slow optical mode for this direction has a polarization of $\mathbf{d}^{(2)} = \{\cos\varphi, \sin\varphi, 0\}$ and the following effective photoelastic constants:

$$p_{\text{ql}}\Big|_{\chi=90°} = p_{\text{ql}}^{(1)} \cos^2\varphi + p_{\text{ql}}^{(2)} \sin^2\varphi + p_{\text{ql}}^{(6)} \sin 2\varphi; \tag{12}$$

$$p_{\text{qs}}\Big|_{\chi=90°} = p_{\text{qs}}^{(1)} \cos^2\varphi + p_{\text{qs}}^{(2)} \sin^2\varphi + p_{\text{qs}}^{(6)} \sin 2\varphi. \tag{13}$$

To the best of our knowledge, the p_{61}, p_{62}, p_{63}, and p_{66} constants have not yet been measured [6], and in all simulations, they were used with zero values.

The maximum values $M_{2\,\text{max}}$ are compared with the calculated data in Figures 7 and 8. In panels (a) and (b), the maximum figure of merit was found for the QL BAW mode and compared to the calculations in Equations (10) and (12). In panels (c) and (g), the maximum figure of merit was found for both QS BAW modes, the fast and the slow, and then compared to the calculations from Equations (11) and (13).

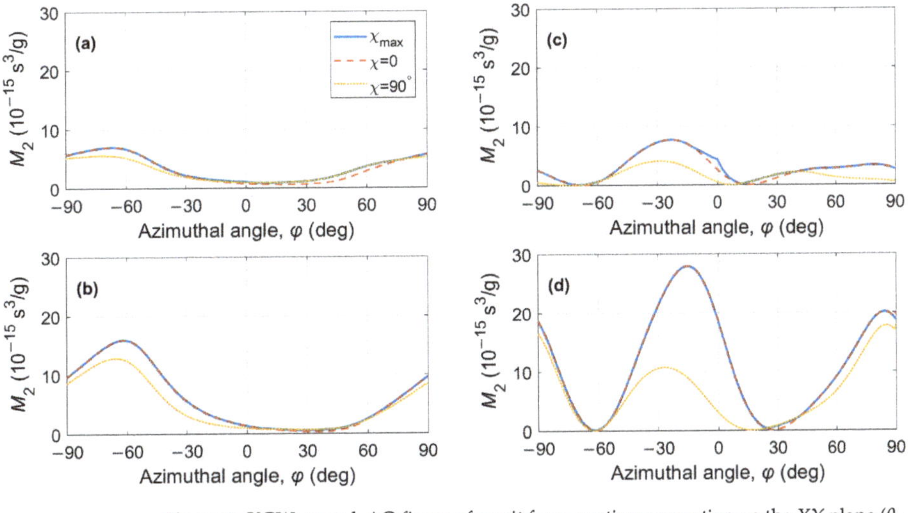

Figure 7. KGW crystal, AO figure of merit for acoustic propagation on the XY plane ($\theta = 90°$): **(a)** QL BAW mode, fast optical mode; **(b)** QL BAW mode, slow optical mode; **(c)** QS BAW mode, fast optical mode; **(d)** QS BAW mode, slow optical mode. Solid lines—data from Figure 4; dashed and dotted lines—calculation according to Equations (10)–(13).

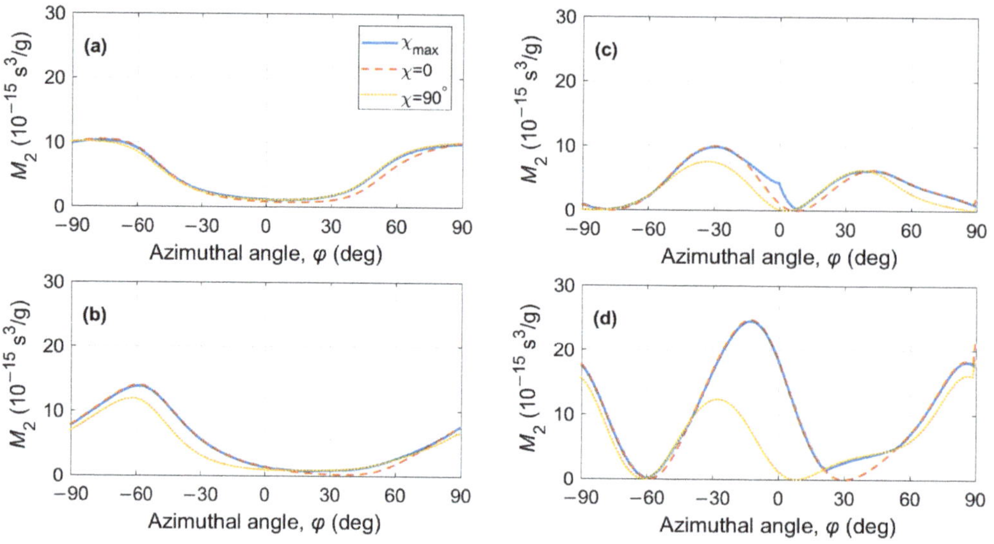

Figure 8. KYW crystal, AO figure of merit for acoustic propagation on the XY plane ($\theta = 90°$): (**a**) QL BAW mode, fast optical mode; (**b**) QL BAW mode, slow optical mode; (**c**) QS BAW mode, fast optical mode; (**d**) QS BAW mode, slow optical mode. Solid lines—data from Figure 6; dashed and dotted lines—calculation according to Equations (10)–(13).

5. Conclusions

We have reported a procedure that can be used to search for the global maxima of the AO figure of merit in crystals of lower symmetry classes. The procedure was able to find the optimal propagation direction of light and the corresponding figure of merit as the functions of the BAW propagation direction in crystals. We demonstrated this procedure with the use of two monoclinic crystals, KGW and KYW, which have previously shown high AO efficiency in certain configurations of AO interaction in the near-IR spectral region.

Our numerical simulations demonstrated that the global maxima of the figure of merit M_2 for isotropic diffraction in KGW and KYW crystals can be obtained for the BAW propagation on the XY plane and the optical propagation along the Z crystal axis. This conclusion is valid both for QL and QS BAW modes, and the highest M_2 values are produced for the slow optical mode polarized along the Y (i.e., N_g) dielectric axis. Nevertheless, we note that numerical data on M_2 in KREW crystals should be refined in the future since 8 out of the 20 photoelastic constants of these crystals have not been measured to date.

Author Contributions: Conceptualization, K.B.Y. and V.Y.M.; methodology, K.B.Y.; software, N.F.N.; investigation, K.B.Y. and N.F.N.; data curation, K.B.Y.; writing—original draft preparation, K.B.Y.; writing—review and editing, N.F.N. and V.Y.M.; project administration, V.Y.M. All authors have read and agreed to the published version of the manuscript.

Funding: This research was funded by the Ministry of Science and Higher Education of the Russian Federation (Priority2030) and the University of Science and Technology MISIS (K7-2022-052).

Institutional Review Board Statement: Not applicable

Informed Consent Statement: Not applicable

Data Availability Statement: The data presented in this study are openly available on Mendeley Data at http://dx.doi.org/10.17632/cg38s35c8d.1, accessed on 24 October 2022.

Conflicts of Interest: The authors declare no conflicts of interest.

Abbreviations

The following abbreviations are used in this manuscript:

AO	Acousto-optic
BAW	Bulk acoustic wave
IR	Infrared
KGW	$KGd(WO_4)_2$, potassium gadolinium tungstate
KREW	Potassium rare-earth tungstate
KYW	$KY(WO_4)_2$, potassium yttrium tungstate
QL	Quasi-longitudinal
QS	Quasi-shear

Appendix A

Appendix A.1

Material constants of the KGW crystal relevant to this research are the following: The crystal density is $\rho = 7.270\ \mathrm{g/cm^3}$. The elastic constants of the KYW crystal are as follows [2]:

$$c_{qr} = \begin{pmatrix} 197.8 & 51.1 & 92.5 & 0 & 0 & -3.7 \\ & 140.0 & 61.3 & 0 & 0 & -0.9 \\ & & 171.7 & 0 & 0 & -3.0 \\ & & & 64.0 & -12.1 & 0 \\ & & & & 45.2 & 0 \\ & & & & & 31.4 \end{pmatrix} \cdot \mathrm{GPa} \quad \text{(A1)}$$

The lower sub-diagonal part of the matrix is not shown because of its symmetry. The photoelastic tensor of the KGW crystal is the following [6]:

$$p_{qr} = \begin{pmatrix} 0.11 & 0.23 & 0.14 & 0 & 0 & -0.053 \\ 0.13 & 0.28 & 0.09 & 0 & 0 & -0.13 \\ 0.13 & 0.23 & 0.04 & 0 & 0 & -0.025 \\ 0 & 0 & 0 & p_{44} & p_{45} & 0 \\ 0 & 0 & 0 & p_{54} & p_{55} & 0 \\ p_{61} & p_{62} & p_{63} & 0 & 0 & p_{66} \end{pmatrix} \quad \text{(A2)}$$

Eight coefficients in the lower half of the matrix are not known. The optical properties of KGW are described with single-term, IR-corrected Sellmeier's equations [31]:

$$\begin{aligned} \varepsilon_{11} = n_\mathrm{m}^2 &= 0.8753 + 3.0976\lambda^2/(\lambda^2 - 0.1554^2) - 0.00042\lambda^2; \\ \varepsilon_{22} = n_\mathrm{g}^2 &= 1.5157 + 2.6141\lambda^2/(\lambda^2 - 0.1733^2) + 0.00183\lambda^2; \\ \varepsilon_{33} = n_\mathrm{p}^2 &= 0.9927 + 2.8661\lambda^2/(\lambda^2 - 0.1523^2) - 0.00139\lambda^2, \end{aligned} \quad \text{(A3)}$$

which correspond to the following refractive indices at $\lambda = 1.06\ \mu\mathrm{m}$: $n_\mathrm{p} = \sqrt{\varepsilon_{33}} = 1.979$, $n_\mathrm{m} = \sqrt{\varepsilon_{11}} = 2.010$, and $n_\mathrm{g} = \sqrt{\varepsilon_{22}} = 2.050$. The angle between the dielectric axis Y and the crystallographic axis $\mathbf{c}$ is $\delta = 21.5^\circ$. This angle was not explicitly used in the calculations, but it is necessary for recalculating the stiffness and photoelastic constants in different frames of reference.

Appendix A.2

Material constants of the KYW crystal relevant to this research are the following: The crystal density is $\rho = 6.565\ \mathrm{g/cm^3}$. The elastic constants of the KYW crystal are the following [2]:

$$c_{qr} = \begin{pmatrix} 202.2 & 51.0 & 90.9 & 0 & 0 & -3.3 \\ & 150.6 & 60.9 & 0 & 0 & -2.7 \\ & & 176.8 & 0 & 0 & -6.2 \\ & & & 61.9 & -15.4 & 0 \\ & & & & 41.8 & 0 \\ & & & & & 33.5 \end{pmatrix} \cdot \mathrm{GPa} \quad \text{(A4)}$$

The photoelastic tensor of the KYW crystal is as follows [6]:

$$p_{qr} = \begin{pmatrix} 0.12 & 0.33 & 0.17 & 0 & 0 & -0.04 \\ 0.14 & 0.27 & 0.10 & 0 & 0 & -0.14 \\ 0.15 & 0.35 & 0.05 & 0 & 0 & -0.02 \\ 0 & 0 & 0 & p_{44} & p_{45} & 0 \\ 0 & 0 & 0 & p_{54} & p_{55} & 0 \\ p_{61} & p_{62} & p_{63} & 0 & 0 & p_{66} \end{pmatrix} \quad \text{(A5)}$$

Eight coefficients in the lower half of the matrix are not known, similar to thet KGW crystal. The optical properties of KYW are described with single-term, IR-corrected Sellmeier's equations [31]:

$$\begin{aligned} \varepsilon_{11} = n_{\mathrm{m}}^2 &= 2.5253 + 1.4638\lambda^2/(\lambda^2 - 0.2083^2) - 0.00237\lambda^2; \\ \varepsilon_{22} = n_{\mathrm{g}}^2 &= 2.3951 + 1.7564\lambda^2/(\lambda^2 - 0.2025^2) - 0.00226\lambda^2; \\ \varepsilon_{33} = n_{\mathrm{p}}^2 &= 2.6986 + 1.1578\lambda^2/(\lambda^2 - 0.2127^2) - 0.00215\lambda^2, \end{aligned} \quad \text{(A6)}$$

which correspond to the following refractive indices at $\lambda = 1.06\ \mu\mathrm{m}$: $n_{\mathrm{p}} = \sqrt{\varepsilon_{33}} = 1.975$, $n_{\mathrm{m}} = \sqrt{\varepsilon_{11}} = 2.011$, and $n_{\mathrm{g}} = \sqrt{\varepsilon_{22}} = 2.053$. The angle between the dielectric axis Y and the crystallographic axis **c** is $\delta = 17.5°$.

References

1. Belyi, V.N.; Khilo, N.A.; Kurilkina, S.N.; Kazak, N.S. Spin-to-orbital angular momentum conversion for Bessel beams propagating along the optical axes of homogeneous uniaxial and biaxial crystals. *J. Opt.* **2013**, *15*, 044018. [CrossRef]
2. Mazur, M.M.; Velikovskiy, D.Y.; Mazur, L.I.; Pavluk, A.A.; Pozhar, V.E.; Pustovoit, V.I. Elastic and photo-elastic characteristics of laser crystals potassium rare-earth tungstates $KRE(WO_4)_2$, where RE = Y, Yb, Gd and Lu. *Ultrasonics* **2014**, *54*, 1311–1317. [CrossRef]
3. Mazur, M.M.; Mazur, L.I.; Pozhar, V.E. Optimum configuration for acousto-optical modulator made of KGW. *Phys. Proc.* **2015**, *70*, 741–744. [CrossRef]
4. Mazur, M.M.; Mazur, L.I.; Pozhar, V.E. Optimal configuration of an acoustooptic modulator based on a KY(WO4)(2) crystal. *Tech. Phys. Lett.* **2015**, *41*, 249–251. [CrossRef]
5. Belyi, V.N.; Kulak, G.V.; Krokh, G.V.; Shakin, O.V. Collinear acousto-optical transformation of Bessel light beams in biaxial gyrotropic crystals. *J. Appl. Spectrosc.* **2016**, *83*, 283–287. [CrossRef]
6. Mazur, M.M.; Mazur, L.I.; Pozhar, V.E. Specific directions of ultrasound propagation in double potassium tungstates for light modulation. *Ultrasonics* **2017**, *73*, 231–235. [CrossRef] [PubMed]
7. Mazur, M.M.; Mazur, L.I.; Pozhar, V.E.; Shorin, V.N.; Konstantinov, Y.P. Acousto-optic modulators based on a KYW crystal. *Quantum Electron.* **2017**, *47*, 661–664. [CrossRef]
8. Martynyuk-Lototska, I.; Trach, I.; Kokhan, O.; Vlokh, R. Efficient acousto-optic crystal, $TlInS_2$: Acoustic and elastic anisotropy. *Appl. Opt.* **2017**, *56*, 3179–3184. [CrossRef]
9. Wu, Q.; Gao, Z.; Tian, X.; Su, X.; Li, G.; Sun, Y.; Xia, S.; He, J.; Tao, X. Biaxial crystal β-$BaTeMo_2O_9$: Theoretical analysis and the feasibility as high-efficiency acousto-optic Q-switch. *Opt. Express* **2017**, *25*, 24893–24900. [CrossRef]
10. Mytsyk, B.; Kryvyy, T.; Demyanyshyn, N.; Mys, O.; Martynyuk-Lototska, I.; Kokhan, O.; Vlokh, R. Piezo-, elasto- and acousto-optic properties of Tl_3AsS_4 crystals. *Appl. Opt.* **2018**, *57*, 3796–3801. [CrossRef]

11. Milkov, M.G.; Voloshinov, V.B.; Isaenko, L.I.; Vedenyapin, V.N. An experimental study of ultra-wide-band and ultra-wide-aperture non-collinear acousto-optic diffraction in an optically biaxial potassium arsenate titanyl crystal. *Moscow Univ. Phys. Bull.* **2018**, *72*, 83–88. [CrossRef]
12. Kupreychik, M.I.; Balakshy, V.I. Peculiarities of acousto-optic interaction in biaxial crystal of alpha-iodic acid. *Appl. Opt.* **2018**, *57*, 5549–5555. [CrossRef] [PubMed]
13. Kupreychik, M.I.; Balakshy, V.I. Investigation of close to collinear anisotropic acousto-optic interaction in a biaxial crystal of alpha-iodic acid. In *Proceedings of the Fourteenth School on Acousto-Optics and Applications*; Grulkowski, I., Linde, B.B.J., Duocastella, M., Eds.; SPIE: Bellingham, WA, USA , 2019; Volume 11210, p. 112100O. [CrossRef]
14. Yushkov, K.B.; Chizhikov, A.I.; Naumenko, N.F.; Molchanov, V.Y.; Pavlyuk, A.A.; Makarevskaya, E.V.; Zakharov, N.G. KYW crystal as a new material for acousto-optical Q-switches. In *Proceedings of the Components and Packaging for Laser Systems V*; Glebov, A.L., Leisher, P.O., Eds.; SPIE: Bellingham, WA, USA , 2019; Volume 10899, p. 1089913. [CrossRef]
15. Pushkin, A.V.; Mazur, M.M.; Sirotkin, A.A.; Firsov, V.V.; Potemkin, F.V. Powerful 3-μm lasers acousto-optically Q-switched with KYW and KGW crystals. *Opt. Lett.* **2019**, *44*, 4837–4840. [CrossRef] [PubMed]
16. Porokhovnichenko, D.L.; Dyakonov, E.A.; Kuznetsov, S.V.; Voronov, V.V.; Fedorov, P.P.; Zaramenskikh, K.S.; Gasanov, A.A.; Zhukova, L.V.; Korsakov, A.S.; Salimgareev, D.D. Indium iodide single crystal: Breakthrough material for infrared acousto-optics. *Opt. Lett.* **2020**, *45*, 3435–3438. [CrossRef] [PubMed]
17. Mazur, M.M.; Mazur, L.I.; Sirotkin, A.A.; Ryabinin, A.V.; Shorin, V.N. Acousto-optic modulators of high-power laser radiation on the basis of KGW and KYW crystals. *Quantum Electron.* **2020**, *50*, 957–961. [CrossRef]
18. Chizhikov, A.I.; Molchanov, V.Y.; Naumenko, N.F.; Yushkov, K.B. Acousto-optic spatial light modulator (SLM) based on KYW crystal. In *Proceedings of the Laser Beam Shaping XX*; Dudley, A., Laskin, A.V., Eds.; SPIE: Bellingham, WA, USA , 2020; Volume 11486, p. 114860B. [CrossRef]
19. Kupreychik, M.I.; Balakshy, V.I.; Pozhar, V.E. Wide-angle acousto-optic devices based on isotropic light scattering in biaxial crystals. *J. Phys. Conf. Ser.* **2021**, *2091*, 012010. [CrossRef]
20. Yushkov, K.B. Noncritical acousto-optic Bragg phase matching: Analysis of orthorhombic and monoclinic crystal systems. *Appl. Opt.* **2021**, *60*, 7113–7121. [CrossRef]
21. Yushkov, K.B.; Naumenko, N.F. Optical beam diffraction tensor in birefringent crystals. *J. Opt.* **2021**, *60*, 095602. [CrossRef]
22. Chizhikov, A.I.; Naumenko, N.F.; Yushkov, K.B.; Molchanov, V.Y.; Pavlyuk, A.A. Acousto-optic modulation of light polarisation in monoclinic crystals. *Quantum Electron.* **2021**, *51*, 343–347. [CrossRef]
23. Chizhikov, A.I.; Mukhin, A.V.; Egorov, N.A.; Gurov, V.V.; Molchanov, V.Y.; Naumenko, N.F.; Vorontsov, K.V.; Yushkov, K.B.; Zakharov, N.G. High-efficient KYW acousto-optic Q-switch for a Ho:YAG laser. *Opt. Lett.* **2022**, *47*, 1085–1088. [CrossRef]
24. Kupreychik, M.I.; Yushkov, K.B. Topological evolution of acousto-optic transfer functions in biaxial crystals. *J. Opt. Soc. Am. B Opt. Phys.* **2022**, *12*. [CrossRef]
25. Ristau, D. *Laser-Induced Damage in Optical Materials*; CRC Press: Boca Raton, FL, USA, 2015.
26. Yariv, A.; Yeh, P. *Optical Waves in Crystals*; Wiley: New York, NY, USA, 1984.
27. Dieulesaint, E.; Royer, D. *Ondes Élastiques Dans les Solides. Application au Traitment de signal*; Mason: Paris, France, 1974.
28. Goutzoulis, A.; Pape, D. *Design and Fabrication of Acousto-Optic Devices*; Marcel Dekker: New York, NY, USA, 1994; p. 497.
29. Yushkov, K.; Naumenko, N. *Acousto-Optic Figure of Merit of KREW Crystals*;Version 1; Published on 24 October 24 2022. Available online: https://data.mendeley.com/datasets/cg38s35c8d/1 (accessed on 24 October 2022). [CrossRef]
30. Naumenko, N.F.; Yushkov, K.B.; Molchanov, V.Y. Extreme acoustic anisotropy in crystals visualized by diffraction tensor. *Eur. Phys. J. Plus* **2021**, *136*, 95. [CrossRef]
31. Loiko, P.; Segonds, P.; Inácio, P.L.; Peña, A.; Debray, J.; Rytz, D.; Filippov, V.; Yumashev, K.; Pujol, M.C.; Mateos, X.; et al. Refined orientation of the optical axes as a function of wavelength in three monoclinic double tungstate crystals $KRE(WO_4)_2$ (RE = Gd, Y or Lu). *Opt. Mater. Express* **2016**, *6*, 2984–2990. [CrossRef]

materials

MDPI

Article

Optically Transparent and Highly Conductive Electrodes for Acousto-Optical Devices

Alexey Osipkov [1,*], Mstislav Makeev [1,*], Elizaveta Konopleva [1], Natalia Kudrina [1], Leonid Gorobinskiy [1], Pavel Mikhalev [1], Dmitriy Ryzhenko [1] and Gleb Yurkov [1,2]

1 Laboratory of EMI Shielding Materials, Bauman Moscow State Technical University, 105005 Moscow, Russia; konoplevaea@student.bmstu.ru (E.K.); natalia.kudrina1@gmail.com (N.K.); lgorobinskiy@bmstu.ru (L.G.); pamikhalev@bmstu.ru (P.M.); dsr@bmstu.ru (D.R.); ygy76@mail.ru (G.Y.)

2 N.N. Semenov Federal Research Center for Chemical Physics Russian Academy of Sciences, 119991 Moscow, Russia

* Correspondence: osipkov@bmstu.ru (A.O.); m.makeev@bmstu.ru (M.M.)

Abstract: The study was devoted to the creation of transparent electrodes based on highly conductive mesh structures. The analysis and reasonable choice of technological approaches to the production of such materials with a high Q factor (the ratio of transparency and electrical conductivity) were carried out. The developed manufacturing technology consists of the formation of grooves in a transparent substrate by photolithography methods, followed by reactive ion plasma etching and their metallization by chemical deposition using the silver mirror reaction. Experimental samples of a transparent electrode fabricated using this technology have a sheet resistance of about 0.1 Ω/sq with a light transmittance in the visible wavelength range of more than 60%.

Keywords: transparent electrode; mesh structure; sheet resistance; transparency; groove; lithography; chemical deposition; shielding efficiency

check for updates

Citation: Osipkov, A.; Makeev, M.; Konopleva, E.; Kudrina, N.; Gorobinskiy, L.; Mikhalev, P.; Ryzhenko, D.; Yurkov, G. Optically Transparent and Highly Conductive Electrodes for Acousto-Optical Devices. *Materials* **2021**, *14*, 7178. https://doi.org/10.3390/ma14237178

Academic Editors: Alexander S. Machikhin and Vitold Pozhar

Received: 1 November 2021
Accepted: 18 November 2021
Published: 25 November 2021

1. Introduction

Electrode element is an indispensable component of modern acousto-optical devices. It defines the shape of the acoustic field induced in the crystalline media [1,2]. Accurate apodization of the electrodes provides diffraction efficiency increase, higher apertures and better acoustic field homogeneity [3].

Transparent electrodes are necessary for acousto-optic devices, where the direction of light propagation coincides with sound wave (Figure 1). Such collinear geometry of acousto-optic interaction is widely used, for example, in acousto-optical tunable filters with high spectral resolution [4], which are necessary for Raman spectroscopy [5] and spectral-domain optical coherence tomography [6]. Propagation of light along ultrasound may be also potentially effective for multi-beam diffraction configurations [7] and other applications. Applying optically transparent electrodes in these devices may simplify its design and reduce its dimensions.

In conventional schemes of such electrodes, an ultrasonic wave may be formed by a transparent piezoelectric transducer based on ceramic materials with a perovskite structure (titanate-zirconate of a divalent metal (for example, lead), etc.) or polymer films (for example, polyvinylidene fluoride and its copolymers) with a thickness up to several tens of micrometers [8–14].

In addition, transparent electrodes are widely used in solar panels, touch screens, organic and inorganic diodes, etc. [15]. Among the approaches to the creation of such electrodes are the usage of transparent conductive oxides [15], metal nanowires [16,17], carbon nanomaterials [18], and metal micro- and nano-grids [19]. Today, thin films based on tin indium oxide (ITO) are the most widely used [20]. Still, their usage is limited by the low ratio of transparency and electrical characteristics for many current tasks; the best samples of ITO coatings have a surface resistance of about 20 Ω/sq with transparency in the visible range of about 90% [21]. In addition, this material is expensive due to the depletion of global indium reserves [22], and requires high-temperature annealing to obtain a high Q factors ϕ_s ($\phi_s = T_{550}{}^{10}/\varrho_s$ [23]), which complicate the production of high-quality transparent electrodes on flexible polymer substrates. The latter is relevant due to the recent intensive development of flexible LCD touch screen technology [24,25].

A decrease in the electrode resistance leads to an increasing the transparent piezoelectric transducer efficiency. Additionally, electrodes with high electrical conductivity and transparency are necessary to increase the efficiency of solar cells or to increase electromagnetic protection and compatibility of electronic devices and equipment.

Among the approaches to creating high-electrical-conductivity and transparent structures, ordered metal mesh structures formed on the surface or inside a transparent substrate have considerable potential. The geometry of such systems can be calculated in advance and specified during manufacture. Electrodes based on these structures have the best characteristics at the moment [21] in terms of the ratio of sheet resistance and transparency (3–5 Ω/sq with transparency of more than 85% [19]).

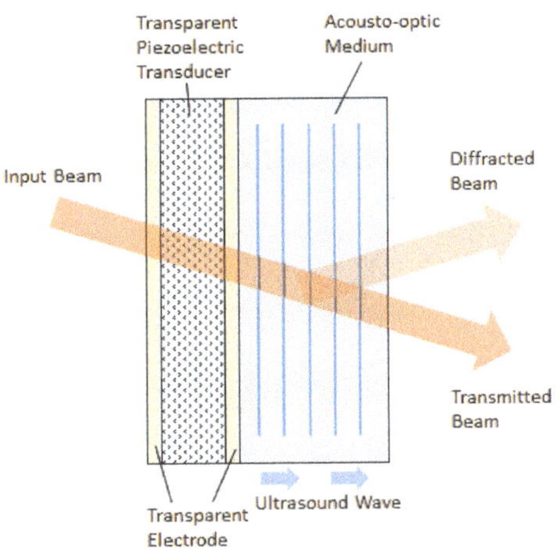

Figure 1. Design of a transparent acousto-optic modulator.

Currently, several approaches are being developed to produce such microgrids [21,26]. One of the general approaches consists of forming microgrooves with a high aspect ratio in an optically transparent material, such as quartz, into which metal, for example, silver or copper, is subsequently deposited [22]. The grooves can be filled with metal by magnetron or electron beam deposition, or chemical deposition from the gas phase. However, due to the high aspect ratio and the small width of the groove, their use is problematic [22]. A rapid blockage of the inlet and the cessation of mental growth inside the grooves may occur during the deposition process. Similar problems are also typical for galvanic metal deposition into structures with a similar aspect ratio [27].

In this paper, the manufacturing technologies for transparent mesh structures are analyzed. A proposed method for filling the grooves produced by lithography methods with silver using the silver mirror reaction is also described. In addition, the article describes the results of studies of the deposited metal morphology and the measurements of transparency, sheet resistance, and the S parameters in the radio and microwave wavelength ranges in the open coaxial waveguide of the obtained structures.

2. The Analysis of the Production Technologies of Transparent Mesh Structures

There are three groups among technological approaches used to obtain ordered mesh structures:

- Approaches based on the formation of grooves of a specified geometry in a substrate or photoresist and their subsequent filling with metal. The grooves are formed using photolithography methods in combination with liquid or ion plasma etching or laser ablation;
- Approaches based on the removal of the preliminarily deposited metal layer using the same etching methods;
- Approaches based on additive processes (electrohydrodynamic printing).

Based on the literature data [27–77], Figure 2 plots of the technological limitations of various approaches to forming grooves in transparent materials or directly forming conductors on the substrate surface. The markers show the values of the width and depth of the groove's or line's height, obtained from the literature sources, and the lines show the technological limitations of the approaches.

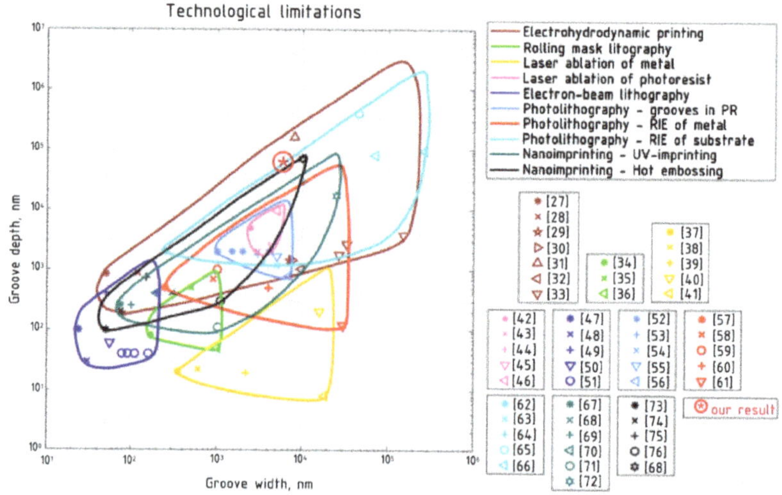

Figure 2. Technological limitations of various approaches to the formation of ordered mesh structures.

Two studies [78,79] provide a description of the calculated model and the results of the conductive mesh structures' calculation, obtaining a screening coefficient SE of more than 50 dB with transparency of more than 90% in the visible range. These calculated values were obtained for a width and height of the mesh conductors of 1.5 and 14 μm, respectively. Per Figure 2, structures with such parameters can be produced by the following methods: forming grooves in a transparent substrate by photolithography followed by reactive ion plasma etching (RIE), nanoimprinting (hot embossing), or multilayer electrohydrodynamic (EHD) printing.

At this stage, the first method was chosen for the practical implementation of the calculated structures due to the wide availability of the necessary technological equipment and verified technological regimes. Notably, the methods of nanoimprinting and EHD printing, with their further development, have considerable potential for creating serial technology for producing large-area, micro-mesh, transparent electrodes.

3. Materials and Methods

Experimental samples were pure quartz wafers with a thickness of 0.5 mm. A system of grooves was formed in the first step of preparation as a network. In the second step, the grooves were filled with silver metal. The topological network was an artificial, disordered mesh with a Voronoi diagram randomly distributed inside a cell of equal probability density (Figure 3).

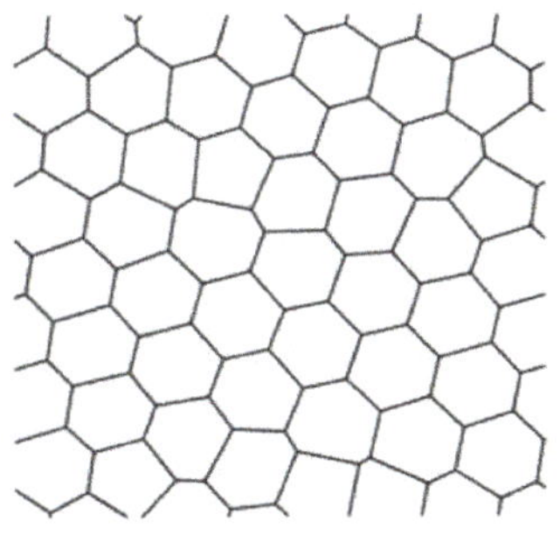

Figure 3. The topological figure of a conductive mesh.

The technological process of groove formation included these steps: the cleaning of a wafer, the coating of an α–Si:H masking layer with plasmochemical deposition using SPTS APM equipment (SPTS Technologies Ltd., Newport, UK); positive photoresist SPR700-1.0 coating by centrifuge, drying, and exposure with a ASML PAS 5500 stepper (ASML, Veldhoven, The Netherlands); followed by the development and plasmochemical etching of the α–Si:H masking layer with the Bosh process; and plasmochemical etching of grooves on the quartz wafer through the windows (holes) at the masking layer with the help of C_4F_8. (Figure 4).

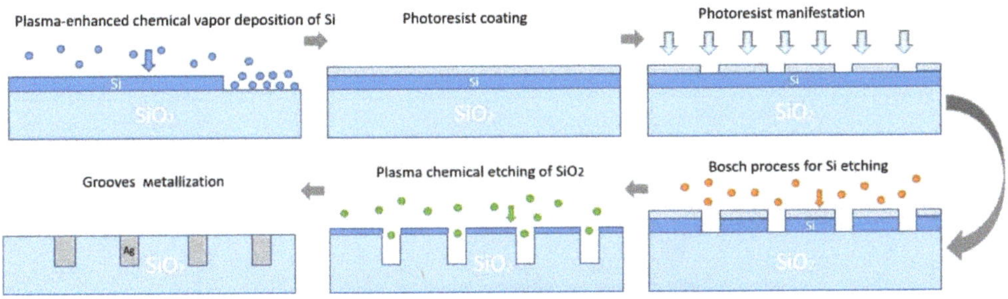

Figure 4. Technological process of micro-mesh formation.

A silver mirror reaction (a process of silver salt reduction, for example, silver nitrate) was used for filling the experimental samples with metal. The formation of a silver layer with few defects is possible only with a low rate of chemical reaction [80]. Glucose was used as a weak reducing agent, which increases silver adhesion to a surface [81]. The synthesis was carried out at 25 °C for the same reason.

Two solutions were used. The first solution was prepared by dissolution of 1.25 g of silver nitrate ($AgNO_3$) in 30 mL of deionized water, the addition of 4 mL of 25% water ammonia (NH_4OH), and the addition of 30 mL of an alkali solution (1.1 g of NaOH was dissolved in water and diluted until 30 mL of solution was formed). The initial rate of the silver mirror reaction decreased with increasing ammonia concentration; the silver nitrate solution stability and the thickness of the silver layer increased simultaneously [81]. The silver alkali ammonia solution was diluted to 100 mL, and the first solution was prepared. The second solution was prepared by dissolution of 1.1 g of glucose in water and dilution to 100 mL. A higher glucose concentration (higher than 1.3 g/L) decreases the maximal thickness of the silver layer [81] and facilitates silver particle agglomeration [24]. The shelf life of the solutions was less than 10 h. The samples for the metallization were placed into a beaker. The first solution was poured, and then the second solution was added in a volume ratio of 1:1. The reaction time was less than 10 min (usually 3–5 min). The long duration of the silver mirror reaction leads to the morphology of silver particles differing from a sphere [82], which can inhibit groove filling. The silver mirror reaction was repeated several times. The excess metal from the surface was removed after each iteration of silver deposition.

The geometric dimensions and surface morphology of the manufactured grooves and the surface morphology and structure of the deposited silver were studied by scanning electron microscopy (SEM, TESCAN ORSAY HOLDING, Brno, Czech Republic) using a VEGA 3. Imaging was performed at an accelerating voltage of 10 kV and a beam current of 10 A in secondary electron detection mode to obtain images with the highest resolution, and in back-reflected electron mode to obtain a compositional contrast.

Sheet resistance was determined by the four-probe method using a Keithley 2000 multimeter (Tektronix, Inc, Beaverton, OR, USA) and a Mill-Max 854-22-004-10-001101 four-probe head (Mill-Max Mfg. Corp., Oyster Bay, NY, USA).

The transmission coefficient in the visible wavelength range of 380 to 780 nm was determined on a Shimadzu UV-3600i Plus spectrophotometer (SHIMADZU CORPORATION, Kyoto, Japan) with a resolution of 1 nm at normal incidence of light on the sample.

Since the materials under development had considerable potential for application in radio engineering, the shielding efficiency (SE) was measured to evaluate the obtained materials' radio engineering properties. The measurements were obtained in the frequency range from 10 MHz to 7 GHz on a specialized measuring stand based on a FieldFox N9916A vector circuit analyzer (Keysight Technologies, Santa Rosa, CA, USA) in a coaxial path (type II). The SE value was determined from the S21 transmission coefficient of the path with the sample in relation to the transmission coefficient S21 of the path without the sample. Electro-sealing gaskets composed of dense metalized fabric were used for better electrical contact between the sample and the walls of the coaxial tract. The SE dynamic measurement range of the stand is 80 dB and the measurement error is ± 2 dB.

4. Results

A photo of an obtained sample is shown in Figure 5. The geometrical dimensions of the sample were an outer diameter of 16 mm and an inner diameter of 6.95 mm. The sample was fabricated in this shape due to the requirements for SE measurements in the coaxial path.

Figure 5. Photo of the manufactured sample.

The grooves in the quartz substrate obtained by lithography and plasma chemical etching were examined by scanning electron microscopy (Figure 6). The groove walls obtained in quartz glass showed vertical deviations up to 10°, which led to distortion of the specified topology (Figure 3). The groove cross-section was trapezoidal with a depth of 16.24 µm and a width of 1.15 (in-depth) to 4.47 µm (at the surface). The deviation from the verticality of the groove walls was caused by lateral subtraction during plasma chemical etching.

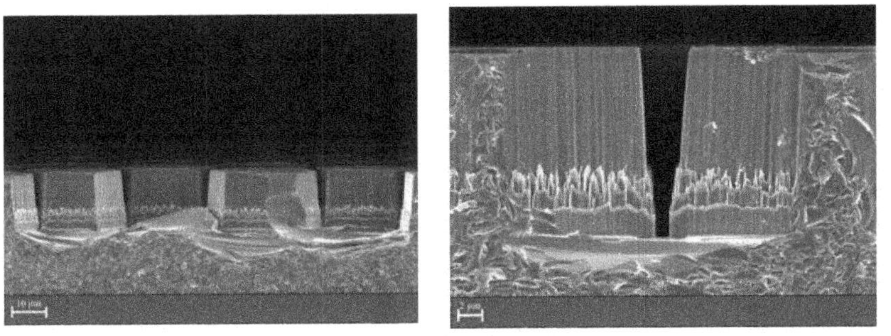

(**a**) (**b**)

Figure 6. SEM images of the sample after the lithography and plasma chemical etching processes: (**a**) 2500× magnification; (**b**) 8000× magnification.

The degree of groove filling with silver was evaluated after the metallization process (Figure 7). We found that the silver filled all the grooves throughout the sample, but the silver structure was porous.

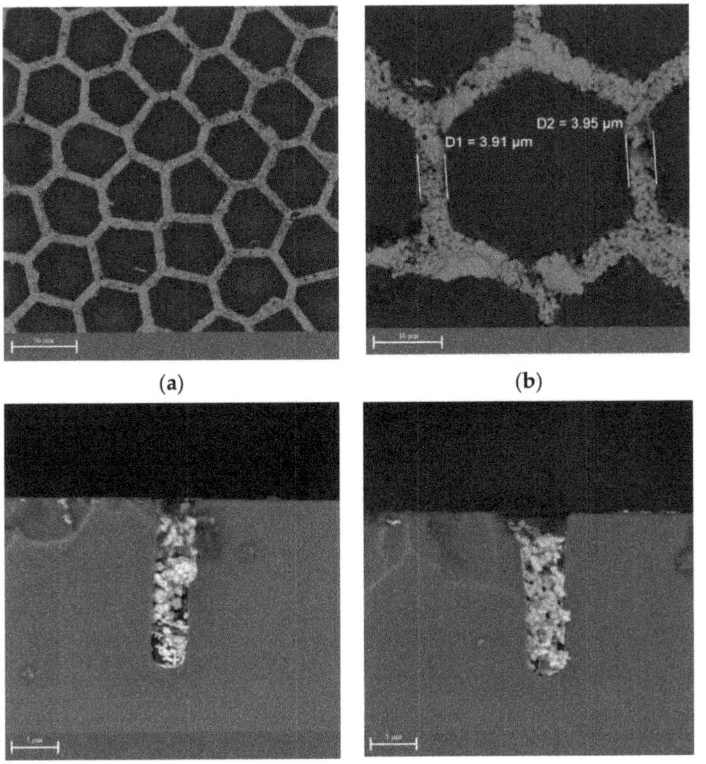

(**a**) (**b**) (**c**) (**d**)

Figure 7. SEM images of the sample after metallization: (**a**) 1000×, (**b**) 4500×, (**c**) 7000×, and (**d**) 7000× magnification.

The sheet resistance of the manufactured sample was (0.084 ± 0.016) Ω/sq. Measurements were taken at 22 different points on the sample surface (Figure 8a).

The frequency dependence of the SE is shown in Figure 8b. The average SE value in the frequency range of 10 MHz to 7 GHz is 54.6 dB. The horizontal behavior of the SE spectrum indicates the absence of significant metallization defects.

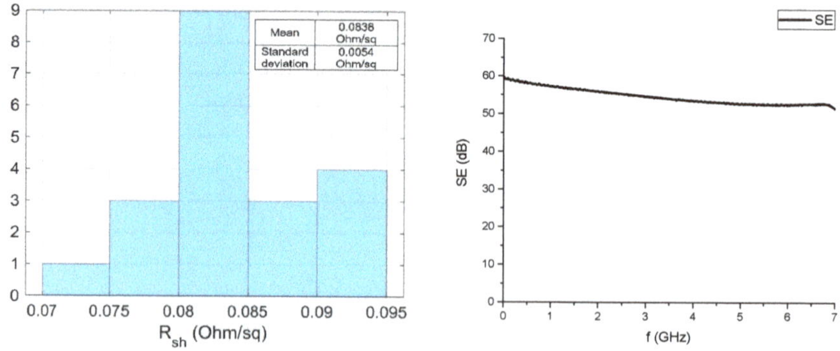

(a) (b)

Figure 8. Sheet resistance distribution (**a**) and frequency dependence of the shielding efficiency (**b**) of the test sample.

The transmission spectrum measurements were taken at eight different points on the sample surface using a specially designed fixture (Figure 9a). The average, maximum and minimum transmission spectrums of the test sample in the visible wavelength range are shown in Figure 9b. The light transmittance coefficient calculated for the wavelength range from 380 to 780 nm following GOST R 54164-2010 is 64.1%.

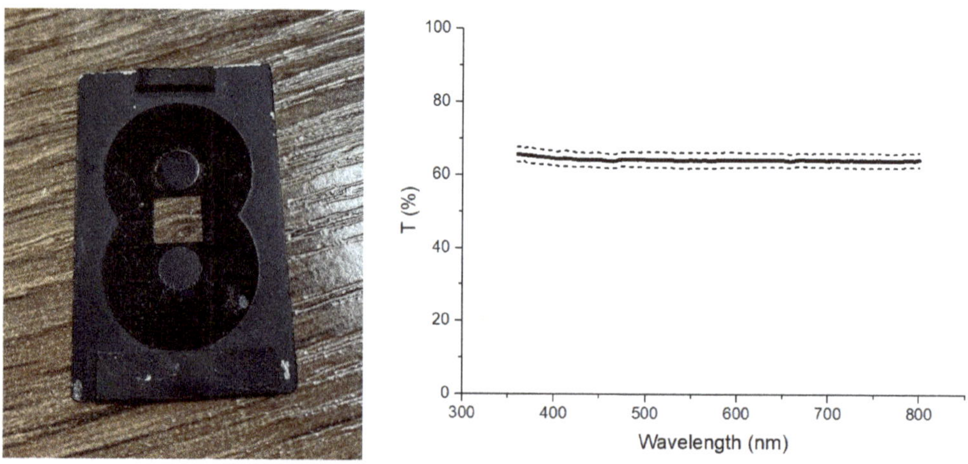

(a) (b)

Figure 9. Photo of measuring equipment (**a**) and transmission spectrum of the test sample in the visible wavelength range (**b**).

5. Discussion

The deviation from the groove walls' geometry due to the technological peculiarities of quartz etching led to a width increase from 1.5 to 4 μm at the groove entrance. This increased the surface area of the conductive grid paths and decreased the light transmission coefficient from 90% (according to the simulation results) to 64%. In order to increase this parameter, further development of the plasma chemical etching process of a transparent substrate is required. Notably, no such problems are encountered, for example, when forming grooves on Si substrates: it is possible to produce grooves with a high aspect ratio without distorting the geometry using the Bosh process, but such samples are only transparent in the IR range (Figure 10).

Transparency can also be increased by applying antireflective coatings based on metal oxides to the sample surface. These coatings will protect the metal mesh structure from oxidation in addition to improving optical properties.

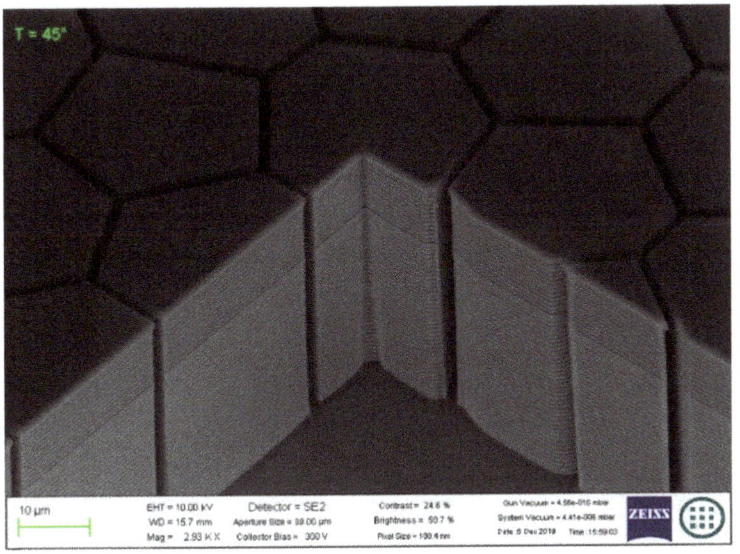

Figure 10. SEM images of the Si sample with grooves formed in its surface using the Bosch process with a high aspect ratio.

The values of SE and sheet resistance obtained on the experimental sample were close to the calculated values. The silver synthesized in the grooves using the silver mirror reaction had a porous structure, and its electrical conductivity differed significantly from that of a monolithic material. The low value of the sheet resistance of the resulting microgrid (0.084 ± 0.016 Ω/sq) and, consequently, the high SE coefficient were caused by an increase in the cross-section of the conductors when the geometry of the conducting lines was distorted. Adjustment of the technology used for filling the grooves accompanied by an increase in the density of the formed channel can significantly increase the values of these parameters.

6. Conclusions

We developed a technology for producing transparent conductive electrodes. The technology consists of forming grooves in a transparent wafer by photolithography methods, followed by reactive ion plasma etching and its metallization by chemical deposition with a silver mirror reaction (Tollen's reagent). This technology was chosen due to the high availability of necessary technological equipment and the verified technological regimes. Notably, methods of nanoimprint lithography and multilayer electrohydrodynamic (EHD) printing have considerable potential with the development of the methods for the creation of serial technology micro-mesh transparent electrodes with a large surface area.

The experimental sample was prepared according to developed technology with a sheet resistance of 0.09 Ω/sq and a visible light transmittance of more than 60%. The development of technology will allow for upgrading the functional characteristics of the prepared samples. In particular, the verticality of the groove wall needs to be improved for transparency improvement. The grade of the metal filling and the density of the metal need to be increased for enhancing the SE and decreasing sheet resistance. The first problem can be solved by reworking the reactive ion plasma etching of transparent wafers. The second task may be achieved using a sintering process for the prepared samples.

The proposed approach to the design of optically transparent electrodes may be effective for multiple scientific and industrial applications including acousto-optics, photoacoustics, lithography, radio shielding applications, etc.

Author Contributions: Conceptualization, M.M. and G.Y.; funding acquisition, D.R. and G.Y.; Investigation, A.O., E.K., N.K. and L.G.; methodology, A.O., M.M., P.M. and D.R.; project administration, P.M.; resources, D.R.; supervision, D.R. and G.Y.; validation, E.K.; writing—original draft, A.O., M.M., E.K. and N.K.; writing—review and editing, A.O., M.M., P.M. and G.Y. All authors have read and agreed to the published version of the manuscript.

Funding: This research work was supported by the Ministry of Education and Science of the Russian Federation under state task No. 0705-2020-0032.

Institutional Review Board Statement: Not applicable.

Informed Consent Statement: Not applicable.

Data Availability Statement: Not applicable.

Acknowledgments: This research work was supported by the Ministry of Education and Science of the Russian Federation under state task No. 0705-2020-0032. The authors also express their gratitude to Bauman Moscow State Technical University staff Bibikov S.B. and Damaratskiy I.A. for the shielding coefficients measurement.

Conflicts of Interest: The authors declare no conflict of interest.

References

1. Maák, P.; Lenk, S.; Jakab, L.; Barócsi, A.; Richter, P. Optimization of transducer configuration for bulk acousto-optic tunable filters. *Opt. Commun.* **2004**, *241*, 87–98. [CrossRef]
2. Dühring, M.B.; Laude, V.; Khelif, A. Improving surface acousto-Optical interaction by high aspect ratio electrodes. *J. Appl. Phys.* **2009**, *106*, 113518. [CrossRef]
3. Pannell, C.N.; Ward, J.D.; Wachman, E.S.; Zhang, B.G.; Reed, M.K. A high-Performance passband-Agile hyperspectral imager using a large aperture acousto-Optic tuneable filter. *Photonic Instrum. Eng. II* **2015**, *9369*, 936906.
4. Tran, C.D.; Huang, G.C. Characterization of the collinear beam acousto-Optic tunable filter and its comparison with the noncollinear and the integrated acousto-Optic tunable filter. *Opt. Eng.* **1999**, *38*, 1143–1148. [CrossRef]
5. Gupta, N.; Fell, N.F., Jr. A compact collinear AOTF Raman spectrometer. *Talanta* **1997**, *45*, 279–284. [CrossRef]
6. Viskovatykh, A.; Machikhin, A.; Pozhar, V.; Pustovoit, V.I.; Viskovatykh, D.A. Combined optical-coherence and spectral microscopy based on tunable acousto-optic filters of images. *Tech. Phys. Lett.* **2014**, *40*, 157–160. [CrossRef]
7. Machikhin, A.; Batshev, V.; Pozhar, V.; Naumov, A.; Gorevoy, A. Acousto-optic tunable spectral filtration of stereoscopic images. *Opt. Lett.* **2018**, *43*, 1087–1090. [CrossRef]
8. Li, Z.; Ilkhechi, A.K.; Zemp, R. Transparent capacitive micromachined ultrasonic transducers (CMUTs) for photoacoustic applications. *Opt. Express* **2019**, *27*, 13204–13218. [CrossRef]
9. Chen, X.; Chen, R.; Chen, Z.; Chen, J.; Shung, K.K.; Zhou, Q. Transparent lead lanthanum zirconate titanate (PLZT) ceramic fibers for high-Frequency ultrasonic transducer applications. *Ceram. Int.* **2016**, *42*, 18554–18559. [CrossRef]
10. Pala, S.; Lin, L. Fully Transparent Piezoelectric Ultrasonic Transducer with 3D Printed Substrate. In Proceedings of the 20th International Conference on Solid-State Sensors, Actuators and Microsystems and Eurosensors XXXIII (Transducers and Eurosensors XXXIII), Berlin, Germany, 23–27 June 2019; pp. 234–237.
11. Sette, D.; Girod, S.; Godard, N.; Adjeroud, N.; Chemin, J.B.; Leturcq, R.; Defay, E. Transparent piezoelectric transducers for large area ultrasonic actuators. In Proceedings of the IEEE 30th International Conference on Micro Electro Mechanical Systems (MEMS), Las Vegas, NV, USA, 22–26 January 2017; pp. 793–796.
12. Luo, G.L.; Fung, S.; Wang, Q.; Kusano, Y.; Lasiter, J.; Kidwell, D.; Horsley, D.A. High fill factor piezoelectric micromachined ultrasonic transducers on transparent substrates. In Proceedings of the 19th International Conference on Solid-State Sensors, Actuators and Microsystems (Transducers), Kaohsiung, Taiwan, 18–22 June 2017; pp. 1053–1056.
13. Ren, D.; Sun, Y.; Shi, J.; Chen, R. A Review of Transparent Sensors for Photoacoustic Imaging Applications. *Photonics. Multidiscip. Digit. Publ. Inst.* **2021**, *8*, 324. [CrossRef]

14. Shehzad, M.; Wang, S.; Wang, Y. Flexible and transparent piezoelectric loudspeaker. *npj Flex. Electron.* **2021**, *5*, 1–6.
15. Pasquarelli, R.M.; Ginley, D.S.; O'Hayre, R. Solution processing of transparent conductors: From flask to film. *Chem. Soc. Rev.* **2011**, *40*, 5406. [CrossRef]
16. Liu, C.H.; Yu, X. Silver nanowire-Based transparent, flexible, and conductive thin film. *Nanoscale Res. Lett.* **2011**, *6*, 1. [CrossRef]
17. Tran, N.H.; Duc, A.N.; Thanh, H.D.; Kim, H.C. Fast and simple fabrication of flexible and transparent electrode based on patterned copper nanowires by mechanical lithography transfer. *Thin Solid Film.* **2019**, *685*, 26. [CrossRef]
18. López-Naranjo, E.J.; González-Ortiz, L.J.; Apátiga, L.M.; Rivera-Muñoz, E.M.; Manzano-Ramírez, A. Transparent electrodes: A review of the use of carbon-Based nanomaterials. *J. Nanomater.* **2016**, *2016*, 4928365. [CrossRef]
19. Voronin, A.S.; Simunin, M.M.; Fadeev, Y.V.; Ivanchenko, F.S.; Karpova, D.V.; Tambasov, I.A.; Khartov, S.V. Technological basis of the formation of micromesh transparent electrodes by means of a self-Organized template and the study of their properties. *Tech. Phys. Lett.* **2019**, *45*, 366. [CrossRef]
20. Lu, H.C.; Mao, J.W.; Chiang, Y.C. Low temperature preparation of ITO thin films by the coating solutions containing solvothermally synthesized ITO nanoparticles. *Surf. Coat. Technol.* **2013**, *231*, 526. [CrossRef]
21. Osipkov, A.; Makeev, M.; Garsiya, E.; Filyaev, A.; Sinyagaeva, A.; Kirillov, D.; Ryzhenko, D.; Yurkov, G. Radio-Shielding metamaterials transparent in the visible spectrum: Approaches to creation. *IOP Conf. Ser. Mater. Sci. Eng. IOP Publ.* **2021**, *1060*, 012007. [CrossRef]
22. Ye, S.; Rathmell, A.R.; Chen, Z.; Stewart, I.E.; Wiley, B.J. Metal nanowire networks: The next generation of transparent conductors. *Adv. Mater.* **2014**, *26*, 6670. [CrossRef]
23. Levy, D.; Castellón, E. *Transparent Conductive Materials: Materials. Synth. Charact. Applications*, 1st ed.; Wiley-VCH: Weinheim, Germany, 2019; pp. 38–42.
24. Jiang, B.; Li, M.; Bai, F.; Yu, H.; Mwenya, T.; Li, Y.; Song, D. Morphology-Controlled synthesis of silver nanoparticles on the silicon substrate by a facile silver mirror reaction. *Aip Adv.* **2013**, *3*, 032119. [CrossRef]
25. Tu, Y.J.; You, C.F.; Chen, Y.R.; Huang, C.P.; Huang, Y.H. Application of recycled iron oxide for adsorptive removal of strontium. *J. Taiwan Inst. Chem. Eng.* **2015**, *53*, 92–97. [CrossRef]
26. Lee, H.B.; Jin, W.Y.; Ovhal, M.M.; Kumar, N.; Kang, J.W. Flexible transparent conducting electrodes based on metal meshes for organic optoelectronic device applications: A review. *J. Mater. Chem. C* **2019**, *7*, 1087–1110. [CrossRef]
27. Wolf, M.; Dretschkow, T.; Wunderle, B.; Jurgensen, N.; Engelmann, G.; Ehrmann, O.; Uhlig, A.; Michel, B.; Reichl, H. High aspect ratio TSV copper filling with different seed layers. In Proceedings of the 58th Electronic Components and Technology Conference, Lake Buena Vista, FL, USA, 27–30 May 2008; IEEE: Piscataway, NJ, USA; pp. 563–570.
28. Galiker, P.; Schneider, J.; Eghlidi, H.; Kress, S.; Sandoghdar, V.; Poulikakos, D. Direct printing of nanostructures by electrostatic autofocussing of ink nanodroplets. *Nat. Commun.* **2012**, *3*, 1–9. [CrossRef]
29. Schneider, J.; Rohner, P.; Thureja, D.; Schmid, M.; Galliker, P.; Poulikakos, D. Electrohydrodynamic nanodrip printing of high aspect ratio metal grid transparent electrodes. *Adv. Funct. Mater.* **2015**, *26*, 833–840. [CrossRef]
30. Seong, B.; Yoo, H.; Nguyen, V.D.; Jang, Y.; Ryu, C.; Byun, D. Metal-Mesh based transparent electrode on a 3-D curved surface by electrohydrodynamic jet printing. *J. Micromech. Microeng.* **2014**, *24*, 097002. [CrossRef]
31. Jang, Y.; Kim, J.; Byun, D. Invisible metal-Grid transparent electrode prepared by electrohydrodynamic (EHD) jet printing. *Appl. Phys.* **2013**, *46*, 155103. [CrossRef]
32. Zhang, B.; Lee, H.; Byun, D. Electrohydrodynamic Jet Printed 3-Dimensional (3-D) Metallic Grid: Towards High-Performance Transparent Electrodes. *Adv. Eng. Mater.* **2020**, *22*, 1901275. [CrossRef]
33. Sujaya, K.; Do-Geun, K.; Jihoon, K. Electromagnetic interference shielding effectiveness of invisible metal-Mesh prepared by electrohydrodynamic jet printing. *Jpn. Soc. Appl. Phys.* **2014**, *53*, 05HB11.
34. Jang, Y.; Hartarto Tambunan, I.; Tak, H.; Dat Nguyen, V.; Kang, T.; Byun, D. Non-Contact printing of high aspect ratio Ag electrodes for polycrystalline silicone solar cell with electrohydrodynamic jet printing. *Appl. Phys. Lett.* **2013**, *102*, 123901. [CrossRef]
35. Kwak, M.K.; Ok, J.G.; Lee, J.Y.; Guo, L.J. Continuous phase-Shift lithography with a roll-Type mask and application to transparent conductor fabrication. *Nanotechnology* **2012**, *23*, 344008. [CrossRef]
36. Chen, J.; Cranton, W.; Fihn, M. *Handbook of Visual Display Technology*; Springer: Berlin/Heidelberg, Germany, 2016.
37. Ok, J.G.; Kwak, M.K.; Huard, C.M.; Youn, H.S.; Guo, L.J. Photo-Roll Lithography (PRL) for Continuous and Scalable Patterning with Application in Flexible Electronics. *Adv. Mater.* **2013**, *25*, 6554–6561. [CrossRef] [PubMed]
38. Eckhardt, S.; Müller-Meskamp, L.; Loeser, M.; Siebold, M.; Lasagni, A.F. Fabrication of highly efficient transparent metal thin film electrodes using Direct Laser Interference Patterning. *Laser-Based Micro-Nanoprocessing IX* **2015**, *9351*, 935116.
39. Müller-Meskamp, L.; Schubert, S.; Roch, T.; Eckhardt, S.; Lasagni, A.F.; Leo, K. Transparent Conductive Metal Thin-Film Electrodes Structured by Direct Laser Interference Patterning. *Adv. Eng. Mater.* **2015**, *17*, 1400454. [CrossRef]
40. Paeng, D.; Yoo, J.H.; Yeo, J.; Lee, D.; Kim, E.; Ko, S.H.; Grigoropoulos, C.P. Low-Cost Facile Fabrication of Flexible Transparent Copper Electrodes by Nanosecond Laser Ablation. *Adv. Mater.* **2015**, *27*, 1500098. [CrossRef] [PubMed]
41. Qin, R.; Hu, M.; Zhang, N.; Guo, Z.; Yan, Z.; Li, J.; Liu, J.; Shan, G.; Yang, J. Flexible Fabrication of Flexible Electronics: A General Laser Ablation Strategy for Robust Large-Area Copper-Based Electronics. *Adv. Electron. Mater.* **2019**, *5*, 1900365. [CrossRef]
42. Wang, Q.; Li, B.J.; Toor, F.; Ding, H. Novel laser-Based metasurface fabrication process for transparent conducting surfaces. *J. Laser Appl.* **2019**, *31*, 022505. [CrossRef]

43. Liu, Y.H.; Xu, J.L.; Gao, X.; Sun, Y.L.; Lv, J.J.; Shen, S.; Chen, L.S.; Wang, S.D. Freestanding transparent metallic network based ultrathin, foldable and designable supercapacitors. *Energy Environ. Sci.* **2017**, *10*, 2534–2543. [CrossRef]
44. Li, M.; Zuo, W.W.; Ricciardulli, A.G.; Yang, Y.G.; Liu, Y.H.; Wang, Q.; Wang, K.L.; Li, G.X.; Saliba, M.; Girolamo, D.D.; et al. Embedded Nickel-Mesh Transparent Electrodes for Highly Efficient and Mechanically Stable Flexible Perovskite Photovoltaics: Toward a Portable Mobile Energy Source. *Adv. Mater.* **2020**, *32*, 2003422. [CrossRef]
45. Xu, J.L.; Liu, Y.H.; Gao, X.; Sun, Y.; Shen, S.; Cai, X.; Chen, L.; Wang, S.D. Embedded Ag Grid Electrodes as Current Collector for Ultra-Flexible Transparent Solid-State Supercapacitor. *ACS Appl. Mater. Interfaces* **2017**, *9*, 27649. [CrossRef]
46. Liu, Y.H.; Xu, J.L.; Shen, S.; Cai, X.L.; Chen, L.S.; Wang, S.D. High-performance, ultra-flexible and transparent embedded metallic mesh electrodes by selective electrodeposition for all-solid-state supercapacitor applications. *J. Mater. Chem. A* **2017**, *5*, 9032–9041. [CrossRef]
47. Jiang, Z.Y.; Huang, W.; Chen, L.S.; Liu, Y.H. Ultrathin, lightweight, and freestanding metallic mesh for transparent electromagnetic interference shielding. *Opt. Express* **2019**, *27*, 24194–24206. [CrossRef]
48. Bilenberg, B.; Jacobsen, S.; Schmidt, M.S.; Skjolding, L.H.D.; Shi, P.; Bøggild, P.; Tegenfeldt, J.O.; Kristensena, A. High resolution 100 kV electron beam lithography in SU-8. *Microelectron. Eng.* **2006**, *83*, 1609–1612. [CrossRef]
49. Taylor, A.B.; Michaux, P.; Mohsin, A.S.; Chon, J.W. Electron-beam lithography of plasmonic nanorod arrays for multilayered optical storage. *Opt. Express* **2014**, *22*, 13234–13243. [CrossRef] [PubMed]
50. Rai-Choudhury, P. *Handbook of Microlithography, Micromachining, and Microfabrication: Micromachining and Microfabrication*; SPIE Press: Washington, DC, USA, 1997; Volume 2, pp. 41–99.
51. Groep, J.; Spinelli, P.; Polman, A. Transparent Conducting Silver Nanowire Networks. *Nano Lett.* **2012**, *12*, 3138–3144. [CrossRef]
52. Wang, L.; Xiong, W.; Nishijima, Y.; Yokota, Y.; Ueno, K.; Misawa, H.; Qiu, J.; Bi, G. Spectral properties of nanoengineered Ag/Au bilayer rods fabricated by electron beam lithography. *Appl. Opt.* **2011**, *50*, 5600–5605. [CrossRef] [PubMed]
53. DONGJIN Photoresist Products DSAM-3000 Series. Available online: https://www.yumpu.com/en/document/read/29811371/dongjin-photoresist-products-dsam-3000-series (accessed on 31 October 2021).
54. Zhang, C.; Khan, A.; Cai, J.; Liang, C.; Liu, Y.; Deng, J.; Huang, S.; Li, G.; Li, W.D. Stretchable Transparent Electrodes with Solution-processed Regular Metal Mesh for Electroluminescent Light-emitting Film. *ACS Appl. Mater. Interfaces* **2018**, *10*, 21009–21017. [CrossRef]
55. Khan, A.; Lee, S.; Jang, T.; Xiong, Z.; Zhang, C.; Tang, J.; Guo, L.J.; Li, W.D. High-Performance Flexible Transparent Electrode with an Embedded Metal Mesh Fabricated by Cost-Effective Solution Process. *Small* **2016**, *12*, 3021–3030. [CrossRef]
56. Wang, W.; Bai, B.; Zhou, Q.; Ni, K.; Lin, H. Petal-shaped metallic mesh with high electromagnetic shielding efficiency and smoothed uniform diffraction. *Opt. Mater. Express* **2018**, *8*, 3485–3493. [CrossRef]
57. Zhao, L.; Yu, S.; Li, X.; Wu, M.; Li, L. High-Performance copper mesh transparent flexible conductors based on electroplating with vacuum-Free processing. *Org. Electron.* **2019**, *82*, 105511. [CrossRef]
58. Xie, R.; Kava, J.; Siegel, M. Aspect ratio dependent etching on metal etch: Modeling and experiment. *J. Vac. Sci. Technol. A Vac. Surf. Film.* **1996**, *14*, 1067–1071. [CrossRef]
59. Blumenstock, K.; Stephani, D. Anisotropic reactive ion etching of titanium. *J. Vac. Sci. Technol. B* **1989**, *7*, 627. [CrossRef]
60. Hu, C.; Small, M.; Schadt, M. Reactive Ion Etching of AI(Cu) Alloys. *MRS Online Proceeding Libr. Arch.* **1986**, *76*, 191–195. [CrossRef]
61. Howard, B.J.; Wolterman, S.K.; Yoo, W.J.; Gittleman, B.; SteinbrÜchel, C.H. Reactive Ion Etching of Copper with SiCl4 and CCl2F2. *MRS Online Proceeding Libr. Arch.* **2011**, *201*, 129–133. [CrossRef]
62. Jin, S.W.; Lee, Y.H.; Yeom, K.M.; Yun, J.; Park, H.; Jeong, Y.R.; Hong, S.Y.; Lee, G.; Oh, S.Y.; Lee, J.H.; et al. Highly Durable and Flexible Transparent Electrode for Flexible Optoelectronic Applications. *ACS Appl. Mater. Interfaces* **2018**, *10*, 30706–30715. [CrossRef]
63. Leech, P.; Reeves, G. Reactive ion etching of quartz and glasses for microfabrication. *Proc. SPIE* **1999**, *3680*, 341281.
64. Holland, A.; Leech, P.; Reeves, G. Reactive ion etching of organic polymers for application in waveguide trench molds. *J. Mater. Sci.* **2004**, *39*, 3505–3508. [CrossRef]
65. Queste, S.; Courjon, E.; Ulliac, G.; Salut, R.; Petrini, V.; Rauch, J.; Besançon Cedex, F. Deep Reactive Ion Etching of quartz, lithium niobate and lead titanate. *JNTE Proc.* **2008**, *1*, 1–2.
66. Zhang, C.; Yang, C.; Ding, D. Deep reactive ion etching of PMMA. *Appl. Surf. Sci.* **2004**, *227*, 139. [CrossRef]
67. Chapellier, P.; Lavenus, P.; Bourgeteau-Verlhac, B.; Gageant, C.; Le Traon, O.; Dulmet, B. Aspect ratio dependent etching in advanced deep reactive ion etching of quartz. In Proceedings of the 2017 Symposium on Design, Test, Integration and Packaging of MEMS/MOEMS (DTIP), Bordeaux, France, 29 May–1 June 2017; pp. 1–6.
68. Zhong, Y.; Yuen, M. Hard mold UV nanoimprint lithography process. In Proceedings of the 2012 14th International Conference on Electronic Materials and Packaging (EMAP), Hong Kong, China, 13–16 December 2012; pp. 1–5.
69. Nanoimprint Lithography (NIL)-SmartNIL®. Available online: https://www.evgroup.com/technologies/nanoimprint-lithography-smartnil/ (accessed on 26 May 2021).
70. Fukuhara, M.; Ono, H.; Hirasawa, T.; Otaguchi, M.; Sakai, N.; Mizuno, J.; Shoji, S. UV Nanoimprint Lithography and Its Application for Nanodevices. *J. Photopolym. Sci. Technol.* **2007**, *20*, 549–554. [CrossRef]
71. Hashmi, S. *Reference Module in Materials Science and Materials Engineering*; Elsevier: Oxford, UK, 2016; pp. 1–8.

72. Kawaguchi, Y.; Nonaka, F.; Sanada, Y. Fluorinated materials for UV nanoimprint lithography. *Microelectron. Eng.* **2007**, *84*, 973–976. [CrossRef]
73. Jianyun, H.; Changsong, Z.; Fan, Y.; Jinghui, Z.; Jinge, T.; Weimin, Y. UV-Curable micro-imprinting method for the fabrication of microstructure arrays. *Microsyst. Technol.* **2019**, *25*, 3311–3316. [CrossRef]
74. Wang, M. *Lithography*; IntechOpen: London, UK, 2010; pp. 457–494.
75. Tang, A.; He, S.; Lee, M. Fabrication of III–V compound nanowires via hot embossing nanoimprint lithography. *Jpn. J. Appl. Phys.* **2016**, *55*, 035001. [CrossRef]
76. Chen, Y. Applications of nanoimprint lithography/hot embossing: A review. *Appl. Phys. A* **2015**, *121*, 451–465. [CrossRef]
77. Roos, N.; Luxbacher, T.; Glinsner, T.; Pfeiffer, K.; Schulz, H.; Scheer, H.C. Nanoimprint lithography with a commercial 4-in. bond system for hot embossing. *Proc. SPIE* **2001**, *4343*, 427–435.
78. Shiriaev, P.P.; Ryzhenko, D.S.; Makeev, M.O. Simulation of transparent electromagnetic interference shielding materials based on periodic conductive networks. *J. Phys. Conf. Series. IOP Publ.* **2019**, *1410*, 012206. [CrossRef]
79. Shiriaev, P.; Makeev, M.; Ryzhenko, D.; Popkov, O. Theoretical study of electromagnetic and optical properties of periodic conductive networks based on Voronoi diagrams. *Mater. Today Proc.* **2019**, *19*, 2179–2182. [CrossRef]
80. Vansovskaya, K.M. Metal coatings applied by chemical method. *Mashinostroenie* **1985**, *2*, 81–85.
81. Shalkauskas, M.I.; Vashkialis, A.I. *Chemical Metallization of Plastics*, 3rd ed.; Chimia: Saint-Petersburg, Russia, 1985; pp. 137–142.
82. Shi, Y.; Li, Y.; Zhang, J.; Yu, Z.; Yang, D. Electrospun polyacrylonitrile nanofibers loaded with silver nanoparticles by silver mirror reaction. *Mater. Sci. Eng. C* **2015**, *51*, 346–355. [CrossRef]

materials

MDPI

Article

Polarization of Acoustic Waves in Two-Dimensional Phononic Crystals Based on Fused Silica

Mikhail V. Marunin and Nataliya V. Polikarpova *

Faculty of Physics Lomonosov, Lomonosov Moscow State University, Moscow 119991, Russia
* Correspondence: polikarpnv@yandex.ru

Abstract: The two-dimensional square-lattice phononic crystal is one of the recently proposed acoustic metamaterials. Strong anisotropic propagation of elastic waves makes the material promising for various potential applications in acoustics and acousto-optics. This paper presents a study of the propagation of elastic waves in two-dimensional phononic crystals based on fused silica. The band structures of a phononic crystal are obtained by solving the wave equation in its variational form by the finite element method. The main phononic crystal acoustic characteristics that are of practical interest in acousto-optics are calculated based on the analysis of the dispersion relations. It is shown that the choice of the phononic crystal geometry makes it possible to control the distributions of both the inverse phase velocities and the energy walk-off angles of acoustic modes. The calculations of the acoustic modes' polarization are in a particular focus. It is demonstrated that under certain conditions, there are exactly three acoustic modes propagating in a phononic crystal, the averaged polarization vectors of which are mutually orthogonal for any directions of the acoustic wave's propagation. It is argued that the acoustic properties of phononic crystals meet the requirements of acousto-optics.

Keywords: two-dimensional phononic crystals; fused silica; finite element method; polarization of acoustic waves

check for updates

Citation: Marunin, M.V.; Polikarpova, N.V. Polarization of Acoustic Waves in Two-Dimensional Phononic Crystals Based on Fused Silica. *Materials* **2022**, *15*, 8315. https://doi.org/10.3390/ma15238315

Academic Editors: Theodore E. Matikas and Tien-Chang Lu

Received: 19 August 2022
Accepted: 16 November 2022
Published: 23 November 2022

1. Introduction

Acousto-optic devices are well known for their broad application in research and technology. To date, more than a dozen acousto-optic devices using different principles and serving various objectives have been introduced and explored [1].

In designing new acousto-optic devices, particular emphasis is put on the choice of a material in which optical radiation interacts with an acoustic wave. The choice of material depends on the acoustic, optic, and acousto-optic properties of the medium [2]. Most commonly, priority is given to the acoustic properties of the material because they play a vital role in defining the diffraction efficiency. It is known that the acousto-optic figure of merit of the material is inversely proportional to the cube of the acoustic wave phase velocity [3]. In the acousto-optic field, it is therefore essential to find and analyze the materials characterized by the low phase velocity of propagating acoustic waves.

A crucial factor in choosing the material is the anisotropy of its acoustic properties. Strong acoustic anisotropy makes it possible to create and implement new configurations of acousto-optic interaction. In particular, a high degree of anisotropy of the acoustic wave phase velocity leads to significantly larger values of the acoustic energy walk-off angle. These high values of the angle between the phase and group velocities allow, for example, efficient implementation of the collinear diffraction mode [4,5].

It is known that due to the photoelastic effect, the acoustic perturbation in the medium leads to the induced phase grating in the material. In this case, the polarization vector of the acoustic wave determines the effective photoelastic constant, a value that directly affects the diffraction efficiency [6]. Information about the polarization composition of the acoustic wave is therefore important for practical applications. It is interesting to consider the conditions under which the acoustic wave becomes purely longitudinal or purely transverse, as well as the conditions under which the wave changes the polarization type.

Monocrystalline materials take a special place in the field of acousto-optics. Among the most common and frequently used media are crystals of tellurium (Te), lithium niobate ($LiNbO_3$), and TAS (Tl_3AsSe_3), as well as various mercury compounds [7–10]. Nowadays though, a paratellurite (TeO_2) crystal is used as a working material in acousto-optic devices most of the time.

A paratellurite crystal is known for its unusual acoustic properties, which distinguish it from other monocrystalline media. As an example, Figure 1 shows the spatial distributions of the inverse phase velocities of acoustic waves and their acoustic energy walk-off angles. It is evident that paratellurite has extremely low values of the phase velocity of acoustic waves propagating along specific directions. Thus, the minimum phase velocity of a slow shear acoustic mode propagating along one direction is 616 m/s [11]. To note is the strong acoustic anisotropy of the medium, which leads to record-high angles of acoustic energy walk-off. The maximum value of the acoustic walk-off angle in the paratellurite crystal reaches 74° [12].

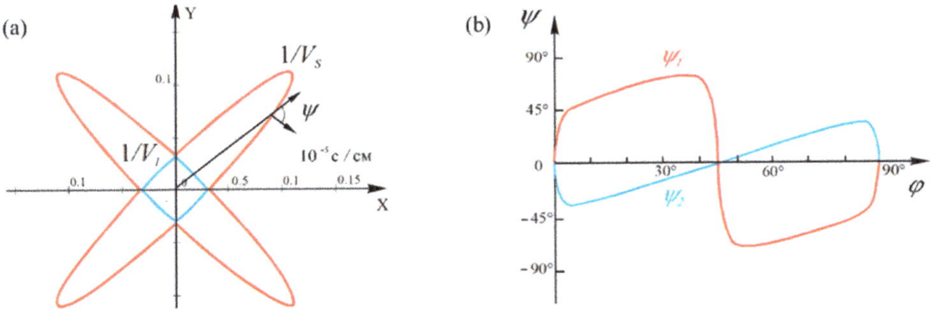

Figure 1. Acoustic characteristics of a paratellurite crystal: (**a**) distributions of the inverse phase velocities of acoustic modes in the XY plane and (**b**) distributions of the walk-off angles of the acoustic modes' energy in the XY plane.

A paratellurite crystal is the standard of modern acousto-optics, and it can be used in creating a wide range of acousto-optic devices. However, with a transparency range of 0.35–6.5 μm [13], paratellurite cannot be used in creating acousto-optic devices operating in the ultraviolet and far-infrared regions of the electromagnetic spectrum. Moreover, the physical properties of paratellurite are determined and they cannot be changed, so the value of the acousto-optic effect in the medium is fixed. Thus, an imperative task of acousto-optics today is not merely the search for acousto-optically effective materials but also the development of innovative artificial media, whose properties could be controlled, depending on the unique challenges in creating devices. In this study, we turn to acoustic metamaterials, namely phononic crystals [14–16].

A phononic crystal is a periodic structure comprising materials with various acoustic properties [17]. The propagation of acoustic waves in such a crystal is described by the dispersion relation, also known as the band structure. An important feature of the band structure is the presence of band gaps, which are frequency ranges in which the propagation of acoustic waves is forbidden [18]. In addition, under certain conditions, the dispersion curves have a negative slope, which results in negative refraction [19,20]. Moreover, the two-dimensional structures may be applied for tunable broadband polarization converters [21], cloaking sensors [22], and plasmonic photodetectors [23]. The characteristics of phononic crystals are key in studying these materials. However, the periodic nature of phononic crystals predetermines a number of other features, such as strong acoustic anisotropy of the material, low phase velocities of acoustic waves, and large angles of acoustic walk-off [24–26]. At the same time, the presence of periodicity and the variation in phononic crystal parameters make it possible to artificially impose various acoustic properties even on isotropic materials.

This paper represents a theoretical study of two-dimensional phononic crystals made of fused silica. The distributions of all the main acoustic characteristics of practical importance in the creation of acousto-optic devices are obtained. The distributions of the inverse phase velocities of acoustic modes of a phononic crystal, the directions of their polarizations, and the acoustic energy walk-off angles are presented. Calculations of the acoustic characteristics are carried out by analyzing the phononic crystal band structure. The dispersion dependences of acoustic waves propagating in a phononic crystal are obtained by solving the elastic wave equation in its variational form, as described in [17]. The work presents the results of studies conducted to assess the possibility of using the acoustic properties of phononic crystals in creating acousto-optic devices.

2. Theoretical Model and Formulations

This paper analyzes a two-dimensional square-lattice phononic crystal based on fused silica. The model of the phononic crystal considered in the work is shown in Figure 2. The phononic crystal represents a sample of fused silica with periodically located cylindrical through holes filled with air (Figure 2a). It is considered an infinite, perfectly periodic medium.

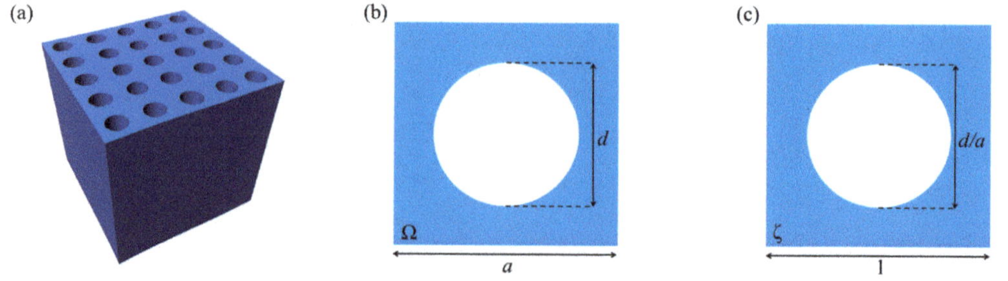

Figure 2. Model of a two-dimensional square-lattice phononic crystal: (**a**) three-dimensional physical model, (**b**) unit cell, and (**c**) unit cell, made dimensionless.

The spatial periodicity of a phononic crystal is characterized by its unit cell, as shown in Figure 2b. The most important geometric parameters of the unit cell are the diameter of its cylindrical holes, d, and the lattice constant, a, which defines the length of the unit cell; the region Ω denotes the part of the unit cell taken up by fused silica.

Fused silica is chosen as a matrix material forming the phononic crystal, due to its wide availability and ease in manufacturing. In addition, fused silica is known for its low attenuation of acoustic waves [27]. Thus, when considering the problem of propagation of elastic waves in a phononic crystal, fused silica is considered an ideal medium and losses due to the propagation of acoustic waves are not considered. A possible approach to accounting for the viscoelastic properties of a phononic crystal, leading to the attenuation of acoustic waves, is presented in [28]. In our work, we also assume that the propagation of acoustic waves in a phononic crystal occurs only in the material of fused silica. The excitation and propagation of acoustic waves in cylindrical holes can be neglected due to the relatively low acoustic impedance of the air.

The theoretical model is based on the elastic wave equation of the form:

$$\frac{\partial T_{ij}}{\partial x_j} + f_i = \rho \frac{\partial^2 u_i}{\partial t^2}, \qquad (1)$$

where u_i is an elastic wave displacement vector, T_{ij} is a Cauchy stress tensor characterizing the reaction of the medium to the applied mechanical stress, f_i is the bulk density of external forces, and ρ is the bulk density of the medium in which the elastic wave propagates. The indices take values of i,j = 1,2,3 and obey the Einstein convention.

The degree of deformation of the medium under the impact of an elastic wave propagating in it is characterized by the strain tensor S_{kl}, which is defined as:

$$S_{kl} = \frac{1}{2}\left(\frac{\partial u_k}{\partial x_l} + \frac{\partial u_l}{\partial x_k}\right). \tag{2}$$

In the limit of the small strains, the stress tensor T_{ij} is proportional to the strain tensor S_{kl} of the medium:

$$T_{ij} = c_{ijkl} S_{kl}, \tag{3}$$

where c_{ijkl} is an elastic constants tensor of the medium. The latest equation represents Hooke's law in the case of an anisotropic medium, in which stresses and strains are related linearly.

Given that the c_{ijkl} tensor is symmetric with respect to the second pair of indices, in the absence of external forces to the bulk material, the wave equation (Equation (1)) can be written as follows:

$$\frac{\partial}{\partial x_j}\left(c_{ijkl}\frac{\partial u_k(r,t)}{\partial x_l}\right) = \rho\frac{\partial^2 u_k(r,t)}{\partial t^2}. \tag{4}$$

In the case of a phononic crystal, the elastic moduli $c_{ijkl}(r)$ and the bulk density $\rho(r)$ are step functions of coordinates. Since the propagation of elastic waves occurs only in the material of fused silica, these functions take on the values of the corresponding constants of fused silica in the Ω region, and they are equal to zero in the region of cylindrical holes.

The spatial periodicity of a phononic crystal may be accounted for using Bloch's theorem. In this case, the general solution of Equation (4) has the form of Bloch waves:

$$u_i(r,t) = \tilde{u}_i(r)\exp[i(\omega t - k \cdot r)], \tag{5}$$

where the amplitude $\tilde{u}_i(r)$ is a periodic function of coordinates, defined in the Ω region of the phononic crystal unit cell, and equal to zero in the region of cylindrical holes; $\omega = 2\pi f$ is the angular frequency of the elastic wave; and k is the wave vector of the elastic wave.

The vector $\tilde{u}(r)$ sets the amplitude of the elastic wave and defines the direction of its polarization. The absolute value of the wave vector $|k| = \omega/V$ sets the phase velocity of acoustic waves propagating in a phononic crystal. Based on the axial symmetry of the unit cell, we limit the analysis to the wave vector in the form $k = (k_x, k_y, 0)$.

The strain tensor of a periodic medium also, according to its definition (Equation (2)), takes the form of a Bloch wave:

$$S_{ij} = \widetilde{S}_{ij}(\tilde{u})\exp[i(\omega t - k \cdot r)], \tag{6}$$

where the amplitudes $\widetilde{S}_{ij}$ equals:

$$\widetilde{S}_{ij}(\tilde{u}) = \frac{1}{2}\left[\left(\frac{\partial \tilde{u}_i}{\partial x_j} + \frac{\partial \tilde{u}_j}{\partial x_i}\right) - i\left(k_i\tilde{u}_j + k_j\tilde{u}_i\right)\right]. \tag{7}$$

The explicit form of solution (Equation (5)) allows obtaining the variational formulation of the wave equation (Equation (4)) for infinite periodic media. The resulting variational problem can be solved by the finite element method (FEM), which makes it possible to find the dispersion dependences of acoustic waves. The core idea of the method is to approximate a solution belonging to some infinite-dimensional function space (e.g., Sobolev space) by a linear combination of the fixed-basis vectors of a subspace of the space. The FEM approximation depends on the choice of a subspace and the associated basis. Usually, a space of piecewise polynomial functions is chosen as such a subspace. In our research, we used the Lagrange P_2-elements as a basis. After choosing the basis functions, the FEM comes down to the optimization problem of minimizing the approximation error of some functional (given by a weak formulation of the original differential equation) on the FEM mesh. An example of the FEM mesh used in our calculcations for a phononic crystal with $d/a = 0.6$ is shown in Figure 3. According to [17], the described variational problem takes the following form:

$$\int_{\Omega} \widetilde{S}_I^*(v) c_{IJ} \widetilde{S}_J(\widetilde{u}) d\Omega = \omega^2 \int_{\Omega} \rho(v^*, \widetilde{u}) d\Omega. \tag{8}$$

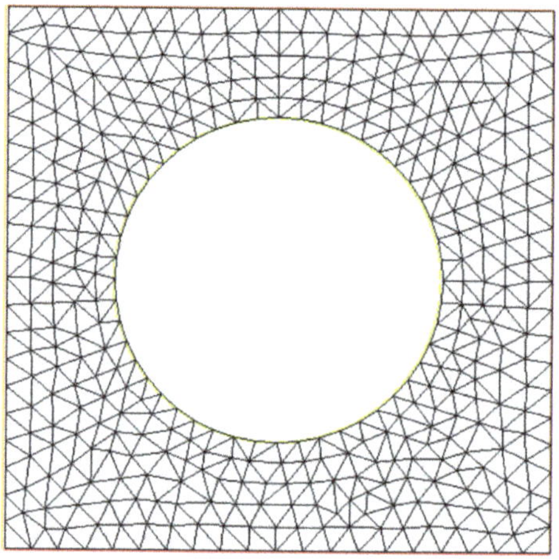

Figure 3. An example of the finite element method mesh used in calculation for a phononic crystal with *d/a = 0.6*.

Here, $v(r)$ is an arbitrary test function that belongs to the same functional space of finite elements as the solution $\tilde{u}(r)$. Introduction of this function is necessary for solving the problem using the finite element method. In this case, the Voigt notation is used for tensor quantities, so the indices I $\leftrightarrow$ *(ij)* and J $\leftrightarrow$ *(kl)* take values from 1 to 6 according to the rule 1 $\leftrightarrow$ (11), 2 $\leftrightarrow$ (22), 3 $\leftrightarrow$ (33), 4 $\leftrightarrow$ (23), 5 $\leftrightarrow$ (13), 6 $\leftrightarrow$ (12). Considering the thermodynamic relations, the components of the six-dimensional deformation amplitudes vector are defined as $\widetilde{S}_I = (2 - \delta_{ij}) \bullet \widetilde{S}_{ij}$, where δ_{ij} is the Kronecker symbol.

The integral equation (Equation (8)) is a generalized eigenfunction and eigenvalue problem. The eigenvalue in this case is $\lambda = \omega^2$. The set of eigenvectors $\tilde{u}_\lambda$ determines the amplitudes and polarization directions of acoustic waves of frequency ω, which can propagate in a phononic crystal. According to Equation (7), the amplitude of the deformation vector is directly affected by the components of the wave vector. Thus, by changing the value of the wave vector within the first zone and solving the integral equation (Equation (8)) for each individual value of the wave vector k, one can obtain the dispersion relation ω(k).

To identify the main features of the solution to the integral equation (Equation (8)), it has to be brought to its dimensionless form. The following dimensionless quantities are introduced:

$$\chi_i = \frac{x_i}{a},\ \xi_i = \frac{\widetilde{u}_i}{a},\ \eta_i = \frac{v_i}{a},\ \kappa_i = k_i \cdot a,\ \rho = \rho' \cdot \rho_0,\ c_{IJ} = c'_{IJ} \cdot c_0, \tag{9}$$

where ρ' and c'_{IJ} are dimensionless bulk density and elastic constants, respectively, carrying the numerical values of the corresponding physical constants of fused silica, and ρ_0 and c_0

are quantities that characterize the dimensions of those physical constants. In dimensionless quantities, the integral equation (Equation (8)) takes the form:

$$\int_{\varsigma} S_I^*(\eta) c'_{IJ} S_J(\xi) d\varsigma = \frac{\rho_0 \omega^2 a^2}{c_0} \int_{\varsigma} \rho'(\eta^*, \xi) d\varsigma, \tag{10}$$

where the integration is carried out over the region ζ of the unit cell, which is made dimensionless, as shown in Figure 2c. The explicit form of the integral equation (Equation (10)) allows making several important observations.

First, the eigenvalues of this equation, and therefore also the eigenvalues of Equation (8), are not as dependent on the specific values of a and d as on their ratio *d/a*. Theoretically, the normalized hole diameter can be infinitely close to unity. However, the maximum *d/a* ratio considered in this work is 0.8. This choice pertains to the technological difficulties in creating phononic crystals, where the normalized hole diameter exceeds this value. It should be noted that increasing the diameter of the holes reduces the space taken up by the matrix material. This, in turn, leads to a decrease in the acousto-optic interaction region, which results in a lower diffraction efficiency. In addition, for a given geometry of a phononic crystal, that is, for a fixed region of integration ζ, the product of ω•a = const. Thus, a change in the unit cell constant *a* of a phononic crystal leads to the corresponding scaling of the dispersion dependences. Further in this work, for the sake of presentation clarity, we consider the solution of the integral equation (Equation (8)) in dimensional quantities. The unit cell constant *a* is considered equal to 10 μm.

3. The Method for Calculations of the Phononic Crystal Acoustic Characteristics

A consistent solution of the generalized eigenfunction and eigenvalue problem (Equation (8)) for different values of the wave vector *k* makes it possible to obtain the dispersion dependence ω(k) of acoustic waves propagating in a phononic crystal. Figure 4 shows the band structure of a phononic crystal with a normalized hole diameter *d/a* = 0.8. Figure 4a shows the dispersion dependences calculated by the finite element method. Only the first three dispersion surfaces are displayed. They correspond to three different acoustic modes with a frequency of up to 200 MHz. Higher-order surfaces are above the indicated frequency. There are no absolute band gaps in the frequency range of up to 200 MHz and this crystal geometry, as evident from Figure 4a.

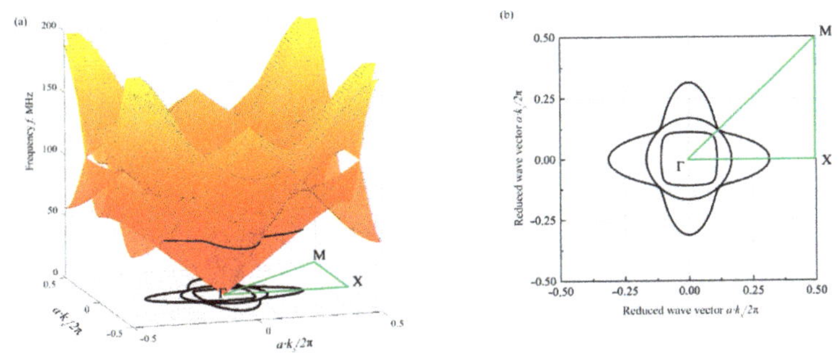

Figure 4. Band structure of a two-dimensional phononic crystal with a normalized hole diameter *d/a* = 0.8: (**a**) the first three dispersion surfaces corresponding to the three lowest acoustic modes in a phononic crystal and (**b**) projection of the dispersion surface contour lines onto the first Brillouin zone at an isofrequency f_0 = 50 MHz.

The dispersion surfaces of the first three acoustic modes of the phononic crystal originate from the center of the first Brillouin zone. In the lower frequency range, up to 50 MHz, these surfaces are mutually exclusive and monotonic and do not intersect. With higher frequencies exceeding 50 MHz, the surfaces are no longer monotonic. The emerging frequency regions with a negative slope may lead to a negative refraction effect. At frequencies above 100 MHz, the second and third surfaces begin to intersect. Such a complex structure is explained by the relatively high *d/a* ratio. It is also evident from Figure 4a that the dispersion surfaces are symmetric with respect to the first irreducible Brillouin zone Γ-X-M-Γ, which is explained by the symmetry of the unit cell.

The dispersion relation $\omega(\mathbf{k})$ connects the frequencies and wave vectors of acoustic waves that can propagate in a medium. By fixing the frequency at $\omega_0 = 2\pi f_0$, it is possible to obtain sets of wavenumbers $k_x(f_0)$ and $k_y(f_0)$ of all acoustic waves of frequency f_0 that can propagate in a phononic crystal. At a given ultrasound frequency f_0, the wavenumbers k_x and k_y are proportional to the components of the inverse phase velocity of acoustic waves, since $k_x = 2\pi f_0/V_x$ and $k_y = 2\pi f_0/V_y$. Thus, the line of the frequency contour f_0 of the dispersion dependence defines the cross sections of the acoustic slowness surface $S(\varphi)$ in the XY plane of the phononic crystal, where φ is the polar angle. In an acousto-optic device, the isofrequency f_0 is the frequency of ultrasound excited by the piezoelectric transducer.

Figure 4a shows the contour lines of the isofrequency f_0 = 50 MHz, while the projection of those contour lines to the first Brillouin zone can be seen in Figure 4b. Similar to the dispersion surfaces, the isofrequency contour lines are symmetric with respect to the first irreducible Brillouin zone Γ-X-M-Γ. For isofrequencies f_0 < 50 MHz, the contour line of each individual acoustic mode is a closed curve. For isofrequencies f_0 > 50 MHz, where the dispersion surfaces become nonmonotonic, the contour lines become piecewise interval lines. The calculation of acoustic characteristics at these higher isofrequencies requires a separate analysis. This work presents the case in which the isofrequency is chosen to be f_0 = 50 MHz.

The spatial periodicity of a phononic crystal leads to anisotropy of the physical characteristics of acoustic waves propagating in a phononic crystal. In particular, the vectors of the phase and group velocities of acoustic waves become non-codirectional. The explicit form of the acoustic slowness curves $S(\varphi)$ allows finding the angle ψ of the acoustic energy walk-off. The acoustic energy walk-off angle between the Pointing vector and the acoustic wave phase velocity, according to [6], can be found as:

$$\psi = \tan^{-1}\left(\frac{1}{S}\frac{dS}{d\varphi}\right). \tag{11}$$

In acousto-optics, one of the most important characteristics of an acoustic wave is the direction of its polarization. The eigenvectors $\tilde{\mathbf{u}}_\lambda(\mathbf{r})$ of the integral problem (Equation (8)) define the amplitudes and polarization directions of ω frequency acoustic waves, which can propagate in a phononic crystal. Since eigenvectors $\tilde{\mathbf{u}}_\lambda(\mathbf{r})$ are periodic functions of coordinates, the polarization direction of each acoustic mode varies within the unit cell. Therefore, from a practical point of view, it seems necessary to carry out the calculation of the polarization vector $\tilde{\mathbf{u}}^0$ averaged over the unit cell region. The components of the averaged polarization vector can be obtained as:

$$\tilde{u}_p^0 = \pm\left(\frac{\int_\Omega |\tilde{u}_{\lambda,p}|^2 d\Omega}{\int_\Omega |\tilde{u}_\lambda|^2 d\Omega}\right)^{1/2}, \tag{12}$$

where the index $p = x,y,z$.

According to Equation (12), the squared values of the integral equation (Equation (8)) solution components are averaged over the unit cell region of the phononic crystal and then normalized. However, only the squared values of the averaged polarization vector components can be obtained this way. The actual components $\tilde{u}_p^0$ of the averaged polarization vector $\tilde{\mathbf{u}}^0$ are defined to within an overall sign, as described in Equation (12). It is clear that in the limit $d/a \to 0$, the averaged solution of Equation (8) should converge to that of solid isotropic fused silica. The requirement of the physical behavior in the mentioned limit makes it possible to choose the sign in Equation (12). We also assume that with a small change in the phononic crystal geometry, that is, with a small change in *d/a*, the change in the polarization vector direction is continuous. With this assumption, the signs of the components of the averaged polarization vector in Equation (12) can be determined for any values of the ratio *d/a*.

4. Results and Discussion

This section presents the results of calculating the main acoustic characteristics of a phononic crystal. The calculations are carried out according to the method described in the previous section. The values of the material constants of fused silica are chosen according to [6]. The elastic constants and density of fused silica are considered as $c_{11} \equiv c_1 = 7.85 \cdot 10^{10}$ N/m^2, $c_{12} \equiv c_6 = 1.61 \cdot 10^{10}$ N/m^2, and $\rho = 2203$ kg/m^3.

4.1. Inverse Phase Velocity Distributions

Before moving on to the various geometries of a phononic crystal, we present the slowness curves of solid isotropic fused silica that can be taken as a reference material. It is well known that in the isotropic case, there are two transverse and one longitudinal acoustic mode propagating in the medium. The polarizations of these modes are mutually orthogonal. One of the ways to obtain the phase velocities of these modes is by solving the Christoffel equation (Equation (6)). The phase velocities of the transverse and longitudinal acoustic modes are:

$$V_\perp = \left(\frac{c_{11} - c_{12}}{2\rho}\right)^{1/2} = 3763 \text{ m/s}, \; V_\parallel = \left(\frac{c_{11}}{\rho}\right)^{1/2} = 5969 \text{ m/s}. \tag{13}$$

In the case of solid isotropic fused silica, the cross sections of the acoustic slowness surfaces are a pair of circles, as shown in Figure 5a. The two transverse shear acoustic modes are shown in blue; the longitudinal acoustic mode is shown in orange.

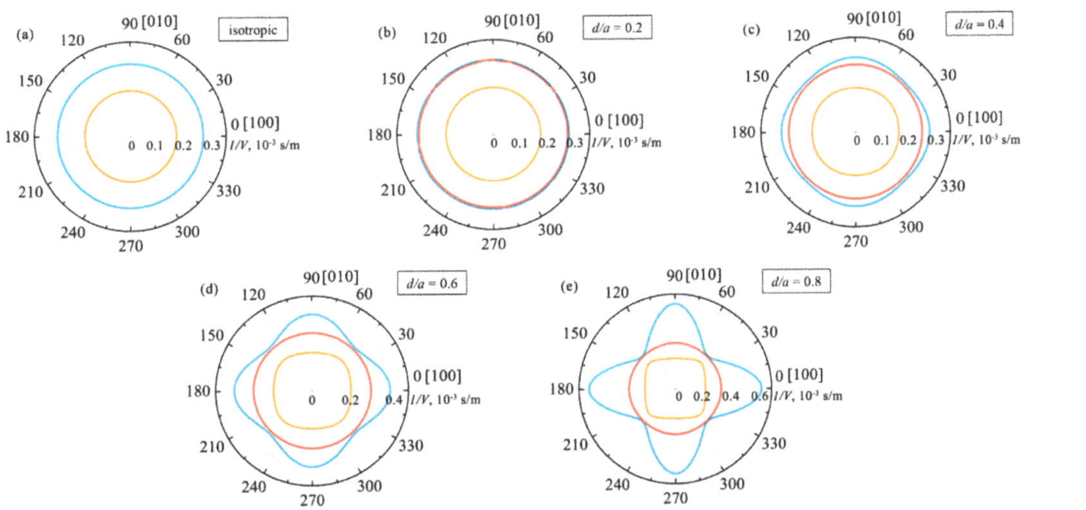

Figure 5. (**a**) Inverse phase velocity curves in the XY plane in an isotropic material of fused silica. Shown in (**b–e**) are inverse phase velocity curves of a phononic crystal with the normalized hole diameter: (**b**) d/a = 0.2; (**c**) d/a = 0.4; (**d**) d/a = 0.6; and (**e**) d/a = 0.8.

Figure 5b shows the cross sections of the slowness surface of a phononic crystal with the normalized hole diameter *d/a* = 0.2. Clearly, the introduction of a spatially periodic inhomogeneity into the material of solid isotropic fused silica leads to the gradual elimination of the degeneracy of transverse acoustic modes, while increasing the normalized hole diameter to $d/a = 0.4$, as shown in Figure 5c, results in the complete removal of the degeneracy. Thus, transverse modes of different polarization directions propagate at different speeds in a phononic crystal. In Figure 5b,c, this is reflected by the appearance of a red curve.

A further increase in the normalized hole diameter leads to notable anisotropy of the phase velocity. One can see that in the case of Figure 5d with the normalized hole diameter $d/a = 0.6$, the longitudinal mode becomes anisotropic, just as well as the slow shear mode. However, the fast shear mode of the phononic crystal shown in red in Figure 5d remains isotropic.

The degree of anisotropy of the acoustic mode phase velocity is characterized by the anisotropy coefficient, which is defined as:

$$\chi = \left(\frac{V_{max}}{V_{min}}^2\right), \tag{14}$$

where V_{max} and V_{min} are the maximum and minimum values of the phase velocity of an individual acoustic mode, respectively. The anisotropy coefficients of the phononic crystal acoustic modes, as well as their minimum and maximum phase velocities, are presented in Table 1.

Table 1. Phase velocities of acoustic modes and their anisotropy coefficients for phononic crystals with various normalized hole diameters.

d/a	**Slow Shear Wave**		**Isotropic Wave**	**Longitudinal Wave**	
	V_{min}, m/s	V_{max}, m/s	V_{iso}, m/s	V_{min}, m/s	V_{max}, m/s
0.2	3707	3708	3709	5808	5816
	$\chi \approx 1$			$\chi \approx 1$	
0.4	3191	3392	3542	5315	5438
	$\chi = 1.13$			$\chi = 1.05$	
0.6	2481	3174	3310	4590	5007
	$\chi = 1.64$			$\chi = 1.19$	
0.8	1594	2976	3004	3759	4536
	$\chi = 3.49$			$\chi = 1.46$	

The phononic crystal acoustic slowness curves with the normalized hole diameter $d/a = 0.8$ are of greatest interest (Figure 5e). It is clear that the slow shear acoustic mode has an extremely strong anisotropy of phase velocity. The anisotropy coefficient of that mode is $\chi = 3.49$. The longitudinal acoustic mode is also anisotropic, with an anisotropy coefficient $\chi = 1.46$. However, the fast shear mode shown in red in Figure 5e remains isotropic even at such a large value of the d/a ratio. Further analysis shows that this is due to the fact that the shear mode of interest is polarized along the cylindrical holes (Z axis).

The results presented in Figure 5 make it possible to observe the dynamics of changes in the distributions of the inverse phase velocities of acoustic modes, depending on the normalized hole diameter d/a. Clearly, an increase in the normalized diameter leads to an increase in the anisotropy of the phase velocities. At the same time, as evident from Table 1, an increase in the normalized hole diameter leads to a decrease in the phase velocities of acoustic modes. The latter is also true for the fast shear mode of a phononic crystal, which is isotropic. Thus, by changing the normalized diameter of the cylindrical holes, it seems possible to impose the necessary distributions and numerical values of the phase velocities of the acoustic modes on the phononic crystal. The obtained results are in qualitative agreement with the slowness curves for the two-dimensional metamaterial based on rutile [29].

4.2. The Distributions of Walk-Off Angles

The explicit form of the acoustic slowness curves makes it possible to calculate the walk-off angle according to Equation (11). For a fixed direction of acoustic wave propagation, the walk-off angle defines the angle between the normal to the slowness curve and the direction of wave propagation. Since in the case of isotropic fused silica, the slowness curves of acoustic modes are a pair of circles (Figure 5a), the walk-off angles of these modes are exactly zero for all directions of their propagation. In the case of small values of the normalized hole diameter, the inverse phase velocity curves are of nearly round shape and the acoustic walk-off angle is close to zero. Thus, in a phononic crystal with the normalized hole diameter d/a= 0.2, the walk-off angles of acoustic modes do not exceed several degrees.

Figure 6 shows the distributions of the walk-off angles ψ (φ) of acoustic modes, depending on the direction of their propagation in phononic crystals of various d/a ratios. The presented walk-off angle distributions of the slow shear and longitudinal acoustic modes correspond to the slowness curves shown in Figure 5c,d. For the fast shear mode, the walk-off angle ψ (φ) distributions are not shown, because the mode is isotropic and its energy walk-off angle equals zero for all propagation directions.

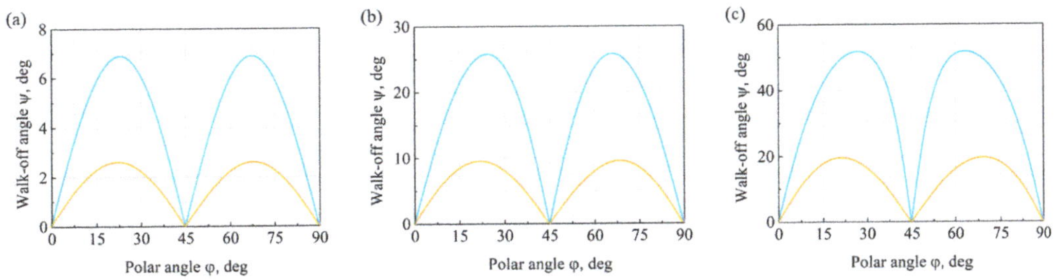

Figure 6. Distribution of acoustic energy walk-off angles in the XY plane of a phononic crystal with a normalized hole diameter: (**a**) $d/a = 0.4$; (**b**) $d/a = 0.6$; and (**c**) $d/a = 0.8$.

With a relatively small normalized hole diameter of $d/a = 0.4$, the maximum walk-off angle of a phononic crystal acoustic mode does not exceed 7°, as is evident from Figure 6a. According to Figure 6b, a further increase in the normalized hole diameter to $d/a = 0.6$ leads to a sharp increase in the acoustic walk-off angle. In particular, the maximum value of the walk-off angle of the slow shear mode exceeds 25°. The maximum values of the energy walk-off angles, ψ_{max}, of acoustic modes in phononic crystals with different normalized hole diameters are presented in Table 2, as well as the propagation directions of the modes, φ *, leading to the maximum walk-off angle.

Table 2. Maximum values of the energy walk-off angles of acoustic modes in phononic crystals with various normalized hole diameters.

d/a	Slow Shear Mode		Longitudinal Mode	
	ψ_{max}, °	φ *, °	χ_{max}, °	ψ *, °
0.4	6.9	23.3	2.62	22.7
0.6	25.8	24.6	9.6	22.2
0.8	51.8	27.4	19.7	21.5

The walk-off angle distributions of acoustic modes shown in Figure 6c are of practical interest. The results suggest that the maximum walk-off angle of the slow shear acoustic mode exceeds 50°. It should be noted that this value is unusually high, and there are only a few monocrystalline structures currently known that exhibit such a large walk-off of acoustic energy. Thus, changing the geometry of a phononic crystal makes it possible to control the energy walk-off angles over a wide range of values.

4.3. The Polarizations of Acoustic Modes

One way of finding the polarization directions of the acoustic modes in isotropic solid fused silica is by solving the Christoffel equation. It has been established that due to the symmetry of the Christoffel tensor, the polarizations of acoustic modes in solid fused silica are mutually orthogonal for any direction of acoustic mode propagation [6]. Calculations indicate that one of the purely shear acoustic modes shown in Figure 5a is polarized in the XY plane of the figure. The other purely shear acoustic mode is polarized along the Z axis. The polarization of the purely longitudinal acoustic mode lies in the XY plane, and it is orthogonal to the polarization directions of the shear acoustic modes.

Figure 7 presents the distributions of the direction of the averaged polarization vector $\tilde{\mathbf{u}}^0(\varphi)$ of acoustic modes, depending on their propagation direction, featuring two different geometries of a phononic crystal. The averaged polarization vector is calculated using the method described in the previous section. The results show that the fast shear mode is polarized along the Z axis of the cylindrical holes of the phononic crystal. The slow shear acoustic mode appears purely transverse for propagation directions $\varphi = \pi n/4,\ n \in \mathbb{N}$ in the XY plane. The longitudinal acoustic mode of a phononic crystal is purely longitudinal when propagating along the indicated directions. This property of the averaged polarization vector is associated with the symmetry of the phononic crystal unit cell.

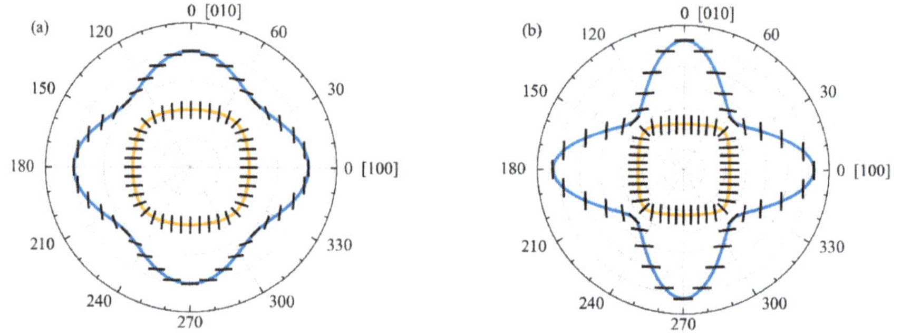

Figure 7. Distribution of the averaged polarization vector of acoustic modes in a phononic crystal with a normalized hole diameter: (**a**) d/a = 0.6 and (**b**) d/a = 0.8.

The degree of polarization of the acoustic mode can be characterized by the angle $\gamma^0(\varphi)$ between the averaged polarization vector $\tilde{\mathbf{u}}^0(\varphi)$ and the direction of the mode φ propagation. Same as the averaged polarization vector, this angle is a function of the acoustic wave propagation direction. By definition of the corresponding terms, for quasi-shear waves, the value of this angle lies in the region $45° \leq \gamma^0 < 90°$; for quasi-longitudinal waves, it lies in the region $0° < \gamma^0 \leq 45°$. The polarization angle γ^0 of the fast shear isotropic mode of a phononic crystal is 90° for all directions of wave propagation. This mode is purely shear, and it is polarized along the Z axis of the cylindrical holes for all phononic crystal geometries.

In accordance with Figure 7a, for the phononic crystal geometry with a normalized hole diameter d/a = 0.6, the polarization angle of the slow shear mode of a phononic crystal belongs to the $69.9° \leq \gamma^0 \leq 90°$ range. The polarization angle of the longitudinal mode takes the values $0° \leq \gamma^0 \leq 20.1°$. For the phononic crystal geometry in which d/a = 0.8, the polarization angle of the slow shear mode lies in the range of $61.6° \leq \gamma^0 \leq 90°$, as seen in Figure 7b. For the longitudinal mode of a phononic crystal, the range is $0° \leq \gamma^0 \leq 28.4°$. Thus, the anisotropic acoustic modes of a phononic crystal are quasi-transverse and quasi-longitudinal. The transformation of the acoustic mode from quasi-transverse to quasi-longitudinal and vice versa is impossible in a phononic crystal. The limitation is associated with the symmetry of the phononic crystal unit cell, as well as with the assumption that the averaged polarization vector $\tilde{\mathbf{u}}^0(\varphi)$ is continuous.

The calculations of averaged polarization vectors help to trace an unusual characteristic of their behavior. It is evident from the results presented in Figure 7 that the averaged polarization vectors of the phononic crystal acoustic modes are mutually orthogonal for any directions of wave propagation at any values of the normalized hole diameter *d/a*. This property is typical for continuous homogeneous media; however, it is not universal for periodic structures in general.

In the case of homogeneous media, the mutual orthogonality of acoustic modes arises from the symmetry of the Christoffel tensor. The kernel of the integral equation (Equation (8)), which makes it possible to identify the acoustic modes of a phononic crystal, is also symmetric with respect to the solution $\tilde{\mathbf{u}}_\lambda$. However, this symmetry is not the reason for the orthogonality of the polarization vectors of the phononic crystal acoustic modes, since for any given propagation direction φ, the solution $\tilde{\mathbf{u}}_\lambda$ is sought at various wavenumbers k_x and k_y. The eigenvalue of the integral equation (Equation (8)) for different acoustic modes matches because these modes are obtained by isocontouring the dispersion surfaces of a fixed isofrequency f_0. However, when searching for the averaged polarization vectors of acoustic modes, different integral equations with different kernels are solved. Thus, in the low-frequency region of excited acoustic modes, up to 50 MHz, a phononic crystal effectively behaves as a homogeneous anisotropic medium in the sense that there are exactly three acoustic modes in it, and their mean polarizations are mutually orthogonal.

5. Conclusions

This work considers the acoustic properties of two-dimensional square-lattice phononic crystals based on fused silica. Distributions of the main characteristics of the phononic crystal acoustic modes are calculated, such as inverse phase velocities, directions of their polarizations, and acoustic energy walk-off angles. It is established that the introduction of spatial periodicity into an initially isotropic material of fused silica leads to strong anisotropy of the acoustic properties. Thus, the energy walk-off angle of the slow shear acoustic mode reaches $\psi = 51.8^\circ$ when the mode propagates along certain directions. It is shown that a change in the phononic crystal geometry makes it possible to control both the spatial distributions and the numerical values of its acoustic characteristics. In particular, the geometry of a phononic crystal is presented, for which the minimum value of the slow shear acoustic mode phase velocity is V = 1594 m/s. This is several times smaller than the phase velocity value $V_\perp = 3763$ m/s of the shear acoustic mode of solid isotropic fused silica. It is demonstrated that under certain conditions, exactly three acoustic modes propagate in a phononic crystal. In this case, the averaged polarization vectors of acoustic modes are mutually orthogonal for any directions of wave propagation. Thus, under certain conditions, a phononic crystal effectively behaves as a homogeneous bulk anisotropic medium.

From the point of view of acoustic properties, phononic crystals satisfy all the necessary requirements for materials in the design of acousto-optic devices. Therefore, the main conclusion of the work is that phononic crystals can be recommended for use in acousto-optic devices, namely filters, deflectors, and modulators, where sound velocity control is required. More practical applications include the use in acoustic filters and delay lines.

Author Contributions: Conceptualization, M.V.M. and N.V.P.; Software, M.V.M.; Validation, N.V.P.; Formal analysis, M.V.M.; Investigation, M.V.M. and N.V.P.; Writing—original draft, M.V.M.; Writing—review & editing, M.V.M. and N.V.P.; Visualization, M.V.M.; Supervision, N.V.P.; Project administration, N.V.P. All authors have read and agreed to the published version of the manuscript.

Funding: This work has been funded in part by the Russian Science Foundation (RSF; grant no. 19-12-00072).

Acknowledgments: This work has been supported in part by the Russian Science Foundation (RSF; grant no. 19-12-00072).

Conflicts of Interest: The authors declare no conflict of interest.

References

1. Xu, J.; Stroud, R. *Acousto-Optic Devices: Principles, Design, and Applications*; John Wiley and Sons. Inc.: New York, NY, USA, 1992.
2. Das, P.K.; DeCusatis, C.M. *Acousto-Optic Signal Processing: Fundamentals and Applications*; Artech House: New York, NY, USA, 1991.
3. Balakshy, V.I.; Parygin, V.N.; Chirkov, L.E. *Physical Principles of Acousto-Optics*; Radio & Svyaz: Moscow, Russia, 1985. (In Russian)
4. Mantsevich, S.N.; Balakshy, V.I.; Molchanov, V.Y.; Yushkov, K.B. Influence of acoustic anisotropy in paratellurite on quasicollinear acousto-optic interaction. *Ultrasonics* **2015**, *63*, 39–46. [CrossRef] [PubMed]
5. Mantsevich, S.N. Investigation of acoustic beam reflection influence on the collinear acousto-optic interaction characteristics. *Ultrasonics* **2016**, *70*, 92–97. [CrossRef] [PubMed]
6. Dieulesaint, E.; Royer, D. *Elastic Waves in Solids: Applications to Signal Processing*; Wiley: Chichester, UK, 1980.
7. Voloshinov, V.B.; Nikitin, P.A.; Trushin, A.S.; Magdich, L.N. Acousto-optic cell based on paratellurite crystal with surface excitation of acoustic waves. *Tech. Phys. Lett.* **2011**, *37*, 754–756. [CrossRef]
8. Voloshinov, V.B.; Khorkin, V.S.; Kulakova, L.A.; Gupta, N. Optic, acoustic and acousto-optic properties of tellurium in close-to-axis regime of diffraction. *J. Phys. Commun.* **2017**, *1*, 025006. [CrossRef]
9. Sarabalis, C.J.; McKenna, T.P.; Patel, R.N.; Van Laer, R.; Safavi-Naeini, A.H. Acousto-optic modulation in lithium niobate on sapphire. *APL Photonics* **2020**, *5*, 086104. [CrossRef]
10. Porokhovnichenko, D.L.; Dyakonov, E.A.; Kuznetsov, S.V.; Voronov, V.V.; Fedorov, P.P.; Zaramenskikh, K.S.; Gasanov, A.A.; Zhukova, L.V.; Korsakov, A.S.; Salimgareev, D.D. Indium iodide single crystal: Breakthrough material for infrared acousto-optics. *Opt. Lett.* **2020**, *45*, 3435–3438. [CrossRef]
11. Declercq, N.F.; Polikarpova, N.V.; Voloshinov, V.B.; Leroy, O.; Degrieck, J. Enhanced anisotropy in paratellurite for inhomogeneous waves and its possible importance in the future development of acouto-optic devices. *Ultrasonics* **2006**, *44*, e833–e837. [CrossRef]
12. Voloshinov, V.B.; Polikarpova, N.V. Acousto-optic investigation of propagation and reflection of acoustic waves in paratellurite crystal. *App. Opt.* **2009**, *48*, C55–C66. [CrossRef]
13. Syrbu, N.N.; Cretu, R.V. The superposition of one- and two-phonon absorption and radiation in TeO_2 crystal. *Infrared Phys. Technol.* **1996**, *37*, 769–775. [CrossRef]
14. Liu, J.; Guo, H.; Wang, T. A review of acoustic metamaterials and phononic crystals. *Crystals* **2020**, *10*, 305. [CrossRef]
15. Olson, R.H.; El-Kady, I. Microfabricated phononic crystal devices and applications. *Meas. Sci. Technol.* **2009**, *20*, 012002. [CrossRef]
16. Maldovan, M. Sound and heat revolutions in phononics. *Nature* **2013**, *7475*, 209–217. [CrossRef] [PubMed]
17. Laude, V. *Phononic Crystals: Artificial Crystals for Sonic, Acoustic and Elastic Waves*; Walter de Gruyter GmbH: Berlin, Germany, 2015.
18. Croenne, C.; Lee, E.J.; Hu, H.; Page, J.H. Band gaps in phononic crystals: Generation mechanisms and interaction effects. *AIP Adv.* **2011**, *4*, 041401. [CrossRef]
19. Craster, R.V.; Guenneau, S. *Acoustic Metamaterials: Negative Refraction, Imaging, Lensing and Cloaking*; Springer: Dordrecht, The Netherlands, 2012.
20. Croënne, C.; Manga, E.D.; Morvan, B.; Tinel, A.; Dubus, B.; Vasseur, J.; Hladky-Hennion, A.C. Negative refraction of longitudinal waves in a two-dimensional solid-solid phononic crystal. *Phys. Rev. B* **2011**, *83*, 054301. [CrossRef]
21. Khajeh, A.; Hamzavi-Zarghani, Z.; Yahaghi, A.; Farmani, A. Tunable broadband polarization converters based on coded graphene metasurfaces. *Sci. Rep.* **2021**, *11*, 1296. [CrossRef]
22. Hamzavi-Zarghani, Z.; Yahaghi, A.; Matekovits, L.; Farmani, A. Tunable mantle cloaking utilizing graphene metasurface for terahertz sensing applications. *Opt. Express* **2019**, *27*, 34824–34837. [CrossRef]
23. Khosravian, E.; Mashayekhi, H.R.; Farmani, A. Tunable plasmonics photodetector in near-infrared wavelengths using graphene chemical doping method. *Int. J. Electron. Commun.* **2020**, *127*, 152472. [CrossRef]
24. Zhou, X.W.; Zou, X.Y.; Wang, T.H.; Cheng, J.C. Effective velocity of 2D phononic crystals with rectangular lattice. *Ultrasonics* **2010**, *50*, 577–582. [CrossRef]
25. Tanaka, Y.; Tamura, S. Surface acoustic waves in two-dimensional periodic elastic structures. *Phys. Rev. B* **1998**, *58*, 7958–7965. [CrossRef]
26. Wu, T.T.; Huang, Z.G.; Lin, S. Surface and bulk acoustic waves in two-dimensional phononic crystal consisting of materials with general anisotropy. *Phys. Rev. B* **2004**, *69*, 094301. [CrossRef]
27. Krischer, C. Optical measurements of ultrasonic attenuation and reflection losses in fused silica. *JASA* **1970**, *48*, 1086–1092. [CrossRef]
28. Wang, Y.F.; Wang, Y.S.; Laude, V. Wave propagation in two-dimensional viscoelastic metamaterials. *Phys. Rev. B* **2015**, *42*, 104110. [CrossRef]
29. Wang, Y.-F.; Maznev, A.A.; Laude, V. Formation of Bragg Band Gaps in Anisotropic Phononic Crystals Analyzed with the Empty Lattice Model. *Crystals* **2016**, *6*, 52. [CrossRef]

materials MDPI

Article

Evaluation of Acoustic Waves in Acousto-Optical Devices by Ultrasonic Imaging

Sergey A. Titov *, Alexander S. Machikhin and Vitold Ed. Pozhar

Scientific and Technological Center of Unique Instrumentation, Russian Academy of Sciences, 15 Butlerova, 117342 Moscow, Russia; np@ntcup.ru (A.S.M.); vitold@ntcup.ru (V.E.P.)
* Correspondence: titov.sa@ntcup.ru; Tel.: +7-(495)333-24-31

Abstract: The structure of the acoustic field defines the key parameters of acousto-optical (AO) devices. To confirm their compliance with the expected values in the presence of multiple real factors, AO crystalline cells require accurate experimental investigation of the acoustic field after being totally assembled. For this purpose, we propose to detect and quantify all the acoustic waves propagating in AO cells using an impulse acoustic microscopy technique. To validate this approach, we have analyzed both theoretically and experimentally the modes, amplitudes, propagation trajectories, and other features of the ultrasonic waves generated inside an AO modulator made of fused quartz. Good correspondence between theoretical and experimental data confirms the effectiveness of the proposed technique.

Keywords: acousto-optic crystals; propagation of acoustic waves; non-destructive testing; impulse acoustic microscopy

Citation: Titov, S.A.; Machikhin, A.S.; Pozhar, V.E. Evaluation of Acoustic Waves in Acousto-Optical Devices by Ultrasonic Imaging. *Materials* **2022**, *15*, 1792. https://doi.org/10.3390/ma15051792

Academic Editor: Alexander N. Obraztsov

Received: 31 December 2021
Accepted: 20 February 2022
Published: 27 February 2022

1. Introduction

Acousto-optic (AO) interaction is a physical effect, which is used for modulation, deflection, and spectral and spatial filtration of electromagnetic radiation [1,2]. As they are compact, monolithic and free of moving components, AO devices are now widespread in industrial, biomedical and scientific applications [3–5]. The physical principle of most AO instruments consists of Bragg diffraction of light by ultrasound in crystalline media. The tuning of ultrasound power and frequency enables smooth and accurate control of amplitude, propagation direction, polarization and other parameters of light waves.

The structure of induced acoustic fields defines the performance of AO devices to a large extent. Attenuation, divergence, walk-off and other features of ultrasonic waves inevitably influence the key characteristics of AO diffraction: efficiency, signal-to-noise ratio, light beam distortions, etc. Precise theoretical consideration and modeling give a chance to predict and protect the characteristic degradation at the design stage, but the real structure of an acoustic field may only be revealed experimentally. It is barely possible to reveal and quantify crystal inhomogeneity and inner defects, heat generation by the piezoelectric transducer, multiple reflections of ultrasound from AO cell facets, and many other factors without experiments [6–8]. Therefore, it is essential to carefully examine each assembled AO cell and certify it in terms of the real acoustic field structure.

The Schlieren method has been adopted as a standard optical technique for imaging acoustic fields in homogeneous transparent media including AO crystals [1,9,10] since it provides a two-dimensional projection image formed due to the diffraction of light on an acoustic beam.

In the interferometric [11,12] and holographic [13] techniques, the spatio-temporal distribution of the probing wide laser beam is recorded by a digital camera. Since the output signal changes at the frequency of the sonic radiator, these techniques are not applicable to high-frequency AO devices. Laser beam scanning also enables visualization of acoustic fields [14–17]. In [14–16], the laser beam diameter is smaller than the sound wavelength;

therefore, this scheme is applicable only to low-frequency fields. In [17], the shape of the diffracted laser beam allows evaluation of high frequency ultrasonic wave in solid media.

In all these methods [9–17], the measurement results are in fact averaged over the entire optical path through the sound field. Therefore, in practice, they are effective for the analysisof simple sound fields that have plane or symmetrical wave fronts. In real AO devices, many waves of various types with different directions and amplitudes can propagate. To obtain three-dimensional distributions of ultrasonic fields, tomographic methods based on the interferometric [18–20] or advanced Schlieren schemes [21–23] are applicable. In this case, data acquisition from different angles is necessary, which is hardly possible for AO devices due to their design features.

The acoustic field structure in crystal influences the wave displacement distribution on the facet of an AO cell located opposite or next to the transducer. Such measurement is available for the laser ultrasonic technique [24] but needs a highly reflective facet, which is normally absent. The gold standard for measuring the acoustic field is the use of needle hydrophones [25,26]. However, they cannot be placed directly on the AO cell facet and have low sensitivity due to their small dimensions. In this paper, we propose to use a focused immersion ultrasonic transducer with the focus placed on the cell surface. This scheme is typical for the receiver of a pulse acoustic scanning microscope [27], which has a high signal-to-noise ratio and spatio-temporal resolution.

In this study, we intend to validate this approach by studying an AO modulator made of fused quartz in order to determine both theoretically and experimentally the modes, amplitudes, propagation trajectories and other features of the ultrasonic waves inside. Below, we describe the experimental setup, the theoretical model of acoustic field structure and the experimental data, then discuss the results and demonstrate good correspondence between theoretical prediction and acoustic microscopy data that confirm the effectiveness of the proposed technique.

2. Experimental Setup and Technique

To confirm the feasibility of this approach, we applied it to testing an AO cell 1 made of optical fused quartz (Figure 1). It has the shape of a straight prism with a thickness of 12 mm. To generate the longitudinal acoustic waves in the cell, a lithium niobate ultrasonic transducer 2 is attached to the bottom plane. To avoid the formation of standing acoustic waves, the upper plane of the cell is inclined at the angle $\alpha = 6.5°$. The transducer consists of two 17.5×3.5 mm^2 sections separated by a 4 mm gap. The central frequency is 50 MHz while the frequency range is 30 MHz.

The upper part of the cell is inserted into a cuvette 3 filled with an immersion liquid (water). The ultrasonic waves generated by the piezoelectrical transducer 2 propagate inside the AO cell 1 and partially penetrate into the immersion liquid. These waves are detected by a focused piezotransducer 4 with a central frequency of 50 MHz and an angular aperture of 15°. The focus of the transducer 4 located at the quartz–water interface is mechanically translated along this plane to record the spatial distribution of the waves.

The ultrasonic data acquisition system of the experimental setup consists of the blocks typical for a scanning acoustic microscope [27,28]. To separate responses of different waves, the pulsed mode is used. The pulser–receiver 5 (5073PR, Panametrics–NDT Inc., Waltham, MA , USA) generates short electrical pulses to feed the AO piezotransducer 2 and amplifies the weak output signals of the transducer 4. The signals are then processed by the analog-to-digital converter 6 (FM412x500M, Insys Inc., Moscow, Russia) at a sampling rate of 500 MHz and a resolution of 12 bits. The mechanical movement of the focused transducer is implemented by the motorized translation stage 7 (8MT167-100, Standa Ltd., Vilnius, Lithuania). At each position of the transducer, the acquired signal is averaged over 16 series in order to increase the signal-to-noise ratio and then is recorded as a function of the wave propagation time t within temporal widows of 4 μs. We acquired the signals at different positions separated by a spatial period of 0.1 mm. The travel distance is set to 44 mm to cover the entire top facet of the AO cell. Recording time of one scan is about 4 s.

To obtain the delayed multiple reflection responses, the scan is repeated several times with different time window settings and signal gain.

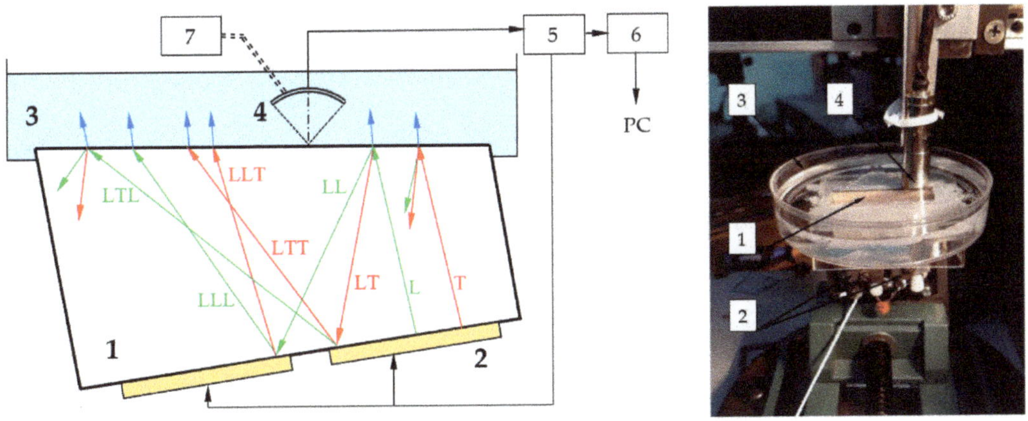

(**a**) (**b**)

Figure 1. Scheme (**a**) and appearance (**b**) of the experimental setup: 1—AO cell; 2—emitting piezotransducer; 3—cuvette with immersion liquid; 4—receiving piezotransducer; 5—pulser-receiver; 6—analog-to-digital converter; 7—motorized translator. Green and red arrows show longitudinal (*L*) and transverse (*T*) waves propagating in AO cell, blue arrows correspond to longitudinal waves in liquid.

There is a variety of ultrasonic waves propagating inside the AO cell (Figure 1), both longitudinal and transverse, while in the immersion liquid there are only longitudinal ones. Normally, the transducer generates a plane longitudinal wave *L*, but a weak transverse wave *T* may also appear. Both waves undergo reflection and mode conversion at the upper solid–liquid interface and partially penetrate the water. Wave *L* produces a longitudinal wave LL and a transverse wave LT, which transform at the bottom plane into two longitudinal waves, LLL and LTL, and two transverse waves, LLT and LTT.

The applied technique presumes detection of the water-penetrating waves by a scanning transducer focused on the upper face of the AO cell. These signals, together with information about the acoustic properties of the cell and immersion liquid are necessary to calculate the amplitudes of upward and downward waves in the AO cell. Then, we may evaluate the amplitudes of the waves inside the AO device using the characteristics of the acoustic absorber. To verify the theoretical data, we detect the waves on the clear part of the upper face partly covered with the absorber.

3. Theoretical Model

In the theoretical analysis described below, we have made the following assumptions. Since the dimensions of the transmitting transducer are much larger than the ultrasound wavelength, the divergence is rather small, the wave fronts in quartz and water are approximately plane and the ray approximation is applicable. As the acoustic attenuation in quartz is small, the amplitudes in the bottom face and near the quartz-water interface are equal. Therefore, the well-known formulas for the reflection, transmission and mode conversion coefficients for plane waves at the solid–liquid and solid–solid interfaces are valid [29]. All the coefficients are real because the angles of incidence do not exceed the critical values. The angles between the wave vectors in water and the axis of the receiving transducer are significantly less than its angular aperture. The distance between the transducer and the interface is constant and, therefore, the sensitivity of the receiver does not depend on the wave propagation direction.

In theoretical analysis, we use the symbols η, μ, ξ, $\kappa = L$, and T to denote waves. Let p_η be the amplitudes of the waves L and T radiated by the transducer (Figure 1). The four waves reflected back at the upper interface have amplitudes $p_{\eta\mu}$. In the next step, eight waves with the amplitudes $p_{\eta\mu\xi}$ propagate in the upward direction.

The incidence angles of the primary waves p_η on the upper surface are equal to the inclination angle α (Figure 2). Incidence angles $\gamma_{\eta\mu}$ and $\theta_{\eta\mu\xi}$ of reflected waves are:

$$\gamma_{\eta\mu} = \beta_{\eta\mu} + \alpha \cdot \theta_{\eta\mu\xi} = \delta_{\eta\mu\xi} + \alpha. \tag{1}$$

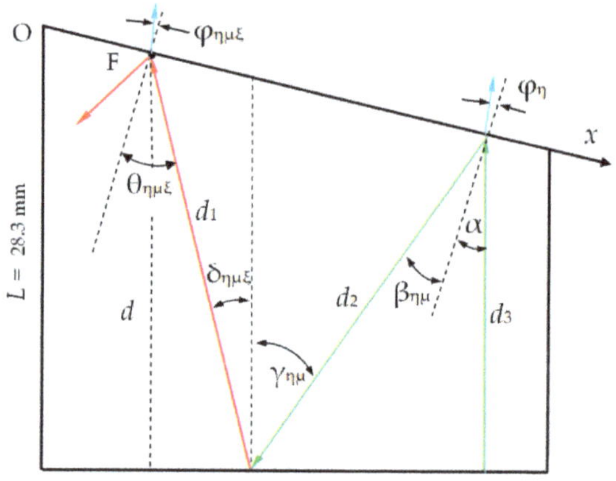

Figure 2. Ray model of the acoustic wave propagation in the AO cell.

The angles of reflection satisfy Snell's law:

$$\sin(\beta_{\eta\mu}) = \frac{C_\mu}{C_\eta}\sin(\alpha),\ \sin(\delta_{\eta\mu\xi}) = \frac{C_\xi}{C_\mu}\sin(\gamma_{\eta\mu}), \tag{2}$$

where C_L = 5960 m/s and C_T = 3760 m/s are velocities of the longitudinal and transverse waves in fused quartz, respectively [27]. Refracted waves in water propagate at angles φ_η and $\varphi_{\eta\mu\xi}$:

$$\sin(\varphi_\eta) = \frac{C_W}{C_\eta}\sin(\alpha),\ \sin(\varphi_{\eta\mu\xi}) = \frac{C_W}{C_\xi}\sin(\theta_{\eta\mu\xi}), \tag{3}$$

where C_W = 1485 m/s is sound velocity in water.

The spatio-temporal signals received by the focused transducer may be described as

$$S_\eta(x,t) = a_\eta w(t - t_\eta(x)),\ S_{\eta\mu\xi}(x,t) = a_{\eta\mu\xi} w(t - t_{\eta\mu\xi}(x)), \tag{4}$$

where x and t are the scanning coordinate and time, $w(t)$ is the impulse response of the experimental setup, a_η and $a_{\eta\mu\xi}$ are the amplitudes of the compression waves in water generated by corresponding modes p_η and $p_{\eta\mu\xi}$, and t_η and $t_{\eta\mu\xi}$ are their delays. The amplitudes a_η and $a_{\eta\mu\xi}$ are proportional to the amplitudes of the waves penetrated into the immersion liquid:

$$a_\eta = p_\eta T_\eta(\alpha) h, \tag{5}$$

where $T_\eta(\alpha)$ are the transmission coefficients at the solid–liquid interface and h is the detector sensitivity coefficient. For $p_L = 1$, this coefficient is equal to

$$h = \frac{a_L}{T_L(\alpha)}. \tag{6}$$

Then, the relative amplitude of the transverse wave T can be found using measured values of a_L and a_T as

$$p_T = \frac{a_T T_L(\alpha)}{a_L T_T(\alpha)}. \tag{7}$$

The amplitudes of the reflected waves may be found as follows:

$$p_{\eta\mu} = p_\eta R_{\eta\mu}(\alpha), \tag{8}$$

where the reflection coefficient $R_{\eta\mu}(\alpha)$ is determined by the acoustic properties of fused quartz and water. Substitution of (5) and (6) in Equation (7) gives

$$p_{\eta\mu} = \frac{a_\eta R_{\eta\mu}(\alpha)}{h T_\eta(\alpha)} = \frac{a_\eta R_{\eta\mu}(\alpha) T_L(\alpha)}{a_L T_\eta(\alpha)}. \tag{9}$$

Since

$$a_{\eta\mu\xi} = p_{\eta\mu\xi} T_\xi(\theta_{\eta\mu\xi}) h, \tag{10}$$

the amplitudes $p_{\eta\mu\xi}$ can be estimated in a similar way using the measured values $a_{\eta\mu\xi}$:

$$p_{\eta\mu\xi} = \frac{a_{\eta\mu\xi}}{T_\xi(\theta_{\eta\mu\xi}) h} = \frac{a_{\eta\mu\xi} T_L(\alpha)}{a_L T_\xi(\theta_{\eta\mu\xi})}. \tag{11}$$

After the absorber is attached, the direct measurement of the wave amplitudes is not feasible. Nevertheless, the amplitudes of the primary waves p_η remain the same, whereas the amplitudes of the reflected waves in the cell covered by absorber $p^*_{\eta\mu}, p^*_{\eta\mu\xi}$ decrease in proportion to the reflection or mode conversion coefficient at the quartz–absorber interface $R^*_{\eta\mu}(\alpha)$:

$$p^*_{\eta\mu} = p_{\eta\mu} \frac{R^*_{\eta\mu}(\alpha)}{R_{\eta\mu}(\alpha)}, \tag{12}$$

$$p^*_{\eta\mu\xi} = p_{\eta\mu\xi} \frac{R^*_{\eta\mu}(\alpha)}{R_{\eta\mu}(\alpha)}. \tag{13}$$

The delays t_η and $t_{\eta\mu\xi}$ of the signals (4) may be calculated as well as the propagation distances of the corresponding waves inside the AO cell (Figure 2). The distance d between the transducer and the receiving focus point F depends linearly

$$d = L - x \tan\alpha, \tag{14}$$

where x is the focus position and L is the length of the left facet. The delays of L and T waves are proportional to the distance x:

$$t_\eta = t_{0\eta} - \varepsilon_\eta x, \tag{15}$$

where

$$t_{0\eta} = \frac{L}{C_\eta}, \; \varepsilon_\eta = \frac{\tan\alpha}{C_\eta}. \tag{16}$$

The delays $t_{\eta\mu\xi}$ depend on the travel distances d_1, d_2 and d_3 (Figure 2) in the following way:

$$t_{\eta\mu\xi} = \frac{d_1}{C_\xi} + \frac{d_2}{C_\mu} + \frac{d_3}{C_\eta}. \tag{17}$$

Since

$$d_1 = \frac{d}{\cos\delta_{\eta\mu\xi}}, \; d_3 = d_2 \cos\gamma_{\eta\mu}, \text{ and } d_2 = d_1 \frac{\cos\theta_{\eta\mu\xi}}{\cos\beta_{\eta\mu}}, \tag{18}$$

$$t_{\eta\mu\xi} = t_{0\eta\mu\xi} - \varepsilon_{\eta\mu\xi} x, \tag{19}$$

where

$$t_{0\eta\mu\xi} = KL,\ \varepsilon_{\eta\mu\xi} = K \cdot \tan\alpha, \tag{20}$$

we derive

$$K = \frac{1}{\cos\delta_{\eta\mu\xi}}\left(\frac{1}{C_\xi} + \frac{\cos\theta_{\eta\mu\xi}}{C_\mu \cos\beta_{\eta\mu}} + \frac{\cos\theta_{\eta\mu\xi}}{C_\eta \cos\beta_{\eta\mu}}\cos\gamma_{\eta\mu}\right). \tag{21}$$

Thus, the values of delay t_η and $t_{\eta\mu\xi}$ are expressed as a linear dependence on the scanning length x with parameters presented in Table 1.

Table 1. Calculated and experimental values of angular, temporal and slowness parameters of wave propagation.

Wave	Calculation						Experiment	
	φ_η, $\varphi_{\eta\mu\xi}$, °	$\gamma_{\eta\mu}$, °	$\delta_{\eta\mu\xi}$, °	$\theta_{\eta\mu\xi}$, °	$t_{0\eta}$, $t_{0\eta\mu\xi}$, µs	ε_η, $\varepsilon_{\eta\mu\xi}$, ns/mm	$t_{0\eta}$, $t_{0\eta\mu\xi}$, µs	ε_η, $\varepsilon_{\eta\mu\xi}$, ns/mm
$\eta = L$	1.62	-	-	-	4.755	19.1	4.75	19.3
$\eta = T$	2.56	-	-	-	7.54	30	7.9	30.8
$\eta\mu\xi$ = LLL	4.77	13	13	19.5	14.02	56.4	13.95	57.5
$\eta\mu\xi$ = LLT	5.74	13	8.16	14.66	16.61	66.8	16.5	67.0
$\eta\mu\xi$ = LTL	5.70	10.6	16.9	23.5	17.2	69.2	16.8	69.1
$\eta\mu\xi$ = LTT	6.67	10.6	10.6	17.1	19.57	78.7	19.6	80.5

4. Experiments

In this section, we present the experimental results of acoustic pulse detection in two different conditions: (1) an AO cell with absorber fully removed and (2) a similar AO cell partly covered with absorber. We analyze the wave patterns and determine their amplitude and delays in order to estimate AO cell characteristics.

The measured ultrasonic signals are shown in Figures 3 and 4 as grayscale images. In these images, the signal value is encoded by the gray levels and is presented as a function of the retarded time τ:

$$\tau = t_0 - \varepsilon x \tag{22}$$

The value of slowness ε is matched to compensate for the spatial dependence on τ. The t_0 values were then measured using the positions of the maxima of the ultrasonic pulse envelopes. As the variable part of the propagation time is compensated, the responses $S(x,\tau)$ look like horizontally oriented patterns. The measured values t_0 and ε are also presented in Table 1. There is a good agreement between the experimental and calculated values that confirms the correctness of the theoretical model.

The response $S_L(x,\tau)$ (Figure 3a) is generated by the main longitudinal wave L. There are two horizontal components, P_1 and P_2, and a set of wavelets E adjacent to P. The size and position of the components P_1 and P_2 correspond to the size and position of the sections of the transmitting transducer indicated by Tr_1 and Tr_2. The wavelets E are radiated by the transducer's edges. This wave pattern is typical for a flat piston transducer [30].

The values of arrival time t_{0T} and coefficient ε_T of the transverse wave T are greater than those for the longitudinal wave L (t_{0L} and ε_L), since $C_T < C_L$. The response $S_T(x,\tau)$ from wave T also contains two components, P_1 and P_2, the sizes and positions of which coincide with the transmitting transducer aperture (Figure 3b). Wave T is much weaker than wave L and, therefore, random noise and artifacts (marked by A in Figure 3b) generated by some unwanted echoes inside the measurement setup are present in the image.

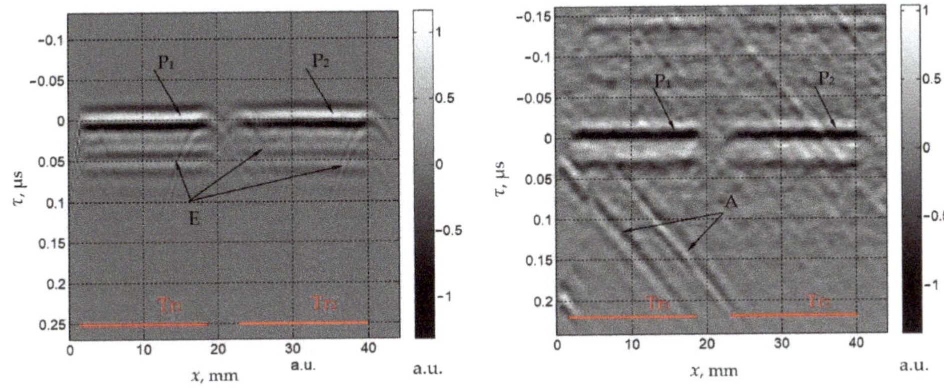

(a) (b)

Figure 3. Measured spatio-temporal signals $S_L(x,\tau)$ (**a**) and $S_T(x,\tau)$ (**b**).

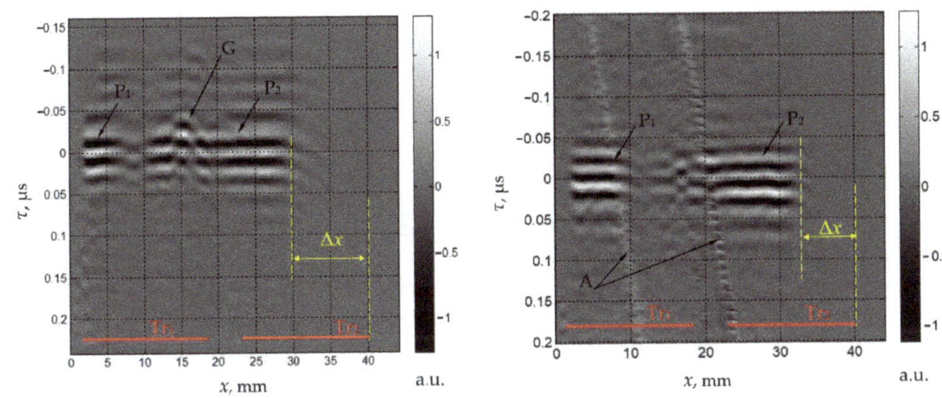

(a) (b)

Figure 4. Measured spatio-temporal signals $S_{LLL}(x,\tau)$ (**a**) and $S_{LTT}(x,\tau)$ (**b**).

The LLL wave formed by triple passes of the longitudinal wave is shown in Figure 4a. There is a lateral displacement Δx of the response due to the slope of the upper facet. The value of Δx may be estimated from the ray model (Figure 2):

$$\Delta x \approx L\left(\tan \delta_{\eta\mu\xi} + \tan \gamma_{\eta\mu}\right) \tag{23}$$

The experimentally evaluated displacement is consistent with the theoretical estimate $\Delta x \approx 12$ mm. The component P_1 is short because the left part of the ultrasonic beam from the section Tr_1 is bounded by the left facet of the AO cell. In the wavelet P_2, there is irregularity G formed by the reflection of the wave LL from the gap between two sections of the transducer on the bottom face instead of the reflection from the transducer's surface.

The measured spatio-temporal signal $S_{LTT}(x,\tau)$ is produced by the mode conversion of the longitudinal wave L at the upper interface (Figure 4b). The structure of this response is similar to $S_{LLL}(x,\tau)$, but the displacement $\Delta x \approx 9.5$ mm is less due to the fact that the angles δ_{LTT} and γ_{LTT} are less than δ_{LLL} and γ_{LLL}, respectively (Table 1).

To determine the amplitudes of the recorded signals, we estimated the maximal values of the envelopes of the ultrasonic pulses. Due to the irregularity of the transducer radiation efficiency, diffraction effects and artifacts, some spatial variations are present in the received responses. To reduce the measurement uncertainty, we calculated the mean and standard deviation of the envelope maximal values within a certain spatial window. The window is set in the range of $30 < x < 40$ mm for L and T waves (Figure 3). For the rest of the waves (Figure 4), a window of $18 < x < 28$ mm was used to compensate for the spatial displacement Δx. The average amplitudes a_η and $a_{\eta\mu\xi}$ are normalized by the amplitude a_L. Their relative standard deviations σ_η and $\sigma_{\eta\mu\xi}$ are presented in Table 2. To assess the measurement error of the experimental setup, the random noise level is estimated as the root mean square of the recorded signal in the areas without wave responses and artifacts. This noise value is 0.3%, which is much less than the standard deviations of the amplitudes.

Table 2. Experimental data for absorber-free AO cell and AO cell covered with ultrasound absorber: refracted angles (ϕ), normalized amplitudes (a, p), transmission (T) and reflection (R) coefficients for various modes and given inclination angle ($\alpha = 6.5°$).

Wave Mode	**Quartz-Water Interface (Absorber-Free AO Cell)**						**Quartz-Epoxy Interface (Absorber-Covered Cell)**	
	Measured							
	a_η, $a_{\eta\mu\xi}$	σ_η, $\sigma_{\eta\mu\xi}$,%	φ_η, $\varphi_{\eta\mu\xi}$,°	T_η (α) $T_\xi(\theta_{\eta\mu\xi})$	$R_{\eta\mu}$ (α)	p_η, $p_{\eta\mu}$, $p_{\eta\mu\xi}$	$R^*_{\eta\mu}$ (α)	p^*_η, $p^*_{\eta\mu}$, $p^*_{\eta\mu\xi}$
$\eta = L$	1	8.2	1.62	0.444	-	1	-	1
$\eta = T$	0.018	13.5	2.56	0.101	-	0.078	-	0.078
$\eta\mu$ = LL	-	-	-	-	0.773	0.77	0.605	0.605
$\eta\mu$ = LT	-	-	-	-	0.161	0.16	0.11	0.11
$\eta\mu$ = TL	-	-	-	-	0.401	0.031	0.27	0.021
$\eta\mu$ = TT	-	-	-	-	0.94	0.073	0.69	0.054
$\eta\mu\xi$ = LLL	0.39	12.5	4.77	0.413	-	0.42	-	0.33
$\eta\mu\xi$ = LLT	0.051	25	5.74	0.222	-	0.10	-	0.068
$\eta\mu\xi$ = LTL	0.074	22	5.70	0.398	-	0.083	-	0.056
$\eta\mu\xi$ = LTT	0.075	22.6	6.67	0.257	-	0.13	-	0.095

To evaluate the influence of the absorber, we used another AO cell of the same design. The cell has the absorber installed on the upper facet. The absorber is made from epoxy resin and covers a part of the facet as shown in Figure 5. The left edge of the absorber is located at $x_0 \approx 30$ mm. Therefore, the signal $S_{LLL}(x,\tau)$ is damped at the interval $[x_1, x_2] \approx [18, 28]$ mm due to the decrease of the reflection coefficient R_L in the presence of the absorber. The ratio of amplitudes measured for the quartz–absorber and quartz–water configurations is 0.75.

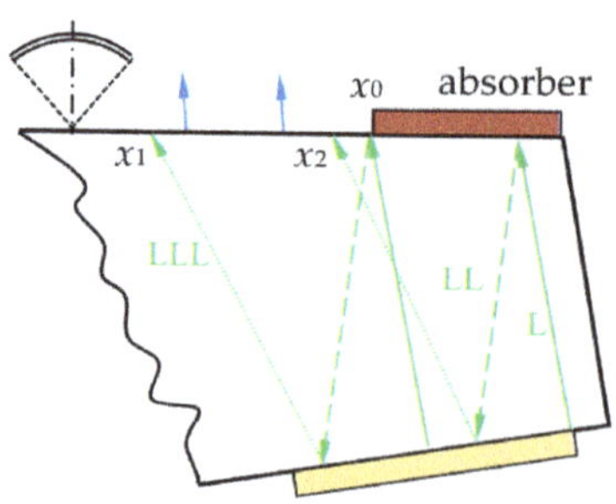

Figure 5. Ray propagation in AO cell with absorber.

5. Discussion

The wave amplitudes in quartz p_T, $p_{\eta\mu}$ and $p_{L\mu\xi}$ are determined from the ones measured in water a_η and $a_{\eta\mu\xi}$ (Table 2) using Equations (7), (9) and (11). Well-known formulas [29] allows calculation of the transmission coefficients $T_\eta(\alpha)$ and $T_\xi(\theta_{\eta\mu\xi})$, reflection and mode conversion coefficients $R_{\eta\mu}(\alpha)$ for the fused quartz–water interface. These coefficients are shown in Figure 6 as functions of the refraction angle φ in water. Their values at the angles φ_η, $\varphi_{\eta\mu\xi}$ are presented in Table 2 as well as the obtained relative amplitudes p_η, $p_{\eta\mu}$ and $p_{\eta\mu\xi}$.

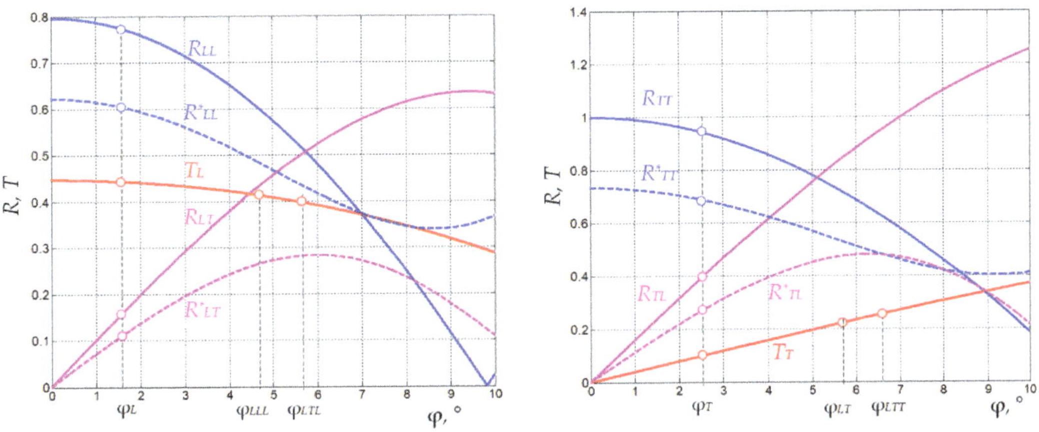

(a) **(b)**

Figure 6. (**a**) Reflection coefficients R_{LL}, R^*_{LL}, R_{TT} and R^*_{TT}, transmission coefficients T_L, and T_T, mode conversion coefficients R_{LT}, R^*_{LT}, R_{TL} and R^*_{TL} for the interfaces fused quartz–water (solid lines) and fused quartz–absorber (dashed lines) vs refraction angle in water φ: longitudinal (**a**) and transverse (**b**) waves in quartz.

We have discovered that the calculated relative amplitude of the direct transverse wave $p_T = 0.078$ is significant. Thus, the ultrasonic transducer generates a non-negligible transverse wave along with the regular longitudinal wave. This unwanted wave propagates in the same direction and cannot be attenuated by the absorber.

Among the reflected waves, the LL wave with an amplitude of $p_{LL} = 0.77$ is the largest. The waves TL and TT, generated at the upper interface by the mode conversion of the wave *L* and reflection of the transverse wave *T*, are much weaker. Therefore, only the propagation of LL and LT waves is considered below.

We should note that the amplitudes $p_{\eta\mu\xi}$ may be calculated using the determined values $p_{\eta\mu}$ and the reflection coefficients at the bottom interface. However, the design of the transmitting transducer is rather complicated and it is difficult to find the reflection coefficients from its surface. In addition, the reflectivity depends on the electrical load of the transducer. Thus, this approach is not reliable and barely practically realizable.

In the presence of the absorber, the amplitudes p^*_η of the direct waves do not change, while the amplitudes $p^*_{\eta\mu}$ and $p^*_{\eta\mu\xi}$ become smaller. These amplitudes are estimated from the amplitudes $p_{\eta\mu}$ and $p_{\eta\mu\xi}$ via Equations (12) and (13) (Table 2). We calculated the required reflection and mode conversion coefficients $R^*_{\eta\mu}(\alpha)$ under the assumption that the absorber is a solid medium with a density of 1150 kg/m^3 and velocities of longitudinal and transvers waves of 2650 m/s and 1100 m/s, respectively, values which are typical for epoxy resin [31]. The coefficients $R^*_{\eta\mu}$ presented in Figure 6 and Table 2 are obtained based on well-established technique [29]. In the experimental data, the ratio p^*_{LLL}/p_{LLL} is 0.75, which is quite close to the value 0.79 calculated from Table 2. We tested the proposed method on a model absorber and confirmed its applicability. Though the estimated amplitudes $p^*_{\eta\mu}$ and $p^*_{\eta\mu\xi}$ are small, they are not negligible and their effect may be significant for the detailed analysis of the AO device characteristics.

6. Conclusions

We have shown that impulse acoustic microscopy is quite informative tool for quantitative characterization of the acoustic field in AO cells. It allows to define the modes, amplitudes, propagation trajectories and other features of the ultrasonic waves propagating in the crystal even after multiple reflections. This information is highly important in practice as it enables the evaluation of the correctness of AO cell design including the cut and facet angles, efficiency of the ultrasound piezotransducer and absorber functioning, etc. Being non-destructive and highly sensitive, impulse acoustic microscopy might be effective for quite fast experimental validation of theoretical estimations and numerical modeling results as well as for accurate certification of AO devices.

Author Contributions: Conceptualization, S.A.T. and V.E.P.; methodology, S.A.T.; validation, V.E.P. and A.S.M.; formal analysis, V.E.P.; resources, S.A.T. and A.S.M.; data curation, S.A.T.; writing—original draft preparation, S.A.T. and A.S.M.; writing—review and editing, V.E.P.; project administration, A.S.M. All authors have read and agreed to the published version of the manuscript.

Funding: This research was funded by Federal State Task Program of Scientific and Technological Center of Unique Instrumentation RAS. This work was performed using the equipment of the Shared Research Facilities of the Scientific and Technological Centre of Unique Instrumentation of the Russian Academy of Sciences.

Institutional Review Board Statement: Not applicable.

Informed Consent Statement: Not applicable.

Data Availability Statement: The data presented in this study are available on request from the corresponding author.

Conflicts of Interest: The authors declare no conflict of interest.

References

1. Goutzoulis, A.P.; Rape, D.R. *Design and Fabrication of Acousto-Optic Devices*; CRC Press: Boca Raton, FL, USA, 2004; 520p.
2. Xu, J.; Stroud, R. *Acousto-Optic Devices: Principles, Design, and Applications*; Wiley: New York, NY, USA, 1992; 672p.
3. Römer, G.R.B.E.; Bechtold, P. Electro-optic and Acousto-optic Laser Beam Scanners. *Phys. Procedia* **2014**, *56*, 29–39. [CrossRef]
4. Hagen, N.; Kudenov, M.W. Review of Snapshot Spectral Imaging Technologies. *Opt. Eng.* **2013**, *52*, 090901. [CrossRef]
5. Duocastella, M.; Surdo, S.; Zunino, A.; Diaspro, A.; Saggau, P. Acousto-optic systems for advanced microscopy. *J. Phys. Photonics* **2020**, *3*, 012004. [CrossRef]
6. Voloshinov, V.; Polikarpova, N. Acousto-optic investigation of propagation and reflection of acoustic waves in paratellurite crystal. *Appl. Opt.* **2009**, *48*, 55–66. [CrossRef] [PubMed]
7. Maak, P.; Takács, T.; Barocsi, A.; Kollar, E.; Richter, P. Thermal behavior of acousto-optic devices: Effects of ultrasound absorption and transducer losses. *Ultrasonics* **2011**, *51*, 441–451. [CrossRef] [PubMed]
8. Balakshy, V.; Voloshin, A.; Molchanov, V. Anisotropic light diffraction in crystals with a large acoustic-energy walk-off. *Opt. Spectrosc.* **2014**, *117*, 801–806. [CrossRef]
9. Voloshinov, V.; Polikarpova, N.; Ivanova, P.; Khorkin, V. Acousto-optic control of internal acoustic reflection in tellurium dioxide crystal in case of strong elastic energy walkoff [Invited]. *Appl. Opt.* **2018**, *57*, C19–C25. [CrossRef]
10. Kudo, N. Optical methods for visualization of ultrasound fields. *Jpn. J. Appl. Phys.* **2015**, *54*, 07HA01. [CrossRef]
11. Ishikawa, K.; Yatabe, K.; Oikawa, Y. Physical-model-based reconstruction of axisymmetric three-dimensional sound field from optical interferometric measurement. *Meas. Sci. Technol.* **2021**, *32*, 045202. [CrossRef]
12. Bertling, K.; Perchoux, J.; Taimre, T.; Malkin, R.; Robert, D.; Rakić, A.D.; Bosch, T. Imaging of acoustic fields using optical feedback interferometry. *Opt. Express* **2014**, *22*, 30346–30356. [CrossRef]
13. Rajput, S.K.; Matoba, O.; Takase, Y.; Inoue, T.; Itaya, K.; Kumar, M.; Quan, X.; Xia, P.; Awatsuji, Y. Multimodal sound field imaging using digital holography [Invited]. *Appl. Opt.* **2021**, *60*, B49–B58. [CrossRef] [PubMed]
14. Malkin, R.; Todd, T.; Robert, D. A simple method for quantitative imaging of 2D acoustic fields using refracto-vibrometry. *J. Sound Vib.* **2014**, *333*, 4473–4482. [CrossRef]
15. Roberto, L.; Vanlanduit, S.; Arroud, G.; Guillaume, P. Underwater Acoustic Wavefront Visualization by Scanning Laser Doppler Vibrometer for the Characterization of Focused Ultrasonic Transducers. *Sensors* **2015**, *15*, 19925–19936. [CrossRef]

16. Ortiz, P.F.U.; Perchoux, J.; Arriaga, A.L.; Jayat, F.; Bosch, T. Visualization of an acoustic stationary wave by optical feedback interferometry. *Opt. Eng.* **2018**, *57*, 051502. [CrossRef]
17. Wu, J.; Xu, Z.; Li, K.; Lv, G.; Li, X.; Wang, C. Analysis of Acoustic Near Field Characteristics in Acousto-Optic Modulator. *IEEE Photonics Technol. Lett.* **2021**, *33*, 201–204. [CrossRef]
18. Yatabe, K.; Ishikawa, K.; Oikawa, Y. Acousto-optic back-projection: Physical-model-based sound field reconstruction from optical projections. *J. Sound Vib.* **2017**, *394*, 171–184. [CrossRef]
19. Torras-Rosell, A.; Barrera-Figueroa, S.; Jacobsen, F. Sound field reconstruction using acousto-optic tomography. *J. Acoust. Soc. Am.* **2012**, *131*, 3786–3793. [CrossRef] [PubMed]
20. Verburg, S.A.; Fernandez-Grande, E. Acousto-Optical Volumetric Sensing of Acoustic Fields. *Phys. Rev. Appl.* **2021**, *16*, 044033. [CrossRef]
21. Koponen, E.; Leskinen, J.; Tarvainen, T.; Pulkkinen, A. Acoustic pressure field estimation methods for synthetic schlieren tomography. *JASA* **2019**, *145*, 2470–2479. [CrossRef]
22. Xu, Z.; Chen, H.; Yan, X.; Qian, M.-L.; Cheng, Q. Three-dimensional reconstruction of nonplanar ultrasound fields using Radon transform and the schlieren imaging method. *JASA* **2017**, *142*, EL82–EL88. [CrossRef]
23. Pulkkinen, A.; Leskinen, J.; Tiihonen, A. Ultrasound field characterization using synthetic schlieren tomography. *JASA* **2017**, *141*, 4600–4609. [CrossRef] [PubMed]
24. Scruby, C.B.; Drain, L.E. *Laser Ultrasonics Techniques and Applications*, 1st ed.; Taylor & Francis Group: New York, NY, USA, 1990; 462p.
25. Hill, C.R.; Bamber, J.C.; ter Haar, G.R. *Physical Principles of Medical Ultrasonics*, 2nd ed.; John Wiley & Sons: West Sussex, UK, 2004; 528p.
26. Harris, G.R. Hydrophone measurements in diagnostic ultrasound fields. *IEEE Trans. Ultrason. Ferroelect. Freq. Contr.* **1988**, *35*, 87–101. [CrossRef] [PubMed]
27. Briggs, G.A.D.; Kolosov, O.V. *Acoustic Microscopy*, 2nd ed.; Oxford university press: New York, NY, USA, 2010; 387p.
28. Vogt, M.; Ermert, H. High-frequency ultrasonic systems for high-resolution ranging and imaging. In *Advances in Acoustic Microscopy and High Resolution Ultrasonic Imaging: From Principles to New Applications*; Maev, R.G., Ed.; Wiley & Son–VCH Verlag & Co, KGaA: Weinheim, Germany, 2013; pp. 93–123.
29. Brekhovshikh, L.M.; Godin, O.A. *Acousics of Layered Media I*; Springer: Heidelberg, Germany, 1990; pp. 87–108.
30. Kino, G.S. *Acoustic Waves: Devices, Imaging and Analog Signal Processing*; Prentice-Hall Inc.: Englewood Cliffs, NJ, USA, 1987; pp. 164–218.
31. Birks, A.S.; Green, R.E.; McIntire, P. *Nondestructive Testing Handbook*, 2nd ed.; The American Society For Nondestructive Testing: Columbus, OH, USA, 1991; Volume 7, pp. 836–841.

MDPI
St. Alban-Anlage 66
4052 Basel
Switzerland
Tel. +41 61 683 77 34
Fax +41 61 302 89 18
www.mdpi.com

Materials Editorial Office
E-mail: materials@mdpi.com
www.mdpi.com/journal/materials

www.ingramcontent.com/pod-product-compliance
Lightning Source LLC
Chambersburg PA
CBHW061446210926
51850CB00008B/1632
* 9 7 8 3 0 3 6 5 9 5 1 9 1 *